Edwin J. Cohn and the Development of Protein Chemistry

EUGENE A. MONTGOMERY

Edwin J. Cohn and the Development of Protein Chemistry

With a Detailed Account of His Work on the Fractionation of Blood

During and After World War II

Douglas M. Surgenor

Published by the Center for Blood Research, Inc.,
Distributed by the Harvard University Press
2002

ISBN 0-674-00962-2

Printed in Canada

To the Physical and Chemical Collaborators

Gordon A. Alles
Eva H. Alameri
John N. Ashworth
Kenneth Bailey
John B. Bateman
Adrien G. Borduas
Norval F. Burk
James B. Conant
Walter B. Dandliker
Maurice D'Hont
David B. Dill
Howard M. Dintzis
Harold Edelhoch
John T. Edsall
Eric Ellenbogen
Drummond Ellis
I. Fankuchen
John D. Ferry
Harry L. Fevold
Marcel Florkin
Joseph F. Foster
Dexter French
Tillman D. Gerlough
Geoffrey A. Gilbert
J. Morton Gillespie
Erland C. Gjessing
Frank H. Gordon
Jesse P. Greenstein
Paul M. Gross, Jr.
Henry E. Guerlac
Frank R. N. Gurd
Selig Hecht
Lawrence J. Henderson
Jessie L. Hendry
Francis F. Heyroth
Norman R. S. Hollies
Walter L. Hughes, Jr.
Margaret J. Hunter
Henry C. Isliker
Frederick F. Johnson
Norman R. Joseph
Frederick W. Kahnt
Ephraim Katchalski
John G. Kirkwood
Bernard A. Koechlin
James B. Lesh
John J. Livingood
Rene Lontie
Barbara W. Low
J. Murray Luck
Thomas L. McMeekin
John W. Mehl
Marshall Melin
Dora G. Mittelman
Naum Mittelman
Herbert Morawetz
Vincent E. Morgan
Robert F. Mouton
Hans Mueller
Dwight J. Mulford
Hans Nitschmann
John L. Oncley
John W. Palmer
Robert B. Pennell
Gertrude E. Perlmann
Louis Pillemer
Jules D. Porsche
Frederic M. Richards
Dan A. Richert
Eugene C. Rochow
Harry A. Saroff
George Scatchard
Harold A. Scheraga
Karl Schmid
Joseph Shack
Richard B. Simpson
Heron O. Singher
Arthur K. Solomon
Christian Sporck
Jacinto Steinhardt
Rudolf Straessle
Danella Straup-Cope
Laurence E. Strong
Douglas M. Surgenor
Charles Tanford
F. H. L. Taylor
Harold L. Taylor
John G. Trump
Johannes van Ormondt
Hubert B. Vickery
Alexander von Muralt
John H. Weare
Philip E. Wilcox
John W. Williams
Kenneth A. Wright
Jeffries Wyman, Jr.

The Clinicians Who, as Part of Their Training Served as Chemical Collaborators

S. Howard Armstrong, Jr.
George Barac
Benjamin A. Barnes
William H. Batchelor
Ray K. Brown
Bernard D. Davis
Robert B. Davis
Gustave Derouaux
David M. Gibson
David Gitlin
Robert S. Gordon
Arda A. Green
Richard W. Greene
James A. Halstead
James T. Heyl
Fritz Koller
David R. Kominz
Edward Leonard
John A. Leutscher
Jacques Lewin
C. H. Liu
Nicholas H. Martin
Benjamin Miller
Bo Norberg
Alexander Rich
William T. Salter
I. H. Scheinberg
Ellsworth Twible

The Immunilogical

Francis G. Blake
William C. Boyd
A. Raymond Dochez
Harry Eagle
Monroe D. Eaton
Geoffrey Edsall
John F. Enders
Ronald M. Ferry
Roger M. Herriott
J. Howard Mueller
Alwin M. Pappenheimer, Jr.
Laszlo L. Reiner
Elliott S. Robinson
Joseph Stokes, Jr.
Eero Uroma
Hans H. Zinsser

Physiological, Pharmacological, Pathological

Orville T. Bailey
James W. Cameron
Walter B. Cannon
John F. Fulton
Clinton van Z. Hawn
Frederick L. Hisaw
Roy G. Hoskins
Otto Krayer
Milton O. Lee
Peter R. Morrison
John R. Pappenheimer
Vladimir A. Pertzoff
A. C. Redfield
Marcus Singer
S. Burt Wolbach
G. Payling Wright

Medical and Surgical Colleagues

Benjamin Alexander
Walter Bauer
William Berenberg
Edgar A. Bering
Edward S. Buckley, Jr.
Allan M. Butler
William B. Castle
Hector Croxatto
Harvey Cushing
Elmer L. DeGowin
Louis K. Diamond
Charles A. Doan
Charles P. Emerson
Clement A. Finch
John G. Gibson, II
Sam T. Gibson
Franc D. Ingraham
Charles A. Janeway
T. Duckett Jones
Roger I. Lee
Walter F. Lever
Robert F. Loeb
George R. Minot
Carl V. Moore
William P. Murphy
Lloyd C. Newhouser
David D. Rutstein
Max M. Strumia
George W. Thorn
James L. Tullis
Bert L. Vallee
Carl W. Walter
Owen H. Wangensteen
Soma Weiss
Maxwell M. Wintrobe
Lorande M. Woodruff

Of Edwin J. Cohn, 1920–1950

Contents

Illustrations

Edwin in the rear. Below, left to right, Alice von Muralt, Edwin Cohn and Marianne Cohn taking lunch in San Martino di Castrozza (left); Edwin J. Cohn, Jr. and Alfred B. Cohn sitting in the Cohn automobile on the dock upon arrival back in Palermo (right). (Courtesy of Edwin J. Cohn, Jr.).

5. Above: At the Brettauer summer home in Lake Placid, New York in 1934. Fred Cohn appears on the left; on the right, left to right, appear Mrs. Brettauer (Marianne's mother), Edwin J. Cohn, Jr., and Marianne Cohn (Courtesy of Edwin J. Cohn, Jr.). Below: George Scatchard, left (courtesy of the MIT Museum); Jeffries Wyman, right (Courtesy of the Harvard University Archives).
6. Above: the Harvard ultracentrifuge installed in the basement of Building E. of the Harvard Medical School in 1938. The machine itself was housed within a concrete structure near the back wall. The long tubes coming forward carried the optical systems for viewing and photographing the sedimentation of proteins during the long measurement period. Viewing the status of the run were Dr. E. G. Pickels of the International Health Division of the Rockefeller Foundation and Mr. Charles Gordon, who built the ultracentrifuge under the direction of Drs. Pickels and J. L. Oncley. Gordon stands near the control panel (Center for Blood Research Archives). Below: The Harvard electrophoresis apparatus being demonstrated by the machine's inventor, Dr. Arne Tiselius of the University of Uppsala, for attendees at the national meeting of the American chemical Society in Boston on September 15, 1939 (Center for Blood Research Archives).

Between pages 198–199

7. Edwin Cohn and a group of young colleagues on the front steps of Building A of the Harvard Medical School in the spring of 1942 at the beginning of the early experiments in the Harvard Pilot Plant directed at isolating human serum albumin from human plasma. Front row, (left to right), L. E. Strong, J. L. Oncley, E. J. Cohn, W. L. Hughes, Jr., J. N. Ashworth; back row, H. L. Taylor, L. Larkin, J. W. Cameron, D. L. Mulford, D. S. Richert, A. H. Sparrow (Center for Blood Research Archives).
8. Above: John F. Enders in his laboratory (Countway Library of Medicine). Below, left to right: A. Baird Hastings, Louis K. Diamond and Carl W. Walter (Countway Library of Medicine).
9. The Harvard Pilot Plant on the first floor of Building E of the Harvard Medical School. It consisted of two connected refrigerated

Figures in text

Foreword

As an editorial in the *New England Journal of Medicine* pointed out,[1] Edwin J. Cohn, who never studied medicine and almost never saw a patient undergoing treatment, exerted an influence on the progress of medicine matched by few contemporary men. He accomplished this feat, first, by creatively broadening the study of protein chemistry in America, and second, through his work on blood fractionation before, during and after World War II. For the first twenty years of his academic career, passed entirely at Harvard, Cohn's studies were primarily devoted to fundamental investigations of the physical chemistry of proteins, amino acids and peptides, whose nature and mechanics were poorly understood at the time. Characteristically, he became the center of a group of investigators, some in his own and some in other laboratories, and inspired scientists of very diverse talents to join him in the attack on these basic problems.

Together with this group of enthusiastic associates, he worked out a pattern of well-defined principles that served to elucidate the relations between the solubility and the structure of proteins, amino acids and peptides, and the significance of the powerful electrical interactions determining them. Not a few chemists and others recognized the fundamental scientific significance of the work, but scarcely anyone could have foreseen that it would prove an indispensable foundation for the later development of fractionation procedures that were to revolutionize the clinical use of products derived from blood. Seldom has the transition been so rapid, within a single laboratory, from the study of very abstract problems of no apparent contemporary practical significance to work on the most urgent problems of practical importance in a time of great crisis. Yet the continuity in the thinking of the workers in Cohn's laboratory was essentially unbroken.

The development of the plasma fractionation program during World War II is a story that has often been told, but the magnitude of the vision that inspired Edwin Cohn, then and later, and the multiplicity of details involved, has perhaps not been fully appreciated. The wartime emergency called fully into play his powers as a great leader and organizer, who drew together chemists, clinicians, pathologists, immunologists and others to achieve the separation and purification of serum albumin, gamma globulin, fibrinogen, thrombin and other fractions of human plasma, and their application to clinical use after a program of careful clinical testing. Dr. Cohn, however, was never satisfied with the immediate achievement of a practical goal. He envisioned a fractionation program that was to become even more complete in the separation of the innumerable protein components of plasma, each to be devoted to its particular use, and the separation of the formed elements of the blood, as well as the plasma proteins, by means so gentle as to preserve them as nearly as possible in their normal state in nature.

As time went on, the wartime fractionation processes were replaced with newer methods involving precipitation by metallic ions such as zinc, which could be removed by ion exchange resins, and the development of a machine of extraordinarily ingenious design made many of the procedures automatic. The revolution Cohn started has still not run its course, and the consequences may be far-reaching for the whole national blood program, for the public health, and for national defense in ways that cannot fully be appreciated.

As the pages that follow fully demonstrate, Edwin J. Cohn truly was what a French Colleague dubbed him in 1952—"Le Roi des Proteins."

ACKNOWLEDGEMENTS

THE INCENTIVE for this book arose in 1991 as I sat in a seldom used office at the Center for Blood Research and discovered that I was surrounded by a trove of books and archival materials dating back to the Cohn period. Previously thought to have been lost, this accumulation could only have been preserved by the original trustees of the Protein Foundation and their successors. The oldest item in this collection was Cohn's 1916 paper with L. J. Henderson on "The Equilibrium Between Acids and Basis in Sea Water," and the newest, an unfinished set of galley proofs entitled "A Collection of Papers Recording an Institutional Growth and Metamorphosis," which Edwin Cohn had been working on at the time of his death. Additional and later resources have included the papers generated by the Protein Foundation and its successor from the time of Cohn's death nearly to the present.

The information provided by these invaluable materials was supplemented through interviews by the author with approximately thirty of Cohn's colleagues still living at the time he commenced research on this work in 1992. John Edsall's input was extremely helpful, for, in addition to contributing information, he read drafts of early chapters that dealt with the prewar period and made many valuable suggestions. Fruitful and important were also the assistance and cooperation of other interviewees such as Bernard Davis, Larry Oncley, Jules Porsche and Sam Gibson who filled in blank spaces by recalling their experiences during World War II. For elucidating facts that occurred in the postwar period, James Tullis, Jack Latham, Robert Pennell and Fred Gilchrist proved equally helpful.

Sizable numbers of Cohn's papers were also found in the Library of the National Academy of Sciences in Washington, in the archives of the Rockefeller Foundation, in The Francis A. Countway Library of Medicine of the Harvard Medical School, and in the Pusey Library in Harvard University. Janice T. Goldblum, archivist at the National Academy of Sciences and Emily Oakhill at the Rockefeller Archives Center were very helpful in making valuable documents available for my use. I particularly want to thank Elin and Richard Wolfe at the Countway Library who helped in innumerable ways in completing this book. Mr. Wolfe provided direction while the manuscript was being compiled and read and edited it through several drafts; afterwards he supervised its publication.

Dr. Saul Benison, formerly Professor of History at the University of Cincinnatti and an historian working with the National Foundation for Poliomyelitis, encouraged me in undertaking this project, citing Cohn's extraordinary relationship with the American Red Cross; his appreciation of the roles of Joseph Stokes and William Hammond in pressing for field tests of Cohn's gamma globulin against polio were important also in filling in details in that aspect of Cohn's work. Periodic discussions with Paul Fremont-Smith, Oglesby Paul, George Thorn, and Ivy DeFriez, whose careers at Harvard partially overlapped that of Cohn, were helpful to me in maintaining my perspective.

In many conversations in Washington, Boston and Cambridge, Edwin J. Cohn, Jr. shared his recollections about his father and mother and their life in Cambridge. Ed's memories of the family's sabbatical year in Germany and the months that followed provided valuable insights in piecing together that important aspect of the Cohn story. Our discussions led quite naturally to subsequent interviews with Alfred Cohn and his wife, Barbara Norfleet Cohn, who were equally open and helpful.

At the Center for Blood Research, Rachelle Rosenbaum, Shirley Lerner Nicholson, Margaret Kramer and Laura Murphy assisted in innumerable ways while I was gathering the material and completing the manuscript. Sandra Taylor's experience was important in processing the notes.

This endeavor gradually engaged the understanding, ability and wisdom of my wife, Lois. As the years passed, we found great satisfaction in working together to complete this book.

Contributions to the Cohn Book Fund at the Center for Blood Research were made by the Alpha Therapeutics Corporation, Los Angeles; the American Red Cross Blood Services, Washington, D.C.; the Baxter Healthcare Corporation, Glendale, California; the Central Laboratory of the Swiss Blood Transfusion Service, Bern; the Central Laboratory of the Netherlands Red Cross Blood Transfusion Service, Amsterdam; the Haemonetics Corporation, Braintree, Massachusetts; Immuno Aktiengesellschaft, Vienna; the New York Blood Center, New York City; Ortho Diagnostic Systems, Raritan, New Jersey; and the Sandoz Pharmaceutical Corporation, East Hanover, New Jersey. The assistance of these organizations was instrumental in completing this project.

Douglas M. Surgenor
August 2001

1 An Auspicious Beginning

Edwin Joseph Cohn was born in New York City on December 17, 1892, the fourth child of Abraham and Mamie Cohn. His father was brought to the United States in the 1850s as a boy when his parents immigrated from Bavaria.[1] The Cohn family settled in New York City and quickly gained a foothold in the New World by dint of hard work and a frugal lifestyle. Abraham had little formal education. He became a member of Temple Emanu-el, but he left the temple in the mid 1870s to follow Felix Adler, the charismatic young leader of the newly established New York Society for Ethical Culture, an avant-garde movement founded to supplant older religions, with an emphasis on the moral factor in human relationships. Later, Abraham Cohn withdrew from the ethical movement as well, either, according to his eldest son, "because it was too much or because it was not enough." Thereafter, Cohn, "a man of powerful intellect, in economics conventional but intellectually uncompromising," was a non-practicing Jew.[2]

Mamie Einstein Cohn came from an old family in Savannah, Georgia where her father had been President of the Synagogue for many years. It was said that he was kept in office because he was the only man in Georgia who could blow a shofar. Reflecting his standing in the community, he was an honorary colonel in the Chatham Artillery, perhaps the oldest military organization in the country. During the financially depressed period after the Civil War, Mamie, like other Southern belles in their teens, was sent to live with friends in New York. It was there that she met Abraham Cohn.[3]

Abraham and Mamie Cohn had four children, Alfred, Leonard,

Myra, and Edwin, born in that order. By the time Edwin appeared in 1892, Abraham had established himself as a successful tobacco merchant, importing Sumatran tobacco. Later he was reportedly one of the first to succeed in growing Sumatran tobacco in Georgia. The family home was a townhouse on East 66th street in Manhattan. There, Edwin grew up in comfortable circumstances in a family with broad interests and a wide circle of acquaintances.[4]

Many years later, he and the author spent a night at the Plaza Hotel in New York. The next morning, on our way to a meeting at the Helen Hay Whitney Foundation in the New York Hospital, we strolled up into Central Park in bright spring sunshine. In a relaxed mood, Cohn reminisced about his childhood and the days when his governess brought him over to the park to play with other children. As we crossed Fifth Avenue at East 70th Street we came upon the mansion that the steel manufacturer and philanthropist Henry Clay Frick had built while Cohn was a student. This is now the home of Frick's collection of paintings, bronzes and enamels that he bequeathed, with the mansion, to the City of New York. On an impulse, Cohn entered and strolled around, whether to see the art or the interior architecture of Frick's home was not clear; he was interested in both.

Alfred Einstein Cohn, thirteen years older than Edwin, received his undergraduate education at Columbia College. He was a brilliant student, going on to the College of Physicians and Surgeons for his M.D. degree, which was awarded in 1904. He trained in pathology at Mt. Sinai Hospital. After postgraduate study in Freiburg, Vienna and London, he was appointed to the staff of the Rockefeller Institute in 1911 and became a Member of the Institute in 1920. At the Hospital of the Rockefeller Institute, Alfred Cohn established a worldwide reputation for his investigations of the physiology and pharmacology of the heart. He was active in the affairs of the New York Heart Association, which became the model for the later establishment of the American Heart Association. He had wide interests. He carried on a prolific correspondence with many prominent people of his day on a wide range of topics. Three collections of his philosophical essays, *Minerva's Progress*, *Medicine, Science and Art*, and *No Retreat from Reason* have been published. In the years following the death of Abraham Cohn in 1911, Alfred Cohn maintained a close relationship with his youngest brother, exerting an important role in matters relating to his education and career aspirations.

Myra Cohn, Edwin's sister, married Edwin M. Berolzheimer, who later was President of the Eagle Pencil Company. The Berolzheimers maintained residences in New York City and Tarrytown, New York, and after 1940, in South Carolina. Edwin enjoyed a close relationship with his sister, visiting her frequently. The Berolzheimers were generous sponsors of medical research projects. Edwin's other brother, Leonard Cohn, had a successful career as a stockbroker with L.M. Rothschild in New York.[5]

Each year it was Abraham Cohn's custom to attend the International Tobacco Market in Amsterdam. Whenever possible he brought his family with him. In this way the family made tours which extended to countries well beyond Amsterdam, where the family indulged their interests in a wide range of cultural and intellectual topics. Those travels in Europe were instrumental in developing Edwin's interests in art and architecture.[6]

Amherst College

Edwin Cohn attended three preparatory schools, the Sachs Institute in New York, the Stearns School in Mt. Vernon, New Hampshire, and the Philips Academy in Andover, Massachusetts. He participated in team sports, playing baseball and tennis. He also liked to ride horseback. In the fall of 1910 he entered Amherst College with the class of 1914. In the freshman chemistry course, he ended up with a grade of D, and said to his adviser, "Thank Heaven, I'll never have to bother with chemistry again." He shied away from chemistry. In his sophomore year he came under the spell of the great humanist, Henry Carrington Lancaster, then a member of the Amherst Department of Romance Languages. He chose French drama as his field of concentration. However, he had difficulty fitting the courses he wanted into his schedule. To fill in the schedule, he enrolled in a survey course on biology offered by "Tip" Tyler, who "taught with the fervor of a revivalist preacher, but not what one would describe as the method of science." Although he continued with French literature, he began to reconsider his earlier judgement.[7]

It was at Amherst that Cohn first encountered George Scatchard, a member of the Amherst Class of 1913. Although they later became close friends and enjoyed a unique association in scientific matters that persisted for the rest of Edwin's life, their encounters as undergraduates were unproductive and rather distant. Although, like Cohn, Scatchard was a fourth child, their backgrounds were otherwise totally different. Scatchard's home was in Oneonta, New York, a village on the Susquehanna River in the northern foothills of the Catskills. He was an avid reader and was rather shy. While in high school, George had worked evenings in his brother's pharmacy. Since his brother wanted to maintain a professional atmosphere in the pharmacy, Scatchard took to reading drug journals, the pharmacopoeia and even the prescriptions. George Scatchard graduated from Amherst in 1913 with the highest marks in his class.[8]

Years later, in a lecture honoring Cohn, Scatchard described his feelings about Cohn in those early days at Amherst:

> I did not like Cohn as an undergraduate. We were as different as our backgrounds. My family had to sacrifice to send me to college though I earned most of my expenses working during the summer. Cohn had no financial problems. If he did not have the best of everything, he had the most expensive when his family did not know that was not the best, which was seldom. He was an esthete. I regarded him as one of the flower children of my generation, less like those of the present than like those of an earlier generation characterized by W.S. Gilbert in *Patience.*
>
> Though the Philistines may jostle,
> you will rank as an apostle
> in the high aesthetic band,
> If you walk down Picadilly
> with a tulip or a lily
> in your medieval hand.[9]

In his third year at Amherst, Cohn began to think seriously about a scientific career in biology. His brother Alfred tried to persuade him to consider medicine, but Edwin would have none of it. On a visit home, Alfred took him to see Jacques Loeb, a colleague who had recently become a Member of the Rockefeller Institute. Although trained as a physician, Loeb's career was devoted to fundamental studies in biology. His interests ranged from the comparative physiology of the brain to organizational growth, the dynamics of living matter, and the mechanistic conception of life. On learning that the young Cohn was then a junior at Amherst College, Loeb immediately suggested that Amherst was not the place to be studying science, commenting, "I don't know the name of a single person on the faculty." He suggested that Cohn secure the catalogs from some leading universities and offered to help him to select a university with a strong program in biology. Acting on Loeb's advice, Cohn transferred to the University of Chicago in 1913 at the end of his junior year at Amherst.[10] Later, in the Amherst yearbook for the class of 1914, he was characterized as "a socialist, anarchist, Ibsenite, Shavian, iconoclast, atheist, agnostic and heretic. He was no backslapper." Years later, a classmate remembered him: "He wore his reasoning straight and penetratingly. Mention of his own accomplishments was shrugged off, he was that modest."[11]

The University of Chicago

In Chicago, Cohn studied chemistry under Julius Stieglitz, mathematics under Arthur Lunn, physics under Robert A. Millikan, and biology, his major field, under Frank R. Lillie. One of his first biology courses was a seminar taught by C.M. Child, a Professor of Biology. Cohn avidly read Child's book, *Senescence and Rejuvenation*. Many years later, he credited Child with exerting a formative influence on his scientific outlook. "It was in Child's seminar that I became acquainted with Arrhenius's idea of quantitative laws in biology, and I knew at once that here was a chemist whose methods I must master." He graduated from the University of Chicago with a B.S. degree in 1914 and continued in the Graduate School as a candidate for the Ph.D degree under Lillie.[12]

At the time, Lillie, influenced by new work of Paul Ehrlich, was pursuing the possibility that sexual behavior involved an underlying immunological mechanism. Lillie's hypothesis was that the fertile egg cell secreted a substance called *fertilizin* that caused the spermatozoa to swim toward the egg. Cohn was given the task of studying the effect of *fertilizin* on sperm cells. The experimental work was done during the summers of 1914 and 1915 at the Marine Biological Laboratory in Woods Hole where Lillie had a laboratory. For a young biology student, Cohn was thus given a unique introduction to biology, for the elite of the biological world gathered in Woods Hole every summer.

Cohn's experimental approach involved adding spermatozoa to a clump of sea urchin eggs suspended in a hanging drop of sea water on the under side of a microscope slide. Observed through the microscope, the fertilizing ability of the sperm was measured crudely by observing the number of eggs that had been fertilized and had reached the four cell stage of growth in a given period of time. The idea was to relate the proportion of fertilized eggs to the activity of the hypothetical *fertilizin*. The first summer at Woods Hole was devoted primarily to standardizing this crude biological system as a basis for more detailed investigations.

Once, while hunched over the microscope observing the spermatozoa actively swimming around the eggs, Cohn noticed that they had stopped moving. Surprised, he paused to consider what could have happened. However, when he once again peered into the microscope, the spermatozoa were again actively swimming! On further observation, he deduced that the observed effect was the result of a change in the sea water caused by exposure to the observer's expired breath. The carbon dioxide in his breath dissolved in the sea water and formed carbonic acid, thus acidifying the sea water and inhibiting the motility of the sperm. When the suspended drop was once again exposed to fresh air, the sperm acted normally. As he pondered the meaning of these observations, Cohn considered the variables within the drop of sea water. These included its content of carbon dioxide, oxygen, and several kinds of salt. He concluded that it seemed clear that before attempting to appraise the effect of *fertilizin*, he needed to know more about the chemistry of sea water. "I had reached this stage in my thinking when, in the summer of 1915, I chanced upon a remarkable book, *Fitness of the Environment* by Lawrence J. Henderson."[13]

Lawrence J. Henderson

"Fitness," a philosophical essay, was based on studies by Henderson of how the balance between acids and bases in living organisms was regu-

lated at or near neutrality. Henderson had discovered that an important chemical mechanism in sea water involved the so-called carbonic acid, bicarbonate buffer system. Henderson's central thesis was that Darwinian fitness is the compound of a mutual relationship between the organism and the environment. "Of this, fitness of the environment is quite as essential a component as the fitness which arises in the process of organic evolution; and in fundamental characteristics, the actual environment is the fittest possible abode of life."[14] *Fitness of the Environment* opened up exciting new horizons for Edwin Cohn, who discovered Henderson's earlier scientific publications in which he had considered the relationship between the living organism and its environment, particularly with respect to environmental water, oxygen and carbon dioxide. He immediately contacted Henderson at Harvard, explaining the problem he was investigating and his status as a beginning graduate student under Lillie, and asking if it would be possible for him to work out the sea water aspects of his problem with Henderson at Harvard.

Henderson was then in his late thirties. Having entered Harvard College at the age of sixteen, he quickly felt the challenge of his new environment. He rose to a level of independence of thought and action which opened doors for him. He became adept with mathematics. He enjoyed physics. From physical chemistry, he mastered the concepts of chemical equilibrium. He won a prize as an undergraduate for an essay on Arrhenius's theory of dissociation. After graduating from Harvard College in 1898, Henderson chose to prepare for an academic career in biological chemistry. However, while that academic discipline was then emerging in some European universities, few formal academic programs in biological chemistry existed in the United States. The only biological chemistry course offered at Harvard was in the Medical School curriculum. Henderson promptly applied and was admitted, graduating four years later with the M.D. degree. However, he never practiced medicine. Instead, he spent the next two years at the University of Strasbourg working under the great German biochemist Franz Hofmeister. On his return to Harvard he obtained a junior appointment as a lecturer. Later, in 1905, Henderson introduced and taught Chemistry 15, the first biochemistry course for undergraduates at Harvard College.[15]

Henderson's response to Cohn's inquiry was encouraging, explaining that it would first be necessary to find a laboratory where Cohn could perform experiments, there being no laboratory available at the Medical School for the work. Henderson turned to Theodore William Richards, then head of the Harvard Chemistry Department. The upshot was that Richards offered Henderson a laboratory bench where Cohn could work in Harvard's Wolcott Gibbs Laboratory in Cambridge. There, Richards treated Cohn as one of his own students, making available for Cohn's use an elegant instrument for the measurement of pH—a measure of acidity or alkalinity—which Richards had just acquired from Germany. Richards was Harvard's first Nobel Laureate, having won the prize in 1914 for his work on the measurement of the atomic weights of several chemical elements. Fortuitously, a Danish scientist, Sven Palitzsch, was also working in the laboratory with Richards on a related problem. Palitzsch had recently been a member of a Danish expedition to the Mediterranean where he had made some measurements of the pH of seawater. While there, Palitzsch offered Cohn some valuable advice in setting up the new experiments. By the end of the 1915 academic year, Cohn had painstakingly completed a careful study of the equilibrium between carbon dioxide and seawater. A paper entitled "The Equilibrium Between Acids and Bases in Sea Water" by Henderson and Cohn, replete with quantitative data, was accepted for publication in the *Proceedings of the National Academy of Sciences* in 1916.[16]

With respect to present day environmental issues, it is noteworthy that Henderson and Cohn predicted that there should be a net escape of carbon dioxide from the warm sea water to the atmosphere in the equatorial regions of the world, while a flow in the reverse direction should occur in the polar regions where carbon dioxide should be absorbed from the atmosphere into the cold ocean water. Among their findings was the discovery that the hydrogen ion concentration in sea water depends upon and can be quantitatively predicted from the carbon dioxide pressure in the atmosphere, the temperature, and the salinity of the water. Cohn was the beneficiary of Henderson's great talent as a teacher. "With graduate students, he practiced a policy that amounted almost to laissez-faire. An enrichment of their ideas was inescapable, as informally he applied the eager clarity of his mind to the innumerable intellectual and personal problems presented by curious young men."[17]

The following summer Cohn returned to Woods Hole to apply what he had learned about sea water to the behavior of sea urchin spermatozoa. He soon demonstrated that the activity of sperm cells that had been attributed to the postulated *fertilizin* was powerfully influenced by changes in the pH of sea water. Indeed, he was able to carry that finding even further. He found that the same ionic variables in sea water which caused the sperm cells of his experiment to behave in certain ways, also affected egg cells but in strikingly different ways. Years later, Cohn stated,

> The problem was to explain this biological specificity—why the same environmental factors (in sea water) made one entity work one way and another work in a diametrically different way. I became convinced that the answer could not be found by studying the responses of the cells. It was as though one expected to understand the automobile engine by watching its response to the fuel mixture, but without looking under the hood. I resolved to look under the hood, to concentrate my search on parts of the cell, and to examine their nature and properties in terms of physical chemistry.[18]

The most profound result of the sea urchin study was the realization that the explanation of biological specificity, as exemplified by the attraction of the egg for the spermatozoa, would not be found in the study of the responses of the cells. Too little was then known even to attack such a problem. Instead, Cohn resolved to concentrate his search on the parts of the cell, and to examine their nature and properties in terms of physical chemistry. Of all the known constituents of cells, the proteins were the most prevalent and most important. However, very little was then known about the chemistry of the proteins. He decided to focus on the study of proteins.

Lawrence Henderson approved, and with Cohn's consent set about making arrangements for Cohn to spend a year in New Haven studying proteins under Thomas B. Osborne at the Connecticut Agricultural Experiment Station there. One might perhaps ask why Osborne, the preeminent American protein chemist at the time, worked at an agricultural experiment station in Connecticut. The best answer: seeds are rich in proteins. In some seeds, the proteins are actually found in the crystalline state. Cohn moved to New Haven early in 1917 and spent an interesting spring in Osborne's laboratory. Two decades later, Cohn wrote:

> Osborne had taken up the difficult problem of protein purification and crystallization where Ritthausen, the discoverer of aspartic and glutamic acids, had left it. Convinced of their respectability as chemical individuals, he prepared beautifully crystalline products, largely from vegetable sources, on which he lavished a meticulous and loving care. He was always convinced of the stoichiometric nature of their combination with acids and bases; i.e. the laws of definite proportions and of the conservation of mass and energy to chemical activity.[19]

Cohn submitted his doctoral thesis to the University of Chicago in two parts: a long and detailed report of the sea urchin work entitled "Studies in the Physiology of Spermatozoa," and a copy of his published paper with Henderson. The sea urchin studies were published in the *Biological Bulletin* in 1918. In its first paragraph, Cohn wrote:

> The changes in the physiological condition of the spermatozoon from the time it is extruded from the genitalia of the male until it undergoes the transformation into a nucleus in the protoplasm of the egg are dependent in rate upon environmental conditions. The germ cells of most marine invertebrates are extruded into sea water, and fertilization of the egg by the sperm follows there. The environment, sea water—or sea water modified by the excretions of the egg or of the sperm—must therefore be studied in order to understand the variations in the physiological condition of spermatozoa that have often been observed.[20]

Years later, Cohn paid tribute to Thomas Hunt Morgan, a man whom he did not know at the time. Morgan, already world famous as a geneticist, had read Cohn's paper that reported the experiments with sea urchin spermatozoa and approved his chemical approach to the solution of cytological problems. In his widely read book *Experimental Embryology*, Morgan quoted Cohn's conclusions and even reproduced many of the tables and figures from Cohn's original paper. "This was one of the most gratifying bits of encouragement I ever received. When a young investigator has finished a painstaking series of experiments, published the results, and then waits through the silence, thinking his report has fallen on deaf ears, it is tremendously stimulating to find his work picked up by one of the great men of science to be quoted and praised. I was a cub when he did that, and I have always felt a keen indebtedness to Dr. Morgan."[21] Edwin Cohn was awarded the Ph.D. degree in Zoology and Chemistry at the 1917 University of Chicago commencement. This established his bona fides as a biologist. However, he never returned to the study of the sea urchin.

Marianne Brettauer

Edwin Cohn was not one to disregard the old "hard work and no play" adage. A handsome young man of strapping vigor, he was a social animal who loved parties and had a good eye for the fairer sex. Nor was he without opportunities for dalliances, particularly during the summers he spent at Woods Hole, and New York was far away. In the end, it was from the set of friends he knew well in New York that he courted Marianne Brettauer and ultimately proposed to her. Marianne, a beautiful young woman, was the only child of Dr. and Mrs. Josef Brettauer of New York City. She was a graduate of the Ethical Culture School in New York, after which she attended finishing school in Paris. On returning to New York she had worked as a volunteer in settlement work. She led an active social life among a wide circle of friends, many of whom Edwin also knew. Marianne's parents were well educated and socially prominent in New York's German Jewish society. Dr. Brettauer, a native of Austria, completed his medical education in Vienna. In 1889 he embarked on a projected trip around the world. While travelling out west, much of it on horseback, he was greatly taken by the United States. He decided to remain in the U.S. and settled in New York where he practiced obstetrics and gynecology, becoming Chief of the Gynecological and Obstetrical Service at Mt. Sinai Hospital.[22]

Marianne and Edwin were married on July 30, 1917 at the Brettauer summer home in Lake Placid. However, there was no time for a honeymoon, for the United States had declared war against Germany. Once again, Edwin Cohn's career path was shaped by Henderson. Cohn was commissioned as a Lieutenant in the U.S. Sanitary Corps and assigned back to Harvard to work once more under Professor

Richards at the Wolcott Gibbs Laboratory. Others there were L.J. Henderson, S. Burt Wolbach and W.O. Fenn. Their mission was to experiment with flours from grains other than wheat, which was in short supply at the time. A particular problem with breads made from other than white flour was the formation of "rope," a sticky glutinous formation caused by bacterial action that rendered the bread inedible. The Harvard group set to work to learn how to make bread. This was a serious scientific study. They published several scientific papers on the physical chemistry of bread making.[23] After the 1918 Armistice, Edwin Cohn, still in uniform, was briefly assigned to the Harriman Research Laboratory at Roosevelt Hospital in New York.

At the Harriman Laboratory, Cohn encountered A. Baird Hastings, who would later become Professor of Biological Chemistry and head of the department at the Harvard Medical School. Hastings was then working across the street in Columbia's old P&S building. In a small (6′ × 6′) darkroom at Harriman, Hastings had an apparatus for generating gaseous hydrogen for measuring pH with a hydrogen electrode. Cohn needed to use the same apparatus, and it was agreed that they would share it. Unfortunately, Hasting's source of hydrogen was "not fancy enough for Edwin." It was not as pure as the hydrogen that Cohn had used in Richard's laboratory. Cohn preferred a hydrogen generator, which used nickel electrodes in a bath of caustic sodium hydroxide (alkali). One morning when they were working together in the darkroom, Cohn turned on the electric current to the generator and there was a loud explosion. Staggering out of the darkroom, they both stripped off their clothes and stood under a nearby emergency shower. A drop of alkali hit Cohn in the eye, but caused no lasting injury. There was no other damage. The cause of the accident was not determined; however, Edwin Cohn was never known for manual dexterity. Hastings later commented that "this was a beginning of sorts of a very close association."[24]

National Research Council Fellowship

On being discharged from military service, and on the advice of Lawrence Henderson, Edwin Cohn set about applying for one of the new fellowships that had been established with Rockefeller Foundation support at the National Research Council. Then, in early 1919, the fellowship application having been submitted, he and Marianne set off for a honeymoon of several months in South America. In September 1919, with a National Research Council Fellowship in hand, Edwin and Marianne Cohn sailed for Europe to study under S.P.L. Sorensen at the Carlsberg Laboratory in Copenhagen. Sorensen, the director of the chemical department, had devoted his career to the study of amino acids, peptides, proteins and enzymes. In the course of studying the effects of acids and bases on proteins, he had discovered the advantages of using the negative logarithm of the hydrogen ion concentration to express acidity. Thus, a hydrogen ion concentration of 10^{-7} molar is expressed as pH 7. In this way a concentrated acid can have a pH as low as 1, and a base as high as 14.

In 1917, Sorensen had completed a landmark chemical investigation of egg albumin, the major protein of egg white. Starting by crystallizing the protein, he had then performed a series of chemical tests to characterize it, estimate its molecular weight, and determine its acid and base binding capacity. He demonstrated with careful measurements that crystallized egg albumin obeyed the classic laws of physical chemistry in its solubility behavior. "A classicist by temperament and training, he began the study of proteins with a consideration of Henry's law and of the phase rule and ended it by demonstrating that even such complex molecules as the proteins could be prepared and studied with such care as had previously not been lavished upon them."[25] This great work profoundly influenced Cohn and served as a paradigm for many of his own later investigations.

Having spent several months in Copenhagen, the Cohns moved on to Sweden to work with Svante Arrhenius, a chemical "giant" who had startled the scientific world by predicting that free sodium ions exist in a solution of sodium chloride, a theory that was subsequently proved. Arrhenius won the 1903 Nobel Prize for developing the theory of dissociation in dilute solutions of salts in water. He was then Director of the Nobel Institute of Physical Chemistry in Stockholm. From Stockholm the Cohns went on to Cambridge, England, where they visited with W.B. Hardy, Joseph Barcroft and other British scientists. While in Cambridge, Barcroft showed Cohn a list of long-range projects that had been planned by the British Medical Research Council. On that list there was a project aimed at using X-ray diffraction to study the structure of the protein hemoglobin. Cohn never forgot this. It was not until much later, in 1948, when the circumstances were favorable, that he initiated X-ray diffraction studies in his own laboratory in Boston.[26]

While the Cohns were in Europe, Henderson, who was still an Associate Professor, was being wooed by Johns Hopkins University. When offered an attractive appointment as a full professor by Hopkins, Henderson informed David Edsall, the Harvard Medical School Dean, who asked what would be needed to persuade Henderson to stay at Harvard. Henderson, then immersed in a study of the equilibrium between oxygen and carbon dioxide in the blood, asked for a laboratory of his own at the Medical School and for an allocation of funds so that Edwin Cohn could rejoin him on his return to Boston. Dean Edsall promptly agreed. Henderson was promoted to Professor and spent the remainder of his career at Harvard.

In September 1920, Cohn's postdoctoral fellowship year came to an end as Edwin and Marianne Cohn attended their first International Physiological Congress in Paris. Although Edwin did not ordinarily identify himself as a physiologist, his primary interest as a scientist lay in the study of the relationships between the structure of proteins and their functions, which made him a physiologist. In the early 1920s, the triennial physiological congresses, usually held in Europe, were the "in" meetings of world-class scientists at the time. To the Cohns' great delight, Lawrence Henderson had come over for the Congress. There was much to share with him. But the big news came from Henderson, who told Edwin and Marianne about his promotion to Professor, and further, about Dean David Edsall's plan to establish a small research laboratory of physical chemistry for Henderson. To cap it all off, Henderson invited Cohn to return to Boston and work in the new laboratory with him.

2 Early Years at Harvard

Edwin Cohn took up his new post in the Laboratory of Physical Chemistry at the Harvard Medical School at the start of the 1920–1921 academic year. Twenty-four years later, George W. Gray, in one of his confidential reports to the Trustees of the Rockefeller Foundation, wrote:

> to create such a department in a medical school was a somewhat unusual and imaginative thing to do. This laboratory was to be free to explore those little-developed aspects of physical chemistry which were considered fundamental for the understanding of biological states and processes. It was quite clear that these researches would not be, at the moment nor for some long time to come, very closely related to the practical problems of sickness and health. But the Harvard authorities had the courage and the wisdom to back a patient, basic, long-range enterprise.[1]

In the broadest sense, the mission of the new laboratory was to introduce the study of proteins at Harvard, and thereby, into American science. To support the new laboratory, Dean Edsall used funds from the De Lamar bequest to the Harvard Medical School that had become available at the end of World War I.[2] In keeping with the terms of the bequest, Cohn and his colleagues in the new laboratory were freed from teaching responsibilities in the Medical School. While providing unusual freedom to those who worked there, it set them somewhat apart from other colleagues who carried teaching and clinical responsibilities. This separation of Cohn from academic colleagues in the school was to lead to difficulties, some small, some larger, in the years ahead. Although Lawrence Henderson was the nominal head of the new laboratory, he gave Cohn a remarkably free hand. Henderson had his own well developed interests to pursue. At the time, he was beginning a series of theoretical studies of the relationships between oxygen and carbon dioxide in the blood. Indeed, shortly afterwards, he was appointed Harvard Exchange Professor for France and spent much of the 1921–1922 academic year in Paris.

While Henderson was in Paris, Cohn and Scatchard met unexpectedly in Amherst. Cohn had driven out to visit some of his former teachers. Scatchard had earned his Ph.D. degree from Columbia University and had just taken a post as instructor in the Amherst Chemistry Department at the invitation of Alexander Meikeljohn, the new President of Amherst, who was launching an effort at educational reform. Years later, in an Edwin J. Cohn Memorial Lecture at a Protein Foundation meeting in Boston, Scatchard commented on that visit with Cohn.

> He then seemed to me to epitomize the outsider's picture of a Harvard man. I learned later that this is usually the result of the insecurity of a spoiled child being thrust into the harsh world without a protector. Cohn's world was very harsh, even though he did have the support of the top men: Richards, Henderson, Cannon and David Edsall. He had just gone to the Medical School in the new Department of Physical Chemistry. There were few Jews in the Medical School especially in the lower ratings, there were fewer Ph.D.'s, and not many of the Faculty believed that physical chemistry had any place in the Medical School.[3]

Scatchard later described his experience at Amherst: "They were exciting years for a young teacher there. Meikeljohn believed that to teach well, one must be learning. He also believed that chemistry and physics are so simple that to learn about them, a man must do research. He brought me to Amherst to prove that research in the physical sciences could be done in a small liberal arts college with the right atmosphere." However, by 1923, sharp disagreements between Meikeljohn and the Amherst trustees came to a head and Meikeljohn resigned. So did Scatchard, who was an ardent Meikeljohn supporter.

With a National Research Council Fellowship, he moved to the MIT Chemistry Department where he had a long and distinguished scientific career.[4]

Shortly thereafter, Cohn and Scatchard drove out to Amherst to attend a reception for the visiting Danish physicist, Niels Bohr, who had just won the Nobel Prize in Physics. This afforded them an opportunity to exchange views and study each other more carefully. In commenting later about this meeting, Scatchard stated, "I was surprised to find that he thought more like I did than anybody I had ever met. His thoughts were more developed and polished than mine, but there was little other difference. From that day we became close friends and frequent collaborators, although we never published a paper together."[5]

Cohn's efforts in those early years were followed with special interest by Theodore William Richards, David Edsall and Walter B. Cannon. Richards was Lawrence Henderson's brother-in-law, and father-in-law of James Bryant Conant, then Professor of Chemistry who later became President of Harvard University.[6] Richards could be counted on to provide strong support for Cohn at the highest levels within the University. From 1910 until 1917, before he became Dean of the Harvard Medical School, David Edsall was the Jackson Professor of Clinical Medicine and Chief of the East Medical Service at the Massachusetts General Hospital. He had recognized the mounting importance of the basic medical sciences for the educational mission of the medical school. Then, and during his tenure as Dean, he launched a number of academic ventures in which he identified able young men and backed them with the needed funding to open up new fields. It was David Edsall who sent Paul Dudley White off to England for advanced training in cardiology, and to establish a program in Boston on his return.[7] In this way, over time, Edsall developed the nucleus of a full-time faculty in the medical school.

Walter B. Cannon, Professor and head of the Medical School Physiology Department, provided the space and the organizational linkage of the new Henderson laboratory to the medical school administration. Cannon had only recently returned to Boston from a tour of duty as a medical officer studying battlefield shock in the front lines in France. He could undoubtedly appreciate more than most of his colleagues in the faculty the wisdom behind Harvard's investment in this new venture to study proteins. Cohn's early scientific papers carried the legend: "From the Department of Physical Chemistry in the Laboratories of Physiology, Harvard Medical School, Boston."[8]

Edwin Cohn had a clear vision of his purpose in science as he set out on his career. From the outset, he lived his daily life according to a deeply held set of standards that he projected to those who worked with him. Clear evidence of those standards can be found in the prolific printed record of his work. However, it was not until 1945 that he found the right forum in which to set down a concise, overarching statement of his philosophy as a scientist. The occasion was a "Laity Lecture" at the New York Academy of Medicine entitled, "Research in the Medical Sciences." Drawing on the sixteenth century examples of Gallileo at Pisa and Vesalius at Padua, cities where universities existed and where it could be supposed that each university wished to become the center of all learning, he argued instead that

> Universities are merely the seats of learning, where past knowledge is accumulated and transmitted. New knowledge is created by men who build new schools of thought which transcend past knowledge. New techniques and new traditions follow and are disseminated from these focal centers. A single center in any one field and at any one time may suffice to open a new era in science, as in art. Those who flock to this center and are trained there carry the new insight and the new techniques to the ends of the earth. The dissemination is not institutional but organic. The conditions for growth in institutions and societies may be barren or fertile with respect to the development of new knowledge in any given field. Institutions can thus inhibit growth or make possible the fullness of a development. Institutions cannot, however, control the rate or the direction of a development. The growth of new knowledge born of the insight of the investigator can follow only directions that are open to it on the basis of contemporary hypotheses and techniques. The importance of the problems that demand solution may be recognized by generations of investigators who nonetheless can make no substantial contribution to them. The pattern of advancing knowledge, usually foreseen only by the creative scientist himself, is readily recognizable in retrospect. The direction of the advance is implicit in the scientific development and in its relation to the scientific attainments of the times.[9]

While it must be conceded that the Edwin Cohn of 1920 might not have phrased the above statement so ably as he did in 1945, it is nonetheless clear that in 1920 Cohn visualized his laboratory as a new center and himself as the individual with vision around which new knowledge would be created and new schools of thought would be formed.

The Cohns chose to reside in Cambridge where Edwin had lived during the sea water studies with Henderson in the Wolcott Gibbs Laboratory. Cambridge was also the home of Henderson and a number of other associates from the earlier days. Although Cohn's laboratory was five miles away at the Medical School in Boston, Edwin and Marianne were convinced that they would be happiest in the environs of the university near other young members of the university family. Although unschooled in science, and unused to the Cambridge academic environment in which she found herself, Marianne Cohn took courses in shorthand and typing and participated eagerly as Edwin's personal assistant. She became proficient in taking down his letters and keeping track of his engagements. Her fluency in German and French was particularly useful in those early days. She came to the laboratory frequently and even travelled with him before the children were born. Marianne quickly became involved with a circle of friends in the activities of other young faculty wives in Cambridge. She was active in the Cambridge Chapter of the League of Women Voters. The Cohn's first child, Edwin Joseph Cohn, Jr., known as Ed, was born in 1921. Their second son, Alfred Brettauer Cohn, called Fred, arrived in 1924.[10]

Early Knowledge About Proteins

Although scientific knowledge about proteins has advanced by leaps and bounds since the end of World War II, surprisingly little was known about proteins at the beginning of the twentieth century. Proteins are ubiquitous in nature. They comprise more than half the dry weight of tissues. Vegetable and animal proteins were long known to be extractable from seeds and tissues. Gliadin, a protein derived from rye or wheat, was important in bread making. Casein from milk was long known for its nutrient value. Fibrinogen, found in blood plasma, was known to be the precursor of the fibrin clot. Hemoglobin, easily recognized by its color, had long roused intense interest. Other proteins serve structural functions. Still others are important in muscular contraction.[11] On the occasion of accepting the Theodore William Richards Medal in Cambridge in 1948, Cohn reflected that

> Protein chemistry has gone through several phases. Unitary theories regarding the composition, the structural pattern and the molecular sizes of proteins have had great attraction for a succession of investigators, each of whom felt that a unifying principle, within the compass of contemporary powers of observation, obviously existed. For Liebig (1803–1873), proteins were alike because their elementary composition was alike. New knowledge has nearly always proved that the deceptively simple unitary principle depended upon observations made with less perfect tools than subsequently became available. The introduction by Dumas (1800–1884) of a better method for analyzing proteins for their nitrogen content demonstrated, over a century ago, that all proteins were not alike, even in this elementary respect. The unitary theories of the latter nineteenth and of the twentieth century have thus depended upon other properties of the protein.[12]

Cohn launched two related tasks: he set out to prepare a scientific review article summarizing current knowledge about the proteins; and he initiated experiments on proteins in his new laboratory. Preparing a review was an appropriate undertaking for anyone aspiring to a place of leadership in a scientific field. In the process of preparing this review, he read more than 360 articles about proteins that had been published in scientific journals since the early 1800s. In organizing the huge body of material to be covered, Cohn exercised his privilege as author by framing the review in terms of his own concepts of the state of knowledge about proteins at the time. The review, "The Physical Chemistry of the Proteins," was published in the American journal *Physiological Reviews* in 1925.[13] Although it is now completely outdated, it serves as a milestone marking the state of knowledge about proteins at the time.

Two distinct theories of protein structure were under intense discussion as Cohn took up his post at the Harvard Medical School in the early fall of 1920. On the one hand there was the colloid theory, proposed in 1876 by Thomas Graham in Edinburgh. Graham and some distinguished chemists at the time believed that proteins were small molecules with molecular weights in the range of 1,000. On the other hand, another group of chemists held that proteins were classical chemical substances, albeit with very high molecular weights, that behaved as typical chemical compounds and obeyed the laws of chemistry. The lack of agreement on this important issue is attributable in large part to the inadequacy of then-existing methods for determining the characteristics of proteins. However, there was a perceptible movement in support of the latter theory. Sorensen, with whom Cohn had worked in Denmark, had begun to use measurements of osmotic pressures of protein solutions to estimate molecular weights. Adair, in England, had startled students of proteins by suggesting that the molecular weight of hemoglobin was approximately 67,000, and that the molecule contained four atoms of iron. Another Cohn mentor, T. B. Osborne, paying no attention to the conflicting schools of thought, continued to pursue chemical studies of plant proteins as chemical entities.

The first amino acid, leucine, was found in a protein digest by the French investigator Proust in 1819. Eleven more amino acids were identified between 1820 and 1889 by investigators in different countries. However, it was not until 1902 that the German chemist, Emil Fischer, realized that the amino acids represented a new family of chemical compounds and demonstrated that amino acids combine with each other to form polypeptides. Fischer thus showed that amino acids are the building blocks of proteins, a discovery which won him the 1902 Nobel prize in chemistry.[14] The last important amino acid, threonine, was discovered in 1935 by W.C. Rose at the University of Illinois.[15] Today, twenty common amino acids are recognized. As a group, they are all closely related structurally. In his review, Cohn stated that

> the demonstration by Fischer of the manner in which amino acids combine with each other has led to the most generally accepted notion of the structure of protein. According to this notion, when two amino acids combine with each other to form a dipeptid [sic], one amino group and one carboxyl group remain free. This dipeptid may combine with still other amino acids or dipeptids to form polypeptids, and this process may be repeated again and again. The polypeptid linkage thus results in molecules consisting of combined amino acids. The free terminal acidic and basic groups of which polypeptides are still possessed enable them to combine with acids

> and bases, or with each other to form the vast molecules upon which the . . . properties of the proteins depend.[16]

Further consequences arose from the existence of individual amino acids in widely differing proportions within proteins, he went on. The chemical nature of the amino acid side chains adds great complexity to protein molecules. Some side chains are nonpolar and aliphatic; they tend to be lipophilic, or fat loving. Charged acidic and basic groups favor solubility in water. The collective influences of these side chain R groups affect the physical properties of proteins, such as their solubility and the number of negative and positive charges on the protein molecule. They also influence the interactions of proteins with other molecules. In this way, the basic polypeptide structure of proteins leads to chemical specificity of proteins, i.e. one protein can be an enzyme, another an antibody. In confirmation of possible complexities in protein structure that were visualized by Cohn at the time, he stated that "the results that are at present available indicate that on the whole proteins that are rich in basic amino acids have high acid combining capacities while those that are poor in this respect have low acid combining capacities . . . With few exceptions, the proteins thus far investigated contain a sufficient concentration of basic amino acids to account for their acid combining capacities."[17]

In 1861, Thomas Graham had observed that proteins cannot pass through the pores of membranes that are permeable to salts.[18] From that observation, he proposed a new classification system for separating the universe of chemical substances into what he saw as two great classes: crystalloids, substances that can be crystallized; and colloids, substances that were unable to penetrate certain membranes, that diffused very slowly and could not be crystallized. Unfortunately, this set off a scientific controversy that persisted well into the twentieth century. Cohn pointed out in his 1925 review that

> the first two of these definitions apply to most proteins, and the last to a few. Egg and serum albumin may be crystallized . . . Certain of the hemocyanins have also been crystallized. Hemoglobin, Bence-Jones protein and many of the vegetable globulins may be crystallized. Proteins thus vary in their crystallizability. Because of the large size of their molecules, proteins have been considered as colloids,

> and laws that had been derived for aggregates of identical, smaller molecules, usually of inorganic origin, have been applied to them.

Nevertheless, he credited Graham for observing that proteins had the ability to scatter light, one of the earliest indications of the large dimensions of protein molecules.[19]

In his 1925 review, Cohn also dealt with the determination of molecular weight by measurement of osmotic pressure, with the estimation of molecular dimensions based on ultrafiltration, and with the measurement of minimal molecular weights estimated from combining weights. One of the earliest estimates of the minimum molecular weight of a protein was provided by the German chemist Hüfner in 1894, employing two classical chemical measurements. He found the molecular weight of hemoglobin that combined with one equivalent weight of gaseous carbon monoxide to be 16,700. Separately, he estimated the minimum molecular weight based on measurements of the iron content of hemoglobin and found it to be 16,619.[20] The agreement between these measures is astonishing. When, much later, it was found that hemoglobin contains 4 atoms of iron per molecule, this meant that the true molecular weight of hemoglobin was 66,800 (4 × 16,700). Cohn's review also included a table of minimum molecular weights measured in his own laboratory as well as osmotic molecular weights and probable molecular weights of nine proteins.

While preparing his long review, Cohn broke the tedium by starting a series of scientific studies in the laboratory. His first published scientific paper, entitled "Studies in the Physical Chemistry of Proteins. I. The Solubility of Certain Proteins at Their Isoelectric Points," was published in the *Journal of General Physiology* in 1922.[21] Cohn was the sole author. As it later turned out, this paper was a rarity in that respect, since almost all of his subsequent publications included one or more coauthors who had contributed substantially to the conduct of the scientific work. However, even in this first case, he acknowledged the aid of Jessie L. Hendry, his first laboratory assistant. The starting point of this study was W. B. Hardy's 1899 observation that "under the influence of a constant electrical current the particles of proteid [sic] in a boiled solution of egg white move with the negative stream if the reaction of the fluid is alkaline; with the positive stream if the reaction is acid." In citing his purpose in the new study, Cohn told that

> As a result of the unsatisfactory nature of the experimental methods that have been employed in determining the isoelectric point of the slightly soluble proteins, and in view of the manifest importance and significance of the isoelectric point for the interpretation of other aspects of the physical chemistry of the proteins, it seems necessary to base these studies upon better criteria of the identity of the proteins under investigation.

On the basis of theoretical considerations summarized in the paper, Cohn went on to show that solubility is constant when the protein under study is uncombined with acid or base. He then found that solubility of a protein at its isoelectric point was independent of the amount of protein in the system.

George Scatchard, then still a bachelor, was a frequent visitor to the Cohn's home. Once at a picnic, Marianne expressed the hope that George would not mind if their son called him "Georgie" rather than Mr. Scatchard because Scatchard was difficult for the boy to pronounce. Although Edwin Cohn joined the children and Marianne in calling Scatchard "Georgie," he never did so publicly. On the countless occasions when George Scatchard attended Cohn's meetings at the Medical School, their interactions, while friendly, tended to be formal. Indeed, the deep friendship between the two men was not generally known to Cohn's associates. Within the laboratory, Cohn was addressed as "Dr. Cohn;" only the senior associates addressed him as Edwin, and then only in private meetings in his office. Cohn generally addressed his close associates by their first names. However, he tended to be formal in addressing younger associates. Scatchard once described Cohn as two people, one of whom occupied the center of the stage, while the other sat in the gallery and watched the first. Cohn was proud of the number of balls he could keep in the air at once in his juggling act. Scatchard said that he used to sit in the gallery with Marianne and Edwin, where they made Edwin laugh at the Edwin on the stage.[22]

Dipolar Ions

Somewhat later, two scientific advances of great importance to the understanding of the physical chemical properties of proteins were reported in Europe. In the first of these, the Danish chemist, Nils Bjerrum provided convincing evidence that amino acids existed in aqueous solution as ions rather than as uncharged molecules.[23] In this view, the carboxyl group of an amino acid loses a proton and is negatively charged while the amino group gains a proton and is positively charged. The result is a dipole, a pair of equal and opposite electric charges separated by a small distance in the same molecule. Bjerrum called them zwitterions, a name that gave way to the term dipolar ion.

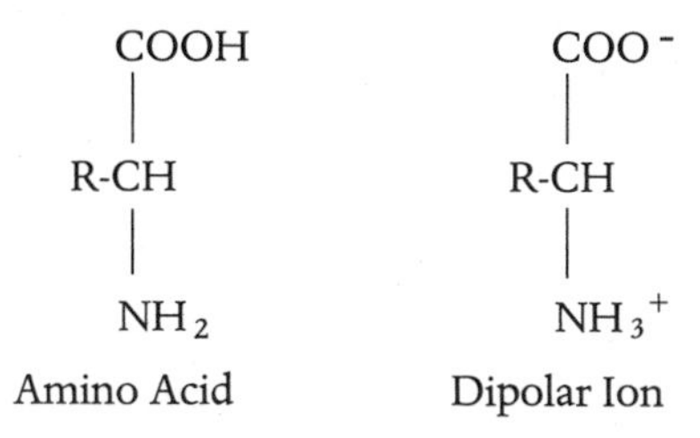

Years later, George Scatchard explained how the term dipolar ion came into use. Professor Keyes, then the head of the Chemistry Department at MIT, objected to the term zwitterion because it was half German and half Greek. Scatchard disliked the connotation of bastard in zwitter. Hybrid ion was ruled out because hybrid to the Greeks meant monster, like the Minotaur. Cohn lightheartedly suggested mule-ion because it would not move in an electric field but would stand still and kick both ways. However, mule-ion was half Latin and half Greek. Scatchard consulted a Greek scholar. The Greek for mule is hemionos, so the ion should be hemion-ion which easily splits in the wrong place to hemi-onion. Otto Folin, head of the Department of Biological Chemistry, took pleasure from the literal translation of hemionos, which is "half-ass." It confirmed his concept of biophysical chemistry. Of course, the worst trouble is that it is not an ion.[24] What is unique about dipolar ions is that they are isomers of the corresponding uncharged molecules; both isomers are electrically neutral. The dipolar ion was an entirely new concept. When applied to polypeptides and proteins it stimulated Cohn and his associates to focus their efforts toward identifying the groups in proteins which could carry electrical charges.

The second important advance was the development of a new theory by Peter Debye, a Dutch physicist then working in Zürich, concerning the electrostatic interactions of ions in solution. Stated simply, Debye and his colleague Hückel proposed that even simple ions such as sodium and chloride ions interact in complex fashion with other ions and with each other in solution.[25] It was immediately obvious to Cohn and his associates that proteins might also be susceptible to electrostatic forces in their interactions with salt ions, dipolar ions and with other proteins in solution. George Scatchard introduced Debye to Cohn one day in 1925. They had some difficulty communicating with each other. Debye didn't know much about proteins, while Cohn was not au courant about Debye's new theory. When Cohn tried to explain what a protein was, Debye interrupted "Never mind that. What's its dielectric constant?" That question took years to answer. Debye later moved to Cornell University and was an occasional visitor to Cohn's laboratory; he won the Nobel Prize in Chemistry in 1936.

Taken together, these two developments opened up a new vista, one in which molecular interactions could be seen to influence the behavior and the physiological actions of proteins in biological environments. They suggested a new range of effects of neutral salts and of non-electrolytes on the solubility of proteins. Even more significant, they opened up the possibility of entirely new effects of dipolar ions on physiological systems. Cohn suddenly found that problems which were expected to take five to ten years to answer could be answered in a few weeks. He was convinced that the functional groups on proteins which combine with acids or bases were the basic and acidic groups on the side chains of certain amino acids. Indeed, experimental techniques had advanced to the point where quantitative statements could be made on this matter.

The first research associate to join Cohn in the new laboratory at Harvard was Ronald M. Ferry. A graduate of Harvard College, Ferry received his M.D. degree in 1916 from Columbia University and finished an internship at Presbyterian Hospital in New York. He arrived at Harvard Medical School in 1920 with a National Research Council Fellowship. His first work in Cohn's laboratory was a study of the preparation of hemoglobin, published in 1923.[26] This was followed by a study of the equilibrium between hemoglobin and oxygen. Ferry became Master of John Winthrop House at Harvard in 1930. His association with Cohn continued until 1943.

Arda A. Green came to work with Cohn in 1924. She left after a year, but returned in 1927 and stayed for another five years. Although trained as a pediatrician, Green quickly became expert at purifying proteins by the methods then available. Her best study was an elegant exploration of the influence of pH on the binding of oxygen by horse hemoglobin, done with Ferry.[27] The hemoglobin used for those studies came from the Massachusetts State Antitoxin and Vaccine Laboratory in Jamaica Plain. During her stay at Harvard, Green published eleven papers. In 1932, she moved on to the Cleveland Clinic. Later, when Carl and Gerti Cori were having problems crystallizing glycogen phosphorylase at Washington University, they brought her to St. Louis, where she taught them what she knew about the crystallization of proteins. She stayed with them for several years. The Coris shared the 1947 Nobel Prize in Chemistry.

Following the publication of his 1925 review article, Cohn began to receive expressions of interest from young scientists seeking training in protein chemistry under his tutelage. Dealing with these requests became a matter of high priority involving communication with the candidates and their sponsors, as well as searching for sources of financial support. In most instances, the applicants had fellowships in hand and needed only Cohn's invitation to come to Boston. In others, he had to help in obtaining support, even to the point of committing funds which he controlled himself. The usual stay for these fellows was two years. However, for those who were particularly productive in research, there was the temptation to keep them on, but this almost always meant the assumption by Cohn of responsibility to find funding for them. Over the years, a number of these postdoctoral fellows were appointed as Tutors in Biochemical Sciences at Harvard College, which provided them with a stipend. The flow of these bright young men and women through the laboratory was critical to its success and yielded some key collaborators who remained with Cohn for long periods and played important roles in the work of the laboratory. Over 110 young investigators worked in the laboratory between 1920 and 1950.

An Important Study With James Conant

Although Cohn had reported minimum molecular weights of proteins between 5,000 and 33,800 in his review article, the concept that proteins might have high molecular weights was not widely accepted. In 1925, Conant returned from a visit to Germany where he had been shown startling new evidence purporting to show that the molecular weights of certain proteins were quite low. This conclusion was based on the results of measurements of the depressing effect of dissolved protein on the freezing point of a solvent, the principle on which the modern use of antifreeze is based. To be sure, the solvent that had been used was somewhat unusual. Nevertheless, the freezing point depression method was a classical physical chemical way of measuring molecular weight. If this new evidence were to be confirmed, it would reopen the old dispute between those who held that proteins were merely colloidal aggregates and those who believed that they were true chemical molecules of great molecular weight. There was no alternative. The German experiments would have to be repeated. The Germans had used liquid phenol as the solvent in their experiments and had reported that proteins dissolved in this solvent had molecular weights between 200 and 600.

Cohn and Conant set to work to measure the molecular weight of zein, a highly purified protein obtained from corn. From the outset, they encountered experimental difficulties; their measurements were not reproducible. On some days the molecular weights were low, on other days they were higher. George Scatchard, who followed the progress of the work on a day-by-day basis, was probably the one who cleared up the puzzle. Suspecting that there was a trace of water in the protein preparations, he suggested a way to eliminate the effect of the water impurity. When this was done, the variability in the experimental measurements disappeared and the molecular weight of zein proved to be uniformly greater than 10,000, which was the upper limit of sensitivity of the method they used. The results of this collaboration, published in 1926, were influential in convincing others that proteins are true macromolecules and not poorly defined aggregates.[28]

In those early years, Cohn and his young group studied a variety of plant and animal proteins which were readily available to them. These included casein from milk, albumin and globulin from animal blood and various tissue proteins. Their objective was to use scientific measurements available at the time to elucidate how protein solubility was influenced by the addition of salt or acids or bases, and to estimate the size of the molecules. Cohn was particularly interested in developing a scientific rationale for understanding the nature of the electrical charges on proteins and determining the behavior of the proteins in solution. These early studies also had an important practical aspect, for they contributed to the development of methods for separating and purifying individual proteins. For example, William T. Salter, a fourth year medical student, working in the laboratory in his spare time, isolated the first crude protein—a globulin—from muscle in 1925. Salter later became Professor of Pharmacology at Yale.

During this period Cohn learned that Theodor Svedberg, a Swedish physical chemist at Uppsala University, was developing a new type of centrifuge called an ultracentrifuge. Svedberg's machine was capable of developing centrifugal fields of such magnitude that dissolved protein molecules would actually be sedimented from solution. By means of a sophisticated optical system, Svedberg could even measure the rate at which the protein sedimented in the centrifugal field. Although the new ultracentrifuge was quite complex, and the calculations to arrive at the experimental results time consuming, it nevertheless provided the only direct means for determining the molecular weight of proteins. Svedberg's early measurements of a few proteins, published in 1926, quickly confirmed that proteins were indeed macromolecules.[29] Cohn resolved to acquire an ultracentrifuge for the laboratory at Harvard. However, in discussions with Svedberg at the 1926 International Physiological Congress in Stockholm, it became apparent that, due to the complexity of the new instrument, it would be some time before another one could be built and installed in Boston.

John T. Edsall, a son of David Edsall, began his association with Edwin Cohn in 1926. After graduating from Harvard College, Edsall entered Harvard Medical School. He interrupted his medical studies to spend two years at Cambridge University in England. There he began an investigation of the effect of oxygen and pH on the activity of heart muscle in turtles. On returning to Harvard in 1926 as a third year medical student, Albert Redfield, his mentor in the Physiology Department, suggested that he talk to Cohn, commenting that "I think the most neglected part of the whole muscle problem lies in the muscle proteins. Edwin Cohn has started some work on muscle proteins. Why don't you go and work with him?"[30] Working in Cohn's laboratory two afternoons a week, Edsall succeeded in isolating the globulin from muscle that was later named myosin. After Edsall received his M.D. degree in 1928, Cohn invited him to stay on and continue his work on muscle. At that time, Alexander von Muralt, a young Swiss physiologist, arrived in Boston hoping to work with Henderson. Von Muralt had been working on the double refraction of light beams by muscle fibres. Since Henderson was not available at the time, Cohn suggested that von Muralt apply his double refraction method to the study of Edsall's myosin. In this way, von Muralt brought the new laboratory face-to-face with the problem of molecular shape, revealing that although many proteins were globular; others were long and filamentous, even in solution. The collaboration between Edsall and von Muralt resulted in a string of important papers on the physical chemistry of muscle globulin that were published in 1929 and 1930.

Cohn Assumes Control of his own Laboratory

In 1927, Lawrence Henderson moved from the Harvard Medical School to the Harvard Business School, thus severing his formal connection with Cohn's new laboratory in Building C. At the Business School, with the support of the Rockefeller Foundation, he established the Fatigue Laboratory. The circumstances which led up to this move were described by Dean Donham of the Business School.[31]

> From about 1922 it was my good fortune to know Henderson well. As I came to appreciate the encyclopedic and imaginative qualities of his mind and his combination of learning with the highest degree of intellectual honesty, I fell into the habit of discussing with him the wider implications of the task facing a school of administration. Up to that time, his intellectual interests had been focussed on science. In 1924–25 his interests in our problems became aroused, and he acquired an understanding of the dangers to organized society which arise from the specialized emphasis of the modern world on technological advance and the relative neglect by men of affairs of human

> problems which arise from such an advance . . . He moved his office here where he could be in continuous contact with, and collaborate in, our work in human problems. This work was important, happy and mutually stimulating.

Despite Henderson's move, his friendship with Cohn remained intact. Their homes in Cambridge were within walking distance of each other. It became Cohn's custom to walk over after dinner occasionally for a visit with Henderson. In this way, the two friends continued the mentor-protege relationship begun while Cohn was a graduate student until Henderson's death in 1942.

In November 1927, the Rockefeller General Education Board made a grant to Harvard Medical School for the development of physiology and physical chemistry. The grant provided $22,500 per year for each of three years. In each year, 60% went to Dr. Walter Cannon in the physiology department, and 40% to Cohn's new physical chemistry laboratory. The annual allocation of medical school funds provided to Edwin Cohn was $15,000 at the time.

3 Characterizing the Active Principle in Pernicious Anemia

On May 4, 1926, at the annual meeting of the Association of American Physicians in Atlantic City, Drs. George Richards Minot and William P. Murphy of Boston announced the discovery of a cure for pernicious anemia, previously considered an intractable, progressive and invariably fatal disease marked by a decrease in the number of circulating red blood cells, increasing weakness and pallor.[1] The treatment announced by Minot and Murphy involved feeding pernicious anemia patients a daily diet of from 120 to 240 grams of cooked calf or beef liver. "They had fed liver intensively and daily to forty-five patients. In many of these patients, symptomatic improvement was obvious within a week . . . Within about sixty days the red blood cell counts had risen on average from low levels to approximately normal."[2]

George Minot, a 1912 graduate of the Harvard Medical School, was a member of a famous Massachusetts medical family. His great grandfather, James Jackson, had been Harvard's second Professor of Medicine in the early nineteenth century. A great uncle, Francis Minot had been Hersey Professor of Theory and Practice of Physic at Harvard. His father, James Jackson Minot, was a distinguished Boston physician. At the time of the announcement in Atlantic City, Minot was Physician at the Peter Bent Brigham Hospital and Physician and Administrator of the Collis P. Huntington Memorial Hospital. Murphy, a 1920 graduate of the Harvard Medical School, held the appointment as Physician at the Peter Bent Brigham Hospital.

Since 1924, there had been occasional anecdotal reports that liver had a beneficial effect in pernicious anemia. However, they had not been convincing. In 1925, George M. Whipple, a pathologist at the University of Rochester, had reported that a liver diet was effective in regenerating hemoglobin in dogs that had been rendered anemic by bleeding.[3] Although the hemorrhagic anemia in Whipple's dogs was distinctly different from pernicious anemia, George Minot thought there might be something of value in it for pernicious anemia. He also had the wisdom to design a rigidly controlled clinical trial in which the dose of liver was deliberately set at a high level. Further, since a daily diet of liver, whether raw or cooked, was decidedly unpalatable for most patients, Minot insisted that the patients in the clinical trial be closely monitored to insure that they ate their liver every day. In the course of their study, Murphy had observed that immature red blood cells, called reticulocytes, appeared in the patient's blood soon after the start of treatment. This provided an early harbinger of a positive response by the patient's bone marrow to the liver diet well before other measures of response were observable.

In preparation for his Atlantic City announcement, George Minot had taken precautions to avoid prior disclosure of the discovery to his colleagues in Boston. Only six days before the Atlantic City meeting, he had given a talk on pernicious anemia at the Boston Medical Library without saying a word about the effect of diet. On the other hand, in anticipation of a rush of requests from practicing physicians, he and Murphy had written an article that provided a complete description of the recommended diet and dealt in some detail with the problem of getting sick patients to comply with it. That article was published in the August 14, 1926 issue of the *Journal of the American Medical Association.*[4]

Isolation and purification of the active principle in liver that cured pernicious anemia was the obvious next step. George Minot turned to Edwin Cohn to see if he would undertake the task. If the active principle proved to be a protein, its successful isolation would bring favorable attention to the Department of Physical Chemistry. Indeed, the successful isolation of the active principle in pernicious anemia would be an achievement fully as significant as the successful isolation of insulin from the pancreas by Banting and Best only five years before. Banting had won the Nobel Prize in Medicine in 1923 for that discovery and was knighted by the King of England. Both Minot and Cohn were familiar with the search for insulin. Indeed, Minot, a newly diagnosed diabetic and patient of Elliott P. Joslin, had been one of the first patients to be treated with insulin.

For Cohn several factors deserved consideration. If the active principle was found to be a protein, he would be in a favorable position in the pursuit of the goal. On the other hand, if the active principle proved not to be a protein, he might face intense competition from other chemists more experienced in the isolation of non-protein substances. Moreover, the attempted isolation of the active principle would require that reliable assays be available in adequate numbers. The clinical assay used by Minot and Murphy in the original work was laborious. It could only be carried out in patients with the disease. It meant feeding or injecting products to be tested to patients with the disease and following their clinical course for days or even weeks. The specificity of the test was unknown.

A more serious question concerned the effect that undertaking a completely new project would have on his own program. At the time, there were only four associates in the laboratory: Ronald Ferry, John Edsall, Francis F. Heyroth, a National Research Council fellow on a one year stay in the laboratory, and John Fulton, a Harvard medical student. Clearly, additional resources—investigators, space, equipment and supplies—would be required. As Cohn discussed these issues with Minot, a working plan emerged. The most urgent need was for a simpler and more rapid test for the active principle. Based on reports in the medical literature, there were some indications that it might be possible to devise such a test. They laid the problem before the Dean.

David Edsall's response was favorable. Using funds from the Farnsworth and DeLamar Mobile Research Fund, he provided a supplementary budget for the project that permitted the recruitment of two additional investigators to assist in the task. Until that time, the space assigned to the Department of Physical Chemistry consisted of the rear half of the fourth floor of Building C-1. The Dean assigned the balance of the space on the fourth floor of Building C-1, including the large oak-paneled front corner office, to Edwin Cohn and the Department of Physical Chemistry.

Without waiting for the arrival of the new personnel, Cohn set to work with the help of John Weare, a laboratory technician. Although progress could be made reasonably rapidly in the laboratory at the medical school, the clinical testing was slow. In order to assay a single liver fraction, Minot had to feed large amounts of the fraction daily for a period of two weeks to pernicious anemia patients. A test was considered positive if there was an increase in circulating reticulocytes, an increase in circulating adult red cells, and a remission in the patient's overall condition. Starting with an aqueous suspension of fresh minced raw liver acquired from a slaughterhouse, Cohn and Weare had succeeded by the spring of 1927 in concentrating the active principle into a fraction they called "G."[5] When fed to pernicious anemia patients in doses of 9.5 grams daily, the response to Fraction G could not be distinguished from the response seen when patients were fed 200 grams of liver. The published report describing the preparation of Fraction G does not provide details about the scale at which raw liver was fractionated; however, considering that the diet of Fraction G to treat a single pernicious anemia patient required about 10 grams of the fraction per day, the amount of liver being fractionated at Harvard in order to obtain Fraction G was quite large, perhaps of the order of twenty or more pounds of fresh liver to test in a few patients.

In testing Cohn's fractions, Minot took pains to assure the validity of the results being obtained. To that end, fractions thought to be inactive were nonetheless tested just as carefully as fractions thought to be active. At one point, Cohn sent Minot "a large bottle full of a reddish, thick, soupy material" to be tested. Minot wrote a note to Herman A. Lawson, the Resident Physician at the Huntington Hospital, explaining that Cohn's "liver mess" contained a residue after extraction of Fraction G and giving careful instructions for its testing. "Count the reticulocytes twice every day, and keep your records carefully. We do not expect a rise, but we might get one if Cohn's extraction has not been complete, or perhaps we will find that the do-good [sic] is present in some other part of the liver as well as in Fraction G."[6] The "liver mess" had no effect in Lawson's patient, a result indicating that Cohn's procedure had concentrated all the activity in Fraction G.

The achievement of such impressive results within a few months was a tribute to the intense efforts of Cohn and Weare as well as of Minot and his associates at the hospital. When the news about the efficacy of Fraction G began to spread among the medical community, another problem arose. Physicians began to call George Minot asking for Fraction G to treat their patients. Minot and Cohn felt a great deal of pressure, which was quickly brought to the attention of Dean Edsall and through him, to President Lowell. Minot's biographer Francis Rackemann commented, "for a moment it looked as though Harvard University, under Dr Cohn, was about to engage in the pharmaceutical business." In the short term, a crisis was alleviated by the appointment of a Harvard Committee on Pernicious Anemia and the development of an arrangement with Dr. G.H.A. Clowes, medical director of the Eli Lilly Company, under which Lilly was given the right to manufacture Fraction G. The Harvard Committee proved to be helpful in dealing with anxious physicians, with the public, and with other organizations in the United States and other countries.

Should Harvard Secure a Patent Covering Cohn's Work?

The Harvard Corporation, seeking to protect the interest of the public by insuring the availability of the product under the most favorable conditions, took the following action on March 12, 1927:

> Whereas Doctors George Minot and Edwin Cohn appear to have discovered a treatment of pernicious anemia by the use of liver, and the method of extracting the effective substance from liver for the purpose,
>
> Voted that with their consent the University apply for a patent with the proviso that the University shall make no profit therefrom; that commercial manufacturers shall be permitted to manufacture and sell the product freely under such supervision as will guarantee its purity, and that excessive profits shall not be charged to the public; it being understood also that the process may be used without restriction for all research and other scientific purposes. The University hereby expresses the intention, if and as soon as this can be done, of depositing this patent and freely giving its privileges to some central national governmental body, such as the United States Public Health Service or the National Research Council, which will assume the administration of the rights and the protection of the public without profit to the holder of the patent.[7]

A few days later, President Lowell, in a letter to Dean Edsall, stated: "The Corporation entirely approved of taking out a patent on the discovery of Drs. George Minot and Edwin Cohn; but Dr. Walcott suggested, and it seemed to us very sensible, that it would be well to do it in consultation with other universities and laboratories, so that it should not be an isolated, but a cooperative action . . . I have asked Mr. Odin Roberts to attend to getting out the patent."[8] Cohn and Roberts set to work to draft the claims to be included in the patent application. During the summer and early autumn of 1927, Cohn had obtained an even more active fraction than Fraction G. However, this new fraction had a mild depressor effect on the patient's blood pressure when administered intravenously. On considering this, Roberts, the patent lawyer, became concerned about a possible conflict with a claim for a hypotensive liver fraction in an existing patent. Although further studies on Cohn's part led him to believe that his pernicious anemia factor was not the same as the patented hypotensive factor, his most active fractions against pernicious anemia at the time retained a mild hypotensive effect. Roberts delayed completing the patent application.

The Eli Lilly Company had by then succeeded in making a product which, though less pure than the Cohn product, was about to be approved by the Harvard Committee on Pernicious Anemia. The Armour and Wilson Companies were also manufacturing a product whose potency was not known. Moreover, the Harvard Committee on Pernicious Anemia had provided information to Professor Best at the Connaught Laboratories in Canada, Professor Dale at the British Medical Research Council, and Professor Krogh in the Insulin Laboratory at Copenhagen that would enable the wider manufacture of Cohn's Fraction G.[9] Taking these into consideration, Cohn began to wonder about the advisability of continuing to seek a patent.

Meanwhile, President Lowell's suggestion had been investigated by looking into other closely related patenting cases. In the case of insulin, Banting's discovery had been patented by the University of Toronto. There the obligation to control the product for safety and efficacy was assigned to the Connaught Laboratory, a manufacturing laboratory within the University that was owned by the Ontario Provincial Government. This solution relieved the University from assuming the role of a regulatory agency. Other influential input was received from F. G. Cottrell, the academic inventor of a commercial process for removing toxic substances from the air. Cottrell forwarded to Harvard a copy of a letter setting down his thinking to an associate at the University of Michigan. The letter began, "The more I see of the attempts to utilize or administer patents on a purely licensing basis, that is, without one's being in the actual manufacturing business and utilizing the patents primarily in a defensive manner, the more I am impressed with the inherent difficulties of the undertaking . . . The rest of the discoveries of our scientific laboratories should be published as promptly and as fully as possible."[10] Cohn later commented, "after a long and serious consideration, it was decided that all of the advantages that were sought for the public could better be obtained without than with a patent."[11] Despite a statement to the contrary by Minot's biographer, Harvard never applied for a patent covering the manufacture of the active principle in pernicious anemia.

Late in 1927, Minot and Murphy reported additional clinical results based on treatment of an additional group of patients using the raw liver diet. This extended the series from forty-five cases, as reported in their first paper, to 105 cases. These new results completely confirmed their original report and stimulated a renewed flow of requests from physicians anxious to obtain the Cohn fraction to treat their patients.[12] At about the same time, Eli Lilly had prepared several lots of Fraction G for clinical testing and general use. Final clinical testing of this product, designated "343," was conducted in thirteen clinics recruited by the Harvard Committee. The test results proved to be satisfactory, and the subsequent availability of this product eased the pressure on Minot and Cohn.

In a paper published in 1930, Cohn, McMeekin and Minot reported substantial further progress in purifying the active principle. Beginning in 1926, when the dose of liver was 200 to 300 grams daily, further stepwise purification efforts had increased the potency and reduced the size of the effective dose to 10 grams of Fraction G by 1927, to 4 grams by 1928, to 0.7 grams by 1929, and to 0.025 grams by 1930.[13] The latter fractions could be administered intravenously without evidence of hypotensive effects. However, the active fractions tested negative when tested for protein. This last paper also contained a hint of an impending new problem: the number of pernicious anemia patients presenting to the hospital with pernicious anemia had fallen sharply since "343" became more generally available, decreasing the possibility of future clinical testing.

In the meantime, the two new investigators authorized by Dean Edsall had been recruited in the hope of developing a simple and more rapid test for the active principle. Gordon Alles, a biologist from California, came in 1927, and George Payling-Wright, an English hematologist, arrived in 1928. By this time, a number of putative *in vitro* assays for the liver factor had been reported in the scientific literature. Alles and Payling-Wright took on the task of investigating each reported test. They concluded that none of the tests were valid indicators of the active factor in pernicious anemia. After only a year in Boston, Alles returned to the California Institute of Technology. Payling-Wright left in 1930 to become Professor of Pathology at Guy's Hospital Medical School in London. Before returning to England, he published two important papers on unrelated factors influencing the respiration of red cells. The possibility of developing an assay based on Murphy's observations on reticulocytes was also explored, but without success.

Earlier, the Harvard Committee on Pernicious Anemia had forwarded a sample of Cohn's Fraction G to Whipple in Rochester who reported back that the fraction was without effect in treating his anemic dogs. This clearly distinguished pernicious anemia from Whipple's secondary anemia brought about by bleeding. By 1931, the availability of commercial liver extracts had solved the problem of therapy for pernicious anemia with the result that fewer patients were available in Boston for use as test subjects for Cohn's work. Without a laboratory test for the active principle, the work came to a halt. For Cohn, the absence of experimental evidence linking the active principle in pernicious anemia to a protein in the liver was a disappointment.

In Rochester, Whipple arranged with the Eli Lilly Company for the preparation of a liver product designated "55" to treat the anemia in dogs. However, "55" was not widely accepted and had to be abandoned. With Whipple's help, Eli Lilly introduced another product, called "lextron," which was compounded of "55" plus iron and the Cohn product "343." Lextron was effective in treating both Whipple's anemic dogs and pernicious anemia in humans. Since Harvard never patented the Cohn process for purifying the active principle, it lacked control over the changes made by Lilly in the preparation of "lextron." Under a direct arrangement with Lilly, Whipple tested every lot of "55" and "lextron" in his anemic dogs. By 1953, the fees paid by Lilly to the University of Rochester for these tests had generated more than $700,000, from which a professorial chair in pathology, a scholarship fund for students, and a visiting lectureship were later established at the University of Rochester.[14]

As might be expected, the discoveries of Minot and Murphy stimulated others to pursue the elusive active principle. One of those was Randolph West in the Department of Medicine at Columbia University. Cohn arranged that the Harvard Pernicious Anemia Commission send West some of the Eli Lilly commercial extract "343." West attempted to further purify the active principle, but without success.[15] West's chief at Columbia was H.D. Dakin, a distinguished English biochemist. Relations between Cohn and Dakin sometimes became very intense. At one point, Cohn showed Hastings an irate letter from Dakin; to which Hastings replied: "if I got a letter like that, I would burn it!" Cohn's reaction was to make a copy of the letter and show it around to others.

George Minot, William Murphy and George Whipple shared the 1934 Nobel Prize in Medicine. In his Nobel Lecture at the Caroline Institute in Stockholm, Minot described his and Murphy's work. In referring to the liver extract, Minot wrote:

> Dr. Edwin J. Cohn, in the Department of Physical Chemistry in the Laboratories of Physiology of the Harvard Medical School, soon made a potent extract suitable for oral use. We tested on patients the preparations he prepared in an attempt to isolate the active principle. Although unsuccessful in this objective, as time passed we demonstrated (1929) that the potent material could be given intravenously and produced maximal effects in very small quantities (0.15 g.). These small experimental preparations were not practical for regular use . . . The exact nature of the potent substance remains unknown, but especially from the studies of E.J. Cohn and R. West it appears to be a relatively small nitrogenous compound.[16]

Further pursuit of the problem was abandoned in 1930.

Although the liver project came to an end, Edwin Cohn continued to use the large front office and his Department occupied the entire fourth floor of Building C-1 thereafter. It happened that Baird Hastings had been offered a professorial appointment at Harvard a few months earlier and had been promised that same corner office plus three laboratories on the fourth floor of C-1 were he to accept the invitation to move to Harvard. However, at the last minute, Hastings had declined the offer. When he later came to Harvard as the Hamilton Kuhn Professor of Biological Chemistry in 1935, Cohn was of course comfortably settled in the C-1 front office, and Hastings was assigned space on the third floor of Building C-2. Much later, in reminiscing about his Harvard career, Hastings said: "When I went to Harvard, there was Edwin Cohn in a full time research professorship. He never saw medical students and very few graduate students. At Harvard he was Mary and I was Martha. I wouldn't trade my career as Martha for that of Mary for anything in this world."[17]

183 Brattle Street, Cambridge

In 1928, buoyed by a promotion to Associate Professor of Biological Chemistry and the favorable publicity his work on the active principle in pernicious anemia had received in the national press, Edwin and Marianne Cohn began to look for a permanent home in Cambridge. The search quickly identified a house that met their requirements. Situated prominently on Brattle Street at the northwest corner of Fayerweather Street, it was an imposing dwelling in one of Cambridge's most elegant neighborhoods, suitably ostentatious to fit Edwin's image of himself and his family. It had ample room for the family as well as quarters for maids. Delighted with their acquisition, the Cohns delayed their occupancy while extensive changes were made in the internal arrangements of the house. These included improvements on the first floor to accommodate dinner parties and other social events. Edwin Cohn played an active role in every stage of the project. When construction was actually underway, he visited the work every day, dealing directly with the carpenters and tradesmen. This was a realm in which he took great delight. He particularly liked detail, such as the addition of egg and dart molding to the living room and the design of small pieces of furniture that were built by one of the carpenters. The family moved into their new home in the autumn of 1929.

Edwin Cohn was meticulous about his personal attire. It may have been while he was in England with Marianne in 1923, or perhaps on a subsequent visit, that he arranged to meet his sartorial needs by having his wardrobe tailored in England. From then on, all of his suits, shirts, hats, overcoats and shoes came from England until the outbreak of the World War II. He favored double-breasted suits, almost always tailored from dark woolen fabrics. Invariably he wore black shoes with leather heels that signaled his approach as he strode down a hallway. He never wore spats. His ties were made of the finest fabrics in conservative designs, and a freshly folded handkerchief always adorned his breast pocket. He wore Homburg hats. In later years, his wardrobe included a few single breasted worsted suits of lighter weight to wear in warm weather. He never took off his jacket in his office, and never gave even a hint of discomfort on the warmest midsummer days in Boston. He invariably cut a handsome figure in dress quite distinct from his academic and medical colleagues. He attracted attention whenever he strolled into a meeting, a lecture, or when he got off the elevator at the fourth floor in Building C-1. His presence was immediately acknowledged by the maitre d's at the Harvard Club in Boston and the Plaza Hotel in New York. The Cohns took a house in Cotuit for the summer of 1929. When there, Edwin usually spent the mornings at his desk writing letters and drafting scientific papers. At lunchtime, he would close up his briefcase and join Marianne. In the afternoon, it was tennis, followed by a swim. This was the only setting in which he wore casual clothes.[18]

On Monday evening, August 19, 1929, the XIIIth International Physiological Congress, the first to be held in the United States, convened at 8:15 p.m. in Sanders Theatre in Cambridge for its opening convocation.[19] Among the dignitaries greeting the members of the Congress were U.S. Surgeon General Hugh S. Cummings, Lieutenant Governor William S. Youngman, and Harvard President A. Lawrence Lowell. William H. Howell, Director of the School of Hygiene and Public Health at Johns Hopkins, presided. After the opening formalities, August Krogh of the University of Copenhagen gave an address on "The Progress of Physiology." Following the convocation, an outdoor reception was held for members of the Congress and their spouses in the Harvard Yard. Edwin Cohn and Alfred Redfield served on the organizing committee.

The scientific sessions of the Congress began the next morning and extended through Friday in the auditoria of the Harvard Medical School, the Boston High School of Commerce, and the Peter Bent Brigham Hospital. The organizing committee put on a splendid social program with tours and teas for the ladies. The high point of the social schedule was a 9 p.m. reception in the court of the Harvard Medical School with a fine concert by the Boston Symphony Orchestra conducted by Arthur Fiedler. There was only one free evening. On that occasion, George Minot, together with Drs. James H. Means and Francis M. Rackemann of the Massachusetts General Hospital, and Soma Weiss of the Boston City Hospital, entertained a large group of visitors at a dinner at the Country Club in Brookline.[20]

Later that year, Dean Edsall casually strolled into Cohn's laboratory one day with a visitor. With accustomed diffidence, the Dean asked if Cohn could interrupt his work and tell the visitor what he was working on. Cohn, preoccupied with an experiment in progress had failed to catch the name of the visitor, and, somewhat embarrassed, did not ask to have it repeated. Nonetheless, he gave the visitor his full attention. A few months later, he learned that the visitor had been Dr. Richard Pierce of the Rockefeller Foundation. Good news was forthcoming!

4 Amino Acids and Peptides as Prototypes of Proteins

More than a decade had passed since Alfred Cohn had assisted Edwin in transferring to the University of Chicago. As their paths diverged, it was perhaps inevitable that they would drift apart. Alfred, rapidly acquiring a reputation as a physician-philosopher, continued his important studies at the Rockefeller Institute, writing essays and pursuing interests in medical education, both in the United States and abroad. He and his wife, Ruth, had no children. There were few occasions when family ties brought the brothers together. On the other hand, despite the differences in their ages, each kept track of the other's growing reputation. During the 1920s, they had found themselves on different sides in the public debate over the Sacco-Vanzetti case. Alfred had sided with Felix Frankfurter, an old friend, in support of the two defendants in the case. This had been a source of embarrassment to Edwin among his Cambridge friends. Thereafter, the brothers took a certain pleasure in trying to score points on each other. After moving into his Brattle Street home, Edwin took umbrage at Alfred for staying with Frankfurter on subsequent visits to Cambridge. This was particularly irritating because Frankfurter lived practically across the street from the Cambridge Cohns.

The rather sudden termination of the efforts to isolate the active principle in pernicious anemia, however disappointing, was not without associated benefits. When it became evident that pursuing the pernicious anemia factor no longer involved the purification of a protein, Cohn had to acknowledge that the task might better be addressed by organic chemists. On the other hand, his successful work on pernicious anemia had suddenly earned him considerable recognition on the part of colleagues in Cambridge as well as in the clinical faculty of the Medical School. Moreover, policy questions concerning the liver work had brought him into contact with President Lowell.

Cohn had been promoted to the rank of Associate Professor in 1928. In 1930, President Lowell notified him that the Rockefeller Foundation had awarded a grant of $175,000 to Harvard to support research in physiology and physical chemistry. The new award provided seven years of support, from 1930 to 1937, at the level of $25,000 per year, of which "approximately three parts shall be devoted to physiology and two to physical chemistry."[1] The total budget of the Department of Physical Chemistry in 1929, including the supplemental funds provided by Dean Edsall for the liver work, had been slightly more than $40,000. These developments bolstered Cohn's self-confidence and fueled his intense ambitions.

The space on the fourth floor of Building C-1 that was occupied by the Department of Physical Chemistry was laid out on both sides of a central corridor. Edwin Cohn occupied the large front corner office that looked out onto the court of the Medical School in the front, and to the Peter Bent Brigham Hospital on the Shattuck Street side. The office was paneled in oak and had built-in bookcases, cabinets and a fireplace. Cohn sat at a handsome oak rolltop desk in an antique Windsor chair. In the center of the room was a solid oak table, at the head of which Cohn sat while chairing meetings and conferences. A door opened directly from his office into his laboratory. The laboratory benches had thick cherry tops. Along one wall, there were built-in refrigerated baths especially made for studying the solubility of proteins. The remaining departmental space consisted of laboratories of different sizes, cold rooms, instrument rooms and a large washroom for cleaning. A few small offices were arrayed around the periphery of the building. The layout of these facilities contributed to the spirit of scientific cooperation and collaboration that was the hallmark of the department. Except for Cohn's front office, the remaining space, including his laboratory, was common space and was never locked. All instruments and special equipment were shared. All dirty glassware was collected promptly by a diener who washed and dried them, and returned them to cupboards for reuse by any member of the staff.

After a decade at Harvard, it was a time of opportunity. Cohn could afford to pause and consider his position. In retrospect, the progress made in his initial studies on proteins had been slow. The molecular weights of only a few proteins were known. The amino acid composition of no protein had yet been worked out. The proteins were too large, too complex, too highly charged electrically, and too poorly characterized for serious study. The results of measurements of the physical chemical properties of proteins at the time defied interpretation because of the inadequacies of current theories.

Drawn back in his thinking to the landmark contributions of Bjerrum and of Debye and Hückel, Cohn began to consider a bold departure from the pathway he had been following. The amino acids were simple prototypes of proteins. Why not go back and pick up the trail from Bjerrum and Debye by investigating the physical chemical properties of the amino acids? Pure amino acids were available. Their chemical structures and molecular weights were known. Except for the amino acid proline, only the side chain R groups differed among the amino acids. For those amino acids which could ionize by dissociating or associating hydrogen ions, the strengths of the dissociating or associating groups could be measured precisely, thus opening the way to defining their electrical charges and dipolar properties in solution. Peptides would be even better prototypes of proteins, although they would have to be synthesized. Nevertheless, if only a few dipeptides and tripeptides could be synthesized, it would be possible to assess how their peptide bonds influenced the dissociation behavior of their C-terminal carboxyl groups and N-terminal amino groups.

Cohn began to visualize acquiring a library of pure amino acids and peptides for use in investigations of their physical chemical properties and reactivities. He was confident that with even a modest library of peptides, knowledge of their electrical charges would open up a new pathway to understanding the nature of the charged groups on protein molecules. It would provide a new level of theoretical insight into these models of the proteins. On the other hand, fully to capitalize on this strategy might require the engagement of most of the scientists working in his laboratory. More than that, it would demand a new level of coordination, communication and interpretation if the results of such studies were to be fully understood. About this time, Jesse P. Greenstein arrived in Cohn's laboratory with a National Research

Council Fellowship in Biochemistry for a year of postdoctoral study. An organic chemist who had just completed his studies for the Ph.D. degree at Brown University, Greenstein's arrival was almost perfectly timed, for he was just the man to undertake the difficult task of synthesizing the library of peptides that Cohn envisioned.

Once again, Cohn undertook an exhaustive review of the scientific literature concerning the physical chemistry of the amino acids and peptides. Although only six years had elapsed since his 1925 review of proteins, he set out to digest the contents of over 300 scientific articles in preparation for the new review. Unlike the earlier review, which had been almost purely descriptive, the new one began with a critical evaluation of the theoretical foundations of physical chemistry as they related to amino acids, peptides and proteins. Two-thirds of the work was devoted to amino acids. It summarized existing knowledge concerning relationships between their chemical structure and the dissociation of their functional groups. Cohn had even assembled from the existing scientific literature a table giving the dissociation constants, isoelectric points and pH of maximum charge of eighteen different amino acids and eleven simple peptides. The manuscript of the new review was submitted to the German journal *Ergebnisse der Physiologie* early in 1931. It bore the same title as the 1925 review: "Die Physikalische Chemie der Eiweisskorper."[2]

Biochemistry at Harvard

Following Lawrence Henderson's move to the Business School, Cohn, sensitive to David Edsall's ability to influence the flow of university resources to the Department of Physical Chemistry, strove to maintain good relations with the Dean. A favorite tactic was to arrange periodic appointments with him. As the frequency of these visits grew, Edsall attempted to restrict their number and duration. In turn, Cohn, undaunted, took to writing letters, some rather lengthy. One such letter, written before his 1931 departure to Europe on sabbatical leave, provides evidence of Cohn's interest in a topic that was destined to become prominent later in his career: the teaching of biochemistry in the University. The topic for discussion was a comparison of the teaching of physiology and biochemistry. Physiology at Harvard was divided into two parts, focussed on different areas, a department of general physiology in Cambridge, born of plant physiology, and a department of physiology in the Medical School, based on animal physiology. In contrast, he noted that "biochemistry, however, cannot be divided into those aspects which are more important in medicine and those which are more important from an academic point of view. There is not one chemistry for the vertebrates, another for the invertebrates. The same substances have been isolated from animals and plants and the same principles define their behavior".

He went on to stress the diversity of issues confronting the biochemists in the University, identifying the two departments in the Medical School, biochemistry and physical chemistry,

> to which must be added three developments in the University, concerned with one or another of the aspects of biological chemistry. One of these is the course in biological chemistry, for years given by Professor Henderson under the auspices of the Department of Chemistry. Professor Henderson's laboratory in the Business School must also be considered as concerned with biological chemistry as it affects certain problems of physiology, especially fatigue. Certain of the courses in organic chemistry, notably Professor Conant's course on the chemistry of natural products, both depend upon and contribute to biochemistry. Finally, it has been the function of the tutors in biochemical sciences to present to their students the diverse material which is the subject of investigation in these different departments of the University.[3]

The Board of Tutors in Biochemical Sciences, he continued, unlike most other Harvard tutorial boards, was not related to a division of the university; there was no division of biochemistry in Cambridge. Instead, the Chairman of the Board of Tutors, John Edsall, and many of the tutors were members of Cohn's own department in the Medical School. "They have functioned without a laboratory in biological chemistry in Harvard College and without the benefit of a strong group of biochemists associated with the University . . . Could this perhaps not best be done by the appointment of a joint committee of the biochemists in the Medical Faculty and the Faculty of Arts and Sciences?"[4] Eliminating this dichotomy was to become an important issue for Cohn in the years ahead.

A Serious Health Problem

When the new studies on amino acids and peptides were well underway, trouble struck. On a weekend skiing party with friends in the winter of 1931, while trying to punch a hole in someone's ski harness, Cohn punctured his own finger. The finger immediately swelled up and the swelling spread rapidly to his hand and arm. This may have been the initial episode of a condition later diagnosed as an allergic angioedema. It was to beset him episodically for the rest of his life. Attacks were marked by rapid onset, accompanied by swelling of his face, lips and tongue, sometimes blurring his speech. These episodes could be controlled by the prompt injection of epinephrine and the use of antihistamines. From then on, he carried a syringe and vial of epinephrine in his brief case, ready for use in case of another episode. The epinephrine altered his moods. He could become hostile and difficult to deal with, although he adamantly denied it. Marianne dreaded these episodes.[5]

While undergoing a thorough medical workup after the edema episode, a second more serious problem was discovered. Joseph Aub, his physician, found that his blood pressure was dangerously high. As related by John Edsall, Aub "viewed his condition very seriously and strongly advised him to lead a relatively quiet and retired life and give up most of his responsibilities."[6] The hypertension had probably existed for some time; it was not thought to be related to the angioedema. The discovery of such a serious condition was alarming to Marianne Cohn who thereafter took fright whenever her husband played tennis, particularly singles. The Cohn boys, then less than ten years old, were not told about this; they first learned about it after their father's death more than twenty years later.

The stresses under which Cohn lived may well have contributed to the hypertension. As related years later by his son Ed, "worries brought by his losses in the stock market crash were added to the tension caused by his immense ambition, his anxiety about being Jewish in a Wasp world, and the underlying sense of being unlovable, instilled by his emotionally deprived childhood. He had been advised to get psychiatric help at this stage, but, of course, he rejected the suggestion indignantly."[7] Aub's advice ran completely counter to Cohn's natural inclinations. If followed, it would require a drastic downward adjustment of his career goals. His response was measured; he would defer completion of the last section of the new review and take a sabbatical leave.

Sabbatical Leave in Munich

The leave had been previously planned although its timing was influenced by the discovery of the hypertension. Cohn had ample scientific reasons for the trip. In his plans to investigate the amino acids and peptides, there were aspects which would benefit from discussion with his friends and scientific peers in Europe. The year would be used for rest, study and renewal. Munich, where they had taken a house, would be the family's base for the six winter months from September to March. They sailed from New York in June 1931. Joseph Aub had advised Cohn to avoid eating fish and eggs, which were suspected of being responsible for the allergic attacks. However, Cohn was never a docile patient. He enjoyed good living. Once, after dining rather well during the Atlantic crossing, he cabled back to Aub: "Minus fish minus eggs equals plus caviar."[8]

The family settled into their Munich house at the end of the summer. It was located on Friedrich Herschel Strasse in the Bogenhausen section of Munich, then a residential district on the south bank of the Isar, across from the City Center. Marianne spoke German with the fluency of a native. Edwin was not as fluent, but made up for it with vigor if not accuracy. Within a short time news of their presence in Munich spread and they were quickly caught up in Munich social life. The household included a maid, a cook, and a young German woman, Marfa v. Gravenitz who served as tutor and companion to the boys. Marfa lived and traveled with the family until they sailed home from Rome in September 1932. Under her tutelage the boys learned German rapidly and became familiar with their Munich surroundings. A good skier, she saw to it that they learned to ski. She also contributed to Edwin's well being by keeping the boys busy.

Cohn had no formal institutional base in Europe. However, several factors had influenced the choice of Munich for his sabbatical. Munich was the home of the organic chemists, Richard M. Willstätter and Heinrich Wieland, both Nobel laureates. They became very interested in Cohn's plan to relate the physical chemistry of the amino acids and simple peptides to their structure. Professor Willstätter, a kind and dignified gentleman, then almost sixty, lived near the Cohn's house and was a frequent visitor there. Cohn had two immediate tasks to be dealt with in Munich. His new review was being translated into German by Frau Else Asher. As published in Munich late in 1931, it ran to more than a hundred pages. Cohn's presence in Germany at the time of the publication of his review generated a favorable measure of publicity for his work among the German scientific community. (A footnote was added by the publisher advising its readers that, due to the illness of the author, a second part of the review had not been completed and would be published at another time.)

A second task involved the preparation and translation into German of an article on the solubility of amino acids and proteins. A young German scientist, H. Ketterl, was responsible for this endeavor, aided by Marianne. This article was also the basis of two lectures which Cohn delivered in November 1931, the first at the Kaiser Wilhelm Institute for Medical Research in Heidelberg, and the second at the University of Würzburg. The substance of these lectures reflected Cohn's developing plan to seek a theoretical basis for the solubility behavior of proteins through the study of amino acids and peptides. Marianne accompanied Edwin to Heidelberg and Würzburg for the lectures and participated in the associated social events. Afterwards they were joined by Ed, Fred and Marfa at Rothenberg and drove back to Munich, visiting other medieval towns on the way.

On weekends, the Cohns often traveled in Bavaria. Their car had British plates, and for some reason it had to be taken out of Germany at least once a month. On those occasions they went to Kitzbühel in Austria, or to Bregenz where a brother of Josef Brettauer still lived. During the Christmas holidays, they visited American friends in Gstaad and relatives in Basel where another branch of the Brettauer family lived. In late winter, the Cohns were swept into a series of balls during Fasching, the German version of Mardi Gras. Writing much later, John Edsall contributed a broader perspective on political conditions in Germany while the Cohns were there.

> The times in Germany were grim, and the Weimar Republic was in the last agonies of its decline. Cohn watched the progress of events

with deep anxiety. He returned to Boston convinced that the Republic was doomed, but still believing that the Nazis would not come to power. That hope was soon shattered, and he watched the events in the following years with a sense of deep horror and outrage and a growing conviction that a general war was inevitable.[9]

In the spring of 1932, after farewells to their friends in Munich, the Cohns took off on an extensive European tour, beginning with several weeks of alpine skiing in Switzerland. There the boys—Ed was then eleven years old while Fred was almost eight—had some excellent alpine skiing with their father, and explored alpine geography and watersheds. During this period, Ed came to realize that his father's enthusiasm for skiing was not matched by his proficiency. On their return to Cambridge, his father's exaggerations of his skiing prowess in social conversation so embarrassed Ed that he lost interest in skiing for several years.[10] Marianne tended to less vigorous exercise, preferring hiking to skiing. On the way down through Italy, the family stopped in interesting places to visit art galleries and view important pieces of architecture. This was a particularly meaningful experience for Ed Cohn, who was beginning to realize the scope of his father's interest in fine arts. More than half a century later, Ed expressed warm memories of that period with his father, although he never understood his father's preference for Romanesque over Baroque architecture. To satisfy a yearning by Marianne, the family spent Easter in Rome. Afterwards, they continued south to Naples where they took a boat to Athens for a prolonged stay in Greece. Returning to Italy, they continued a slow return northward, ending up in Heidelberg for a reunion with Alex von Muralt and his family. Then, in two cars, the families drove back to Rome to attend the Physiological Congress in mid September.

While the program of the Congress offered formal opportunities for the presentation and discussion of new scientific discoveries, the Italian hosts had also arranged an elaborate social program, with banquets, luncheons and visits to Roman antiquities. On such occasions Cohn was always a charming host, sparing no effort to entertain guests, and full of exuberance and enthusiasm. He had a "con brio" vitality that was especially appealing to women. When the Cohns found that the Hastings were also attending the Congress, they graciously swept Margaret and Baird into a round of parties with the von Muralts and other European colleagues. Baird later reminisced about the good times they had with the Cohns in Rome, remarking that Edwin "was very-well to do, and his wife was even more so."[11] Alfred and Ruth Cohn were also in Rome for the Congress. In a conciliatory mood, Edwin and Marianne had arranged to return home on the same ship with them. However, the attempted rapprochement was a failure; Alfred and Ruth announced that they always took their meals alone. Nevertheless, the boys had breakfast most mornings with their uncle and aunt.

Theoretical Studies on Amino Acids and Peptides

On his return to Boston, Edwin Cohn found that a good start had been made in the study of the physical chemical properties of the amino acids and peptides. Greenstein had by then synthesized six new dipeptides and had measured their titration curves. Some results from his study of aspartic acid and its peptide, aspartylaspartic acid give an indication of the potential importance of Cohn's strategy in initiating these studies. Inspection of the chemical formulae of aspartic acid and aspartylaspartic acid reveal that aspartic acid has three charged groups—one positively charged and two negatively charged, while aspartylaspartic acid has four charged groups, one positively charged and three negatively charged.

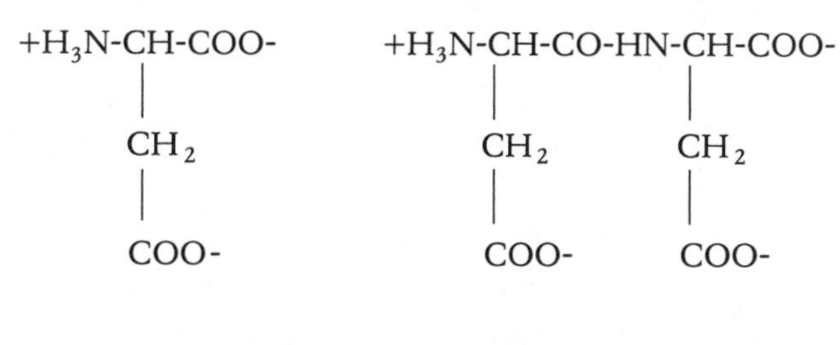

Aspartic Acid Aspartylaspartic Acid

Working with solutions of aspartic acid (ASP-open circles) and aspartylaspartic acid (ASPASP-solid circles), Greenstein confirmed[12] the zwitterion theory of Bjerrum and revealed the state of association or dissociation of each functional group as a function of pH (see Fig-

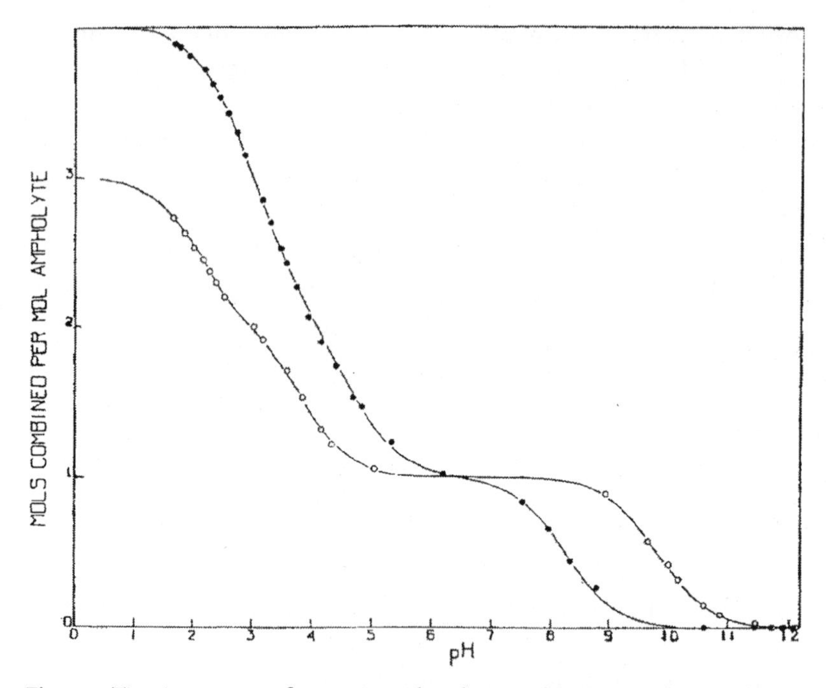

Fig. 1. Titration curves of aspartic acid and aspartyl-aspartic acid. • indicates aspartyl-aspartic acid; o, aspartic acid. These confirm that aspartyl-aspartic acid has four titratable groups per molecule while aspartic acid has only three.

ure 1). He also identified the dissociation constants for each group. Additional information was derived from the relative positions of the titration curves along the horizontal (pH) axis. Below pH 7 the curve for ASP can be seen to lie to the left of the curve for ASPASP. The reverse is true above pH 7, where the curve for ASP lies to the right of that for ASPASP. These results were attributed to the effect of the peptide bond in aspartylaspartic acid on the strength of the acidic and basic groups.

Overall, some seventy scientific studies on amino acids and peptides by Edwin Cohn and his associates were published between 1931 and 1938. Although most of these papers were couched in terse esoteric physical chemical terms scarcely intelligible to any but expert readers, it soon became apparent that, taken together, they constituted an intellectual breakthrough of considerable importance. They represented Edwin Cohn's most important contributions to basic scientific knowledge about proteins. The Rockefeller Foundation, in appraising Cohn's work in 1938, described it as "painstaking, abstruse, and likely only slowly to come to widespread recognition for its essential importance."[13] Indeed, in only a few years, these studies were to provide the conceptual scientific framework for Cohn's great practical achievements with human serum albumin, gamma globulin and other blood products during the war.

In directing these landmark scientific studies on amino acids and peptides, Edwin Cohn displayed extraordinary leadership qualities. The backbone of the enterprise consisted of associates in the Department of Physical Chemistry, John Edsall, R.M. Ferry, Jesse Greenstein, and T.L. McMeekin, each of whom had worked with Cohn for several years. They were supported by the group of postdoctoral fellows who passed through the laboratory during the period. In discussing his handling of students and fellows, Scatchard characterized Cohn as a benevolent despot: "he thought he knew better than they what they should do." He added that "the benevolence far outweighed the despotism."[14] Nevertheless, some students resented Cohn's tendency to micromanage their work. If an experiment that was suggested by Cohn did not come out the way he expected, reporting the outcome could require a dash of diplomacy. With Muriel Blanchard, his longstanding assistant, Cohn was tipped off more by her eyebrows than by her words if her results had not gone as he expected. Once, a new postdoctoral fellow reported an experimental result that was the opposite of what Cohn expected. Cohn immediately arranged to have the young man work under someone else. (Later it emerged that the young man had been right all along. His experiment had been done on bovine hemoglobin, while Cohn's expectations were based on experience with equine hemoglobin. The divergent results were attributable to the species difference.)

Jeffries Wyman, Jr. first worked in Cohn's laboratory under a National Research Council postdoctoral fellowship between 1927 and 1929. When his fellowship expired, Wyman joined the Harvard Biology Department but he continued his studies with Cohn on the amino acids. As Cohn pondered the possible effects of electrical charges on the behavior of amino acids and proteins and on their interactions, he became interested in the dielectric constant, a measure of the force acting between electrically charged bodies in solution; the higher the dielectric constant, the lower the force. But measuring the dielectric constant was not easy, so he arranged for Wyman to work with George W. Pierce, who had been measuring dielectric constants of liquids in the Harvard Physical Laboratories in Cambridge.

The apparatus used by Pierce required gallon quantities of the liquids being measured, far more than the amounts of amino acids and peptides that could be made available. Wyman succeeded in modifying the Pierce apparatus so that it could be used to study very small volumes. Wyman and McMeekin went on to discover that the dielectric constants of amino acid and peptide solutions are higher than any other substances.[15] Moreover, the dielectric constant increased in proportion to concentration, and, more interesting, to the distance separating the positively charged ammonium from the negatively charged carboxyl groups.[16] Later, Edsall and Wyman showed that the increment in dielectric constant increased with the dipole length of the amino acid, suggesting that the molecules existed as relatively rigid rods, with their charged groups at either end.[17]

Wyman's important contributions to the developing knowledge about the physical chemistry of amino acids and peptides exemplified Cohn's ability to attract able collaborators. Their association was voluntary and without any agreement in advance. For Wyman, it was a pathway of research in which he developed a strong personal interest; Cohn's problem was challenging to him. Wyman also became deeply interested in the broader dimensions of Cohn's investigations. In return, Wyman generated a rich flow of new information which contributed to the achievement of Cohn's broad objectives. Wyman was the author of ten scientific papers that grew out of this work.

Early on, Cohn had sought to identify a mathematical physicist who could develop a general theory defining the behavior of dipolar ions in solutions of varying dielectric constant and ionic strength. It was not until 1933 that this task was undertaken by Kirkwood and Scatchard at MIT. Their work introduced important extensions and modifications to the ion interaction theory of Debye and Hückel in order to apply it to the special case of dipolar ions. Their contributions to the enterprise were of great importance in providing a theoretical rationale that resulted in a unified set of concepts.

The scientific publications resulting from these studies reflect Cohn's policy for sharing with his colleagues the responsibility and the credit for the scientific findings emerging from the Department of

Physical Chemistry. Although the policy was never reduced to writing, the responsibility fell to the person who was the de facto leader, or was in the best position to prepare the first draft of a paper. Thus, for example, Wyman was usually the senior author of papers on dielectric constants, Cohn on solubility or molal volumes of amino acids, Greenstein on titration curves. Cohn's name appears as an author on less than a fifth of the papers in this long series.

At one point in the investigation of the physical chemistry of the amino acids and peptides, Edwin Cohn's musings led him to suspect that the compressibility of solutions of the dipolar ions would be interesting to investigate. Characteristically, he made this suggestion to Professor Percy Bridgman in the Harvard Physics Department, and provided Bridgman with samples for study. The findings were indeed interesting to Bridgman but they proved to be peripheral to the study of the physical chemistry of the amino acids and peptides.[18]

Life in Cambridge

On their return from Europe, the Cohn's settled once again into life in Cambridge. The boys attended Shady Hill School which meant that they were at school six days a week. Marianne Cohn had a close circle of friends, among them Molly Powell, Alice Shurtliff and Pearl Wise. She continued her involvement in the activities of the Cambridge League of Women Voters. Still concerned about Edwin's health, she tried to persuade Edwin to avoid resuming his old pace, and succeeded to a degree. Cohn breakfasted in bed every morning, and then worked in his study on manuscripts of scientific papers, preparing lectures and writing letters. He took to using the telephone more frequently. He rarely arrived at his office in the Medical School before 11:15 AM. When at home alone, he often had supper in bed. If he awoke during the night, he sometimes played solitaire, but he was not a good card player. His nighttime reading consisted mostly of biographies of powerful and successful men.

The Cohns led an active social life, hosting formal dinner parties every two or three weeks. Their guests were drawn from a wide circle of friends and associates that included the Conants, Cutters, Redfields, Churchills, Scatchards, Alex Forbes, Wislockis, David Edsalls and Ronald Ferrys. Later, in a National Academy of Sciences Memoir, John Edsall provided a description of the social intercourse at their home.

> The conversation ranged widely over politics, literature, art, history and science. During these earlier years, before the Second World War, Edwin Cohn in such gatherings was full of gaiety and humor. He was a brilliant conversationalist, with incisive, penetrating, and frequently controversial, ideas on very diverse topics. He read widely, and thought deeply, for example, on the functions of universities both in past history and today. His knowledge of art and architecture was unusual; he could, for instance, in a conversation at luncheon, write on the back of a card the locations of all the finest Romanesque churches in France, and characterize almost any one of them from memory in considerable detail. His knowledge of many aspects of Italian art and architecture was at a similar level. With his energy, his outspokenness and his intensity of feeling he sometimes tended to dominate the conversation, but the discussions that went on in the evenings at his house were never dull.[19]

Beginning in 1933, the Cohns spent their summers in Cotuit. There, after spending the morning at his desk, Cohn had lunch with Marianne, followed by a nap. He liked to play tennis in the afternoon, although Marianne usually succeeded in restricting him to doubles. He enjoyed swimming. Their neighbors at Cotuit included John Northrop and his family from the Rockefeller Institute, and the Fremont-Smiths and the DeFriezes from the Boston area. Each summer, one or two Harvard students lived with the family, giving the boys tennis and sailing lessons and supervising swimming activities.

Before the sabbatical leave, Cohn had very little direct involvement with the boys, offering them neither encouragement nor support. This became a source of tension for the boys. Later, when they were attending Shady Hill, they felt strong paternal pressure to conform to what they considered to be absurd standards of scholastic and athletic achievement. Criticism by their father extended to matters of dress within their home. Ed Cohn attempted to conform to this discipline, which included dressing for Sunday lunch even when, later, as a Harvard undergraduate, he and his father ate lunch by themselves. Ed felt abused, although not physically, by his overbearing father. On the other hand, Fred Cohn rebelled and became a nonconformist, taking to wearing suntan pants or shorts year round. At Shady Hill, Fred experienced serious academic difficulties, which Marianne Cohn viewed with gravity, although she was unable to deal with it constructively. Edwin Cohn interpreted Fred's problems as behavioral in nature, even attributing them to laziness.[20]

The boys also felt considerable paternal pressure to conform to high standards of achievement in athletic activities. When recruited to play doubles tennis with his father, Ed Cohn strove to play well so as to merit his father's approval. However, he soon realized that his father's paramount emphasis was not so much focussed on proficiency, that is, on winning, but rather on displaying proper gentleman's behavior on the court. Still later, however, Ed came to feel that the standards set by his father, while ostensibly for the benefit of the boys, actually stemmed from a concern that the boys' performance in school and in their subsequent careers reflect favorably on him. In contrast to his older brother, Fred Cohn rejected tennis altogether. The boys were not aware that they were Jewish. Among their friends, they had no idea who was or who was not Jewish. The matter arose for the first time in a perfunctory way in the fall of 1935 when Ed was getting ready to go off to Philips Exeter Academy as a boarding student. At that point his father advised him to avoid involvement in groups of Jewish students.[21]

Changes at Harvard

Between 1933 and 1935, turnover in three top Harvard posts portended changes that lay ahead for Edwin Cohn. In 1933, James B. Conant succeeded A. Lawrence Lowell as President of Harvard. In 1934, Otto Folin, the Hamilton Kuhn Professor of Biological Chemistry and Chairman of the Medical School Biology Department, died. And in 1935, David Edsall retired as Dean of the Harvard Medical School and was replaced by C. Sidney Burwell. In keeping with academic tradition, a faculty committee was appointed to search for a new Professor to succeed Folin. With a new class of freshman medical students about to arrive in the fall, the faculty search committee felt some pressure to complete their task before another academic year began. Although qualified candidates from outside had been considered, the only serious internal candidates were Cyrus Fiske and Edwin

Cohn, both associate professors. Fiske enjoyed an international reputation as the discoverer of phosphocreatine and ATP. He had considerable experience in teaching, having taught for several years in Folin's biochemistry course for first year Harvard medical students. Cohn had earned renown for his work on amino acids and peptides and for the partial isolation of the factor in liver that cured pernicious anemia. On the other hand, although one or two medical students had worked in his laboratory each year, questions could be raised about his abilities to take over the responsibility for teaching the Medical School course.

On the completion of its search, the members of the committee arrived at an unusual recommendation. They recommended that Edwin Cohn and Cyrus Fiske be appointed cochairmen of the Department of Biological Chemistry. Their recommendation was forwarded to the Medical School Committee of Professors, then to Dean Edsall and thence to the President and the Harvard Corporation. The Committee of Professors in the Medical School expressed strong objections and advised against the proposed plan. The reasons for its position were not made clear. Among the obstacles, the sharing of leadership in an important academic department may have been seen as a disadvantage, weakening the department in relation to other medical school departments, most of which had strong single chairmen. In the end, the recommendation was disposed of by President Conant, who turned it down. The outcome was a bitter disappointment to Edwin Cohn. Had the Committee's recommendation been approved, it would have resulted in his promotion to the rank of full professor.

President Conant took the appointment of a successor to Otto Folin into his own hands. He immediately contacted Baird Hastings, indicating that he wished to meet with him and discuss the possibility of Hastings taking over the headship of the Biological Chemistry Department with an appointment to the Hamilton Kuhn chair. Presumably their talks went smoothly. Harvard had offered the chair to Hastings in 1928, and Conant, then still a Professor of Chemistry, had been party to those discussions. Hastings knew Cohn and although he had never met Fiske, was familiar with his work. In his first meeting with Conant, Hastings insisted that if he were to accept the invitation to Harvard, both "E.J. Cohn and Cyrus Fiske be elevated to full professorships, because I didn't want to start with that sort of millstone around my neck. They were both more distinguished scientists than I was."[22]

Hastings took up the post at Harvard in the fall of 1935 and Cohn became Professor of Biological Chemistry in Hastings' department. The tying of Cohn's appointment to Hastings's department marked the beginning of a relationship between the two men which was never smooth. The two differed in personality, in management style and in background; they were never destined to develop close associations. As a senior Harvard official put it, "they never hit it off." Nevertheless they maintained a relationship that was outwardly cordial and friendly over the years. The budget of the Department of Physical Chemistry continued to be provided directly by the Dean of the Medical School.

Baird Hastings considered the members of his department to be family, that is, his family. In speaking of them much later, he said, "If they didn't fit in the family, I suppose they didn't stay." The teaching of the biochemistry course for the medical students was the major preoccupation of Hastings and his department. Weekly department meetings were held at lunchtime on Mondays. Each year, beginning in October, the meetings were devoted primarily to developing the lecture schedule and planning the laboratory exercises for more than 100 students. The research interests of the members of the department were dealt with in department seminars held at another time. The medical school course occupied the period from early in February until the end of May, during which time it was customary for all members of the department to attend each lecture and assist during long exercises in the laboratory. There was little time for anything else.[23]

In the Department of Physical Chemistry, the weekly luncheons were also held on Mondays. As a result, only rarely did members of one department attend a meeting of the other. Attending Cohn's meetings was the "in" thing for medical students to do in that period. With masterly skill, Cohn could play more than one role in these meetings. He was particularly perceptive and sensitive to subtle currents that might be operating below the surface. He seldom missed an innuendo. His demeanor could range from gentle to tough. He did not tolerate mediocrity. For a young man or woman giving a talk in the meeting for the first time, taking part could be a difficult experience. On the other hand, the exercises were seldom climactic. The agenda for each meeting was usually developed no more than a day in advance. Several short reports were usually heard at each conference. On occasion, a conference might be devoted to a series of related reports on the same general topic. Thus, scientists were more likely to be called on periodically for short reports rather than for infrequent long reports. Most neophytes took these experiences in stride. One important advantage of this way of doing things was that individual scientists were oriented to the problems being worked on by others. In turn, this led to increased interactions, corridor consultations and discussions. It was not necessary to wait for an annual or semiannual seminar to learn about a colleague's research.

Cohn was skillful in eliciting the strengths and weaknesses of the presenter's work. Although his comments and suggestions were taken seriously, he seldom dominated the discussion. He was particularly adept at involving a guest who might be present. Cohn's senior associates were thoroughly accustomed to these meetings and participated actively. Visits by investigators from other institutions were usually scheduled to coincide with these laboratory conferences. In those days when travel was by train, even from New York, visits were likely to involve much or all of a day. Cohn saw to it that visitors were "passed around" to his associates; in that way, even junior postdoctoral fellows came to meet distinguished visitors from other institutions. For Edwin Cohn, the introduction of a visitor was always done in a gracious manner and included some small talk. But this was never allowed to intrude for long in the scientific discourse. It was during this period that Cohn fine-tuned his policy of discouraging visits that extended more than one day. He liked to state that a visitor could stay for a day or for a year, but not for any period in between.

The most important function of these weekly meetings was the broad discussion and integration of the scientific findings about the physical chemical characteristics of amino acids and peptides, and about their meaning within the context of the theoretical framework emerging from the studies being carried out by Scatchard and Kirkwood. The result was the reasoned organization and integration of the several facets of the amino acid and peptide investigations. In 1935, Edwin Cohn had been approached by the American Chemical Society to prepare a monograph about the amino acid and peptide studies. He agreed to do so, but when the time came, he was caught up in a new set of responsibilities occasioned by the mounting international crisis in Europe. When that happened, John Edsall stepped into the breach and completed the task, which had grown to major proportions by then. The manuscripts were completed in 1941 while Edsall was on a sabbatical leave with Linus Pauling at the California Institute of Technology.

Proteins, Amino Acids and Peptides, a 650 page book, was published in 1943. In the preface, Edwin Cohn and John Edsall concluded that "Something approaching an adequate theory of the physical chemical behavior of amino acids and peptides . . . appears to have been achieved by the cooperative efforts of workers in several different laboratories. These studies on the simpler molecules have laid a foundation for the study of some aspects of the chemistry of proteins.[24] Though never destined to be a best seller, "Cohn and Edsall," as the book came to be known, surprised its authors by going through five printings before the demand for it in the worldwide scientific community subsided.

In 1936, President Conant appointed Edwin Cohn to succeed Otto Folin as Chairman of the Division of Medical Sciences of the Harvard Faculty of Arts and Sciences. At Harvard, all Ph.D. degrees are administered by the Faculty of Arts and Sciences. The Division was composed of those members of the Faculty of Medicine who had graduate students registered in the program. It thus functioned as an important bridge between the Faculty of Medicine and the Faculty of Arts and Sciences. As Chairman of the Division, Edwin Cohn became a voting member of the Faculty of Arts and Sciences and administered the Ph.D. program in the medical sciences.

Cohn and the Fourth Estate

In the spring of 1936, Edwin Cohn had a bizarre encounter with the press. He had accepted an invitation to give a major lecture on the "Significance of the Size, Shape, and Electric Moments of Amino Acids and Proteins for Biology and Medicine" at the annual meeting of the American Chemical Society to be held in Kansas City in April 1936. Shortly before the meeting, a public relations officer of the Society asked him for an advance statement that could be used in a press release. Cohn discussed this request with C. Sidney Burwell, the newly appointed HMS dean and it was agreed that Cohn should provide the Society with a statement that was both adequate and accurate. This followed the practice that Cohn had adopted in dealing with the press in 1929 at the International Physiological Congress in Boston. Cohn submitted a two page statement, the lead paragraph of which read:

> It has long been known that bodily functions are accompanied by electrical manifestations. The contraction of muscles, the secretion of glands, the conduction of nervous impulses are all associated with electrical changes. The passage of electricity along the nerve has occasionally been compared to that along a telegraph wire, although it is certain that the conduction tissue is in no sense a metallic wire, but rather composed like other organs of the body of proteins and lipoids, acids and bases, salts and water. Of these substances, only acids, bases and salts are electrolytes, and conduct electric current, but the rate at which they move in an electric field is far too slow to account for the electrical properties of body tissues.[25]

The public relations officer returned a draft news release that contained what Cohn termed "unwarranted expansions." Corrections by Cohn were followed by further unacceptable drafts. As the date of the lecture was fast approaching, Cohn insisted that final copy be forwarded to him. When it was received, he went over it with Ronald Ferry and John Edsall, who removed more objectionable text and prepared a revised draft that was acceptable to them. A primary source of alarm was the running head, "New Electric Units Discovered," suggested by the Society. In a strongly worded letter, Cohn pointed out that

> What has been discovered is not new molecules, nor new electrical forces, but hitherto unsuspected electrical properties of molecules that have previously been known. If there are further changes in this statement that you would like to make, I trust that you will let me know so that we can decide whether they retain the accuracy of statement that has now been achieved.

There was no response from the Society. Even before he arrived in Kansas City, the newspapers were carrying stories that were completely unacceptable to Cohn. One such story appeared on the front page of *The New York Times* under the byline of William L. Laurence:

> Vast Network of "Living Dynamos' Supplies Electric Power to the Body." Prof. E.J. Cohn of Harvard Discovers Giant Molecules Generating Currents That Transmit Messages on Nerve Fibres. Ten Year Research to be Told to Chemists Session.[26]

The story was carried on the following day by *The Times* (London) under the headline: "Living Dynamos of the Body U.S. Scientist's Theory."[27]

Before his lecture, Cohn flatly turned down a *New York Times* request for permission to take photographs of his laboratory for its rotogravure section, and angrily refused to have anything further to do with the Chemical Society's news service. Much later, Laurence came to see Cohn and said he was sorry to hear that Cohn had not liked the article. Cohn vigorously protested against the headline that had been used, as well as the substance of the text, emphasizing the precautions he had taken in writing the article. In the course of their discussion, Laurence revealed that in preparing to write about Cohn's work, he had interviewed Professors Harold Urey and Hans Clarke at Columbia University. He even gave Cohn a copy of his notes from those meetings. According to the notes, Urey, speaking in general terms, interpreted his understanding of the electrical effects described by Cohn. The notes showed Urey stating:

> Dr. Cohn's findings represent a beginning in the understanding of how such behavior can happen – how muscles move, how nerve impulses discharge. The molecules themselves are shown not to travel. The mechanism of the process is being explained . . . When the circulation of the blood was discovered, there could have been no answer to the question: What use is this discovery to medicine? No one at that time could have foreseen the outcome of this discovery to the science of medicine, or named any definite applications. A parallel can be drawn between the discovery of the circulation of the blood and discoveries such as this concerning electrical phenomena in the body.

Clarke told Laurence that "Dr. Cohn clarifies the general picture of the electric condition in body fluids and tissues. He has made accurate determinations of the electric forces of the amino acids, protein, and phospholipoids. . . . His work aids in explaining the conduction of nerve impulses. . . . His is one piece in a great field of research. It is a "germ." The way in which electrical forces are initiated is not yet known—nor is it known what happens when an impulse reaches its destination."[28]

The publicity surrounding Cohn and his work did not escape the attention of the officers of the Rockefeller Foundation. Cohn stopped off to see Alan Gregg at the Foundation offices in New York on his way home from Kansas City and carefully raised the possibility of requesting supplemental funding from the Foundation in order to vigorously pursue studies of the chemistry of nervous tissue. His audacious proposal met with a lukewarm reception from Gregg who pointed out that a request for new funds before the expiration of the term of his present grant might disturb his relationship with the Foundation. On April 30, 1936, Cohn wrote to Gregg stating that "I am entirely willing to leave to your discretion the wisdom of requesting funds in addition to those that have already been granted by the Rockefeller Foundation in order to pursue vigorously studies on the chemistry of nervous tissue . . . Though disappointed, I am quite prepared to abide by your judgement."[29] There the issue was dropped. Cohn never initiated studies on the chemistry of nervous tissue.

It is difficult to escape the conclusion that Edwin Cohn was at least partially responsible for the contretemps over the "Living Dynamos." Speaking as a chemist to a national meeting of chemists, he was free to deal with arcane concepts such as dipolar ions and dielectric constants. On the other hand, in attempting to speak to the public, he risked being misunderstood. In so doing, he fell victim to his own characteristic tendency to generalize from his special field of science to the broader universe of biology. Although he did not coin the phrase "living dynamos," the lead paragraph in his statement to the Society press office clearly introduced the association between bodily functions and electrical manifestations. It is hardly surprising that William Laurence and other newsmen, trained in their craft to frame the "lead" for their story, seized on Cohn's first paragraph for their headline. Where did the term "living dynamos" originate? It was not found in Cohn's draft press release, nor was it characteristic of his usage. Since it was used in the headlines of both the *New York Times* and the London *Times*, it is likely that the term was included in the Chemical Society's press release on the Cohn story. Thereafter, Edwin Cohn took pains to avoid direct contacts with the press.

5 Return to the Study of Proteins

While the studies on the amino acids and peptides continued at a diminishing rate until 1939, studies on proteins were reintroduced as early as 1935. In one of the first of these, horse hemoglobin was substituted for aspartylaspartic acid in the titration experiment cited in the previous chapter. The results with hemoglobin are exemplified in figure 2, the coordinates of which are the same as those used in the figure in the previous chapter except that the scale of the vertical axis has been increased. The data plotted in the figure reveal that there were 255 groups capable of reacting with acid or base per 100,000 grams of hemoglobin. A few years later, when the molecular weight of hemoglobin was confirmed to be 68,000, Cohn and his colleagues estimated that the isoelectric hemoglobin molecule contained seventy-five dipole pairs, i.e. seventy-five positively charged groups and seventy-five negatively charged groups per molecule.

The beauty of the prototype studies lay in knowing the molecular weights and chemical structures of the amino acids and peptides being studied. When applied to the proteins, despite the fact that their chemical structures were not known, they began to reveal important new information. For example, the studies strengthened the hypothesis that it is the nature and the number of reactive side chain groups of amino acids in the protein which are responsible for the electrical characteristics of the protein molecule. In addition, the shape of the titration curve at specific regions of pH led to tentative conclusions about

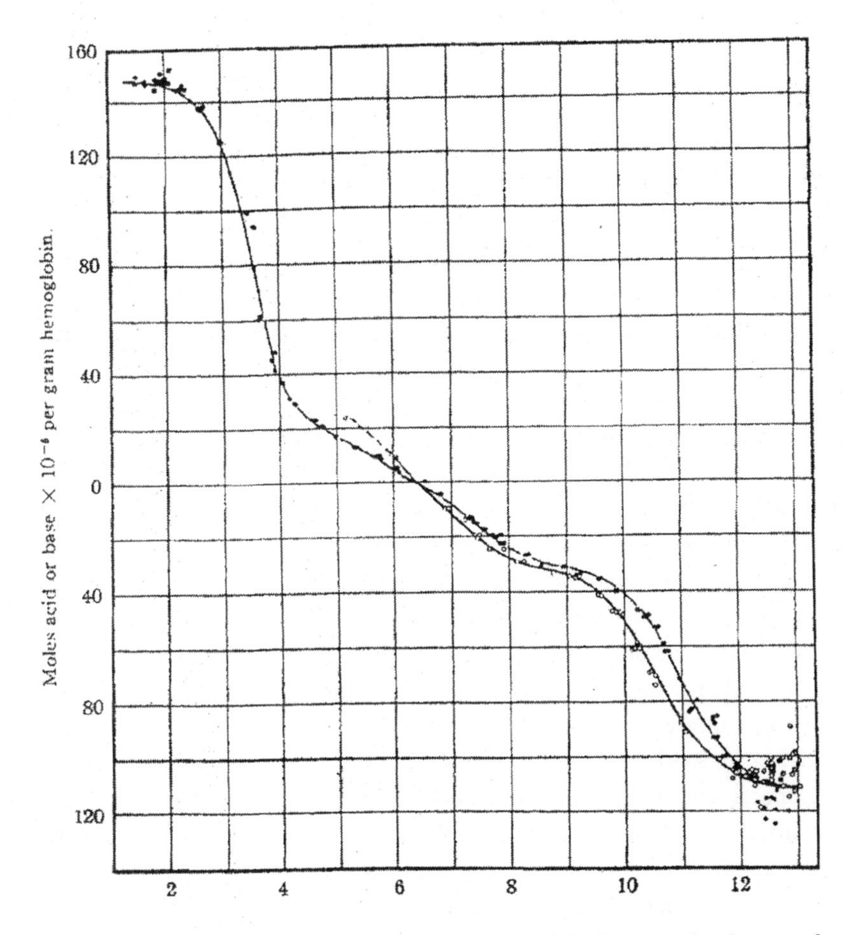

Fig. 2. The titration curve of carboxyhemoglobin of the horse in the absence of added salt and in the presence of 1 *M* NaCl.

which amino acid side chain R groups were dissociating or associating acid or base at any given pH. The hemoglobin titration study also explained the buffering property of hemoglobin against the effects of acids and bases in maintaining the physiological neutrality of the internal environment. This important property depends upon the concentration of hemoglobin in the blood, which is high, and upon the slope of the titration curve at physiological pH, which is positive.

Powerful New Tools

New technical advances contributed significantly to the growth in knowledge about proteins. In 1935, glass electrodes became available for the measurement of pH. They would gradually replace the cumbersome hydrogen electrodes that Cohn had used since his days at Columbia after the first war. Some of the new instrumentation developed for the amino acid studies was adaptable to the study of proteins. Just as Wyman's instrument for measuring dielectric constants provided critically important new information about amino acids and peptides, John L. (Larry) Oncley's new radio frequency bridge extended the measurement of dipole moments and dielectric dispersion to provide important new information about proteins.

It was during this period that Edwin Cohn first encountered Larry Oncley. While still a graduate student at the University of Wisconsin, Oncley had become interested in measuring the rotation of chemical molecules in solution. He had succeeded in detecting some evidence of rotation by small molecules, but needed larger molecules where the effects might be more pronounced; that is how he became interested in proteins. With a NRC postdoctoral fellowship, he came to work with Fred Keyes, the Chairman of the Chemistry Department at MIT in 1936. While at MIT, Oncley met Edwin Cohn who offered to supply some proteins for study and help purchase the apparatus he would need to begin studying the dielectric properties of proteins. The beautiful studies of Wyman on amino acids and peptides had revealed that the amino acids and peptides behaved as rigid rod- like structures with their charges at either end. Would it be possible to make similar studies on proteins? This proved to be a much more complex problem. Oncley's first publication, "Studies of the Dielectric Properties of Protein Solutions. I. Carboxyhemoglobin," appeared in 1938 as a joint contribution from the Research Laboratory of Physical Chemistry, Massachusetts Institute of Technology, and the Department of Physical Chemistry, Harvard Medical School.[1]

For that study, he used a newly developed apparatus called a radio frequency bridge to study the electrical symmetry of hemoglobin, i.e. the state of distribution of the seventy-five positive and seventy-five negative charges within the isoelectric hemoglobin molecule. The principle of the method was similar to that used earlier by Wyman to measure the dielectric constant of amino acids, except that Oncley's apparatus made possible the study of more highly charged molecules. Were the effective center of all the positive charges on the hemoglobin molecule to be separated from the effective center of the negative charges by even a small distance, the isoelectric hemoglobin molecule would act like a huge dipole. That proved to be true of hemoglobin. Oncley found that it has a dipole moment of approximately 500 Debye units—about twenty times greater than that of the simplest amino acid, glycine. This suggested that even at its isoelectric point, there is an asymmetry in the distribution of electric charges in hemoglobin, i.e. it is a dipole. This study was the first of its kind on proteins. A second study was conducted by John Ferry and Oncley using several water-soluble proteins of horse serum. After receiving his Ph.D. in chemistry from Stanford University, Ferry had come to Harvard in 1936 with an appointment as Instructor in Biochemical Sciences. He was a Member of the Society of Fellows between 1938 and 1941, after which he moved over to the Department of Physical Chemistry.

Ever since his 1926 visit with Svedberg in Stockholm, Cohn had resolved to obtain an ultracentrifuge for his laboratory. No other instrument could enable the direct measurement of the molecular weight of proteins and provide insight into their size and shape. However, because of their technical complexity, few Svedberg ultracentrifuges had been built outside Sweden. In 1936, Cohn learned that a new ultracentrifuge was being built by J.H. Bauer and E.G. Pickels at the International Health Division of the Rockefeller Foundation, using a new air turbine drive designed by J.W. Beams, a professor of physics at the University of Virginia. Cohn decided to build a similar instrument for his laboratory. As a first step, President Conant appointed a Harvard Ultracentrifuge Committee and empowered it to develop the project. It consisted of George Minot, Hans Zinsser, Baird Hastings, Jeffries Wyman, Jr. and William T. Salter, with Edwin Cohn as its chairman. Early on, it emerged that a qualified scientist would be required to oversee the construction of the instrument and serve as principal investigator in future research projects that would be undertaken. Oncley was given this responsibility. However, since his academic appointment was then at MIT, steps were instituted by Cohn and the Ultracentrifuge Committee to obtain the approval of President Conant and MIT President Karl Compton for an unusual joint faculty appointment for Oncley at the two institutions.[2]

When the committee came to the choice of instrument to build, Cohn summed up the discussion by stating that

> there is no doubt that further improvements will be made in both types of centrifuge. My consultations have satisfied me that the simplicity and relative inexpensiveness of the air-driven type of ultracentrifuge indicate its adoption for our purposes at the present time. It is surely adequate for the study of viruses and all but the smallest protein molecules. It is even probable that further improvements can be made in the method of driving or suspending the rotor, or in the shape or dimensions of the latter. The convenience, moreover, of building an instrument developed in America, parts and repairs for which can be made in America, should not be overlooked.[3]

Arrangements were made with Beams, Bauer and Pickels to obtain the blueprints as well as certain critical components of the machine. The committee also identified an expert instrument maker, and arranged for space to house the ultracentrifuge and a new machine shop where many of the parts could be fabricated. Preliminary discussions were initiated with a foundation regarding the possibility of grant support for the ultracentrifuge laboratory.

It had been known for many years that dissolved proteins migrated in an electric field, a mechanism known as electrophoresis. Recent advances by Arne Tiselius at Uppsala University in Sweden materially improved the precision and general utility of the method. These included the design and fabrication of a special optical glass cell with parallel sides, a schlieren optical system, and performing the electrophoretic analyses at a temperature of 4°C, where water has its maximum density. While on sabbatical leave in 1937, Ronald Ferry spent some time in Tiselius' laboratory in Sweden, and on his return to Boston, he brought with him one of the new optical cells together with the latest specifications for the complete electrophoresis apparatus. This made it possible to assess the purity of protein mixtures, e.g. albumin, and alpha, beta and gamma globulins, according to their mobility in an electrical field. The Harvard ultracentrifuge and the Tiselius electrophoresis apparatus were demonstrated in September 1939 during a Boston meeting of the American Chemical Society.

Application for Continued Rockefeller Foundation Support

Meanwhile, the seven-year Rockefeller Foundation Grant to Cannon and Cohn was due to expire in 1937. Since Walter Cannon was then approaching retirement age, the officers of the Foundation decided to separate any subsequent support for Cannon and Cohn. They initiated an in-depth review of Cohn's scientific contributions and extended a grant to Cohn for one additional year of support at the level of $12,500. This provided time for reviewing a new grant application without interrupting Cohn's scientific work. (Walter Cannon received continued support through a separate grant.) As the first step in the renewal process, a formal application to the Foundation for a new grant to support Cohn's work was submitted by Dean Burwell to Warren Weaver, Director of the Division of Natural Sciences of the Rockefeller Foundation in March 1937. In a brief letter of transmittal, President Conant described Edwin Cohn's efforts as "one of the important research activities now going on in the University." In justifying the need for continued support of the work in the Department of Physical Chemistry, he indicated that other funds to fully support the work were not available. However, he raised the possibility that Harvard might be able to interest "certain donors in capitalizing" Cohn's venture in the future. The Harvard application requested a grant of $15,000 per year for a period of from five to ten years.[4]

In a separate letter to Warren Weaver, Dean Burwell expressed confidence that Cohn's department, lying between physics and chemistry on the one hand, and physiology on the other, would yield important translations into physiology of the great advances in physics and chemistry. He stated that Cohn "has attacked his problems with energy and insight, he has not been deflected from his major plans, but has gone steadily ahead. He has established a group which is certainly one of the leading groups engaged in the investigation of this important field, and there seems every reason to expect a continuation of both the energy and the quality of his attack on the subject."[5]

In discussing the recent work of the laboratory, Cohn cited several milestones in protein research, beginning with the electrostatic force theory of Debye and Hückel. He recounted how the Harvard studies had led to the development of a theory concerning the behavior of dipolar ions in solution, stressing the relationships between their free energy, electric moments and changes in ionic strength, in dipole moment and the dielectric constant of the medium. He attributed great importance to the work of Jeffries Wyman on the dielectric constant of solutions of amino acids and peptides. Another important step was the development by Oncley of ways to measure the dielectric constants of protein solutions, revealing the importance of defining their size, as well as their number and distribution of electrically charged groups. In referring to the recent studies on amino acids and peptides, Cohn stressed the opportunities they provided to relate physical chemical behavior to their chemical structures. As one example, he cited the influence of glycine on the interaction between the amino acid cystine and sodium chloride as the prototype of interactions of proteins with each other and with electrolytes in biological systems.

The grant application provided a picture of the financial support for the Department of Physical Chemistry at the time. Fifteen persons were being paid from funds totaling $51,000, including $24,475 from the De Lamar Bequest of the Medical School, $16,530 from Harvard, and $10,000 from the expiring Rockefeller Foundation Grant. Separate Harvard funds provided support for Ronald Ferry as Master of John Winthrop House as well as for John Edsall, John Ferry, Jesse Greenstein and John W. Mehl, who served as Tutors in Biochemical Science in Cambridge. Oncley, although holding an appointment at the Medical School at the time, received all his support from MIT.

The officers of the Rockefeller Foundation conducted their own evaluation of Cohn's accomplishments by consulting an international group of experts familiar with his work. Of particular interest were the following comments. John H. Northrop at the Rockefeller Institute in Princeton wrote that "the results of the work in his laboratory are of great interest both theoretically and practically. His laboratory is one of the few in the world where reliable and accurate data are being compiled on proteins and related compounds."[6] Michael Heidelberger, the immunologist, hesitated to appraise Cohn's qualities as a physical chemist, stating that "I am, however, able to appreciate the care and accuracy with which he works, and to see the importance of a thorough knowledge of the physical chemistry of protein structure and behavior." Heidelberger added that he had talked with Professor Svedberg about Cohn's hemoglobin titration curves and reported Svedberg's view that no other physical method could provide that information.[7] Linus Pauling, in a conversation with Warren Weaver, rated John Edsall's work on the Raman effect as "most interesting and promising," but stated that "while the accumulation of so much data was necessary and interesting, it had in fact led to very little as regards our fundamental understanding of the proteins."[8]

W.T. Astbury, the British crystallographer who had recently visited Cohn in Boston, offered a colorful assessment.

> Judging by the way they worked me—besides a formal lecture I gave a colloquium lasting the best part of a day—there can be no doubt of their enthusiasm there for proteins, and I gathered the impression of a live band of workers. As for Dr. Cohn himself, I had not met him before and I could hardly say just what part he plays in the papers he publishes with various collaborators, but he seemed full of energy and his subject.[9]

Max Bergmann, a professor of organic chemistry in Germany who had recently become a member of the Rockefeller Institute in New York, wrote:

> As objects of chemical and physico-chemical investigation, the proteins confront the scientist with two main problems. The first is to ascertain the general structural principle by which the building stones of a protein molecule are linked to each other. This branch of investigation is the aim of the organic chemist . . . The second problem originates in the fact that the proteins must contain definite accessory linkages which transform the peptide chain into either extended micelles or sphere-like structures. These accessory linkages, which originate in the many electropolar groups of each protein molecule, determine the physico-chemical and biological properties of proteins in colloidal suspension and in solution. It is the study of these accessory linkages and forces that is the field of the physical chemist, and the object of the many sided investigations of Dr. Cohn and his collaborators. Due to the extreme complexity of the protein molecules, experimental investigations of this kind and their theoret-

> ical interpretation belong to the most difficult tasks of physico-chemistry. They have to be attacked on a broad experimental basis and demand a wide knowledge and great skill...The work of Dr. Cohn, which fulfills these requirements, is and gives promise of continuing to be of very great importance to the development of physico-chemistry and to many fields of biology.[10]

C.R. Harington of University College Hospital Medical School in London stated that the Department at Harvard

> has been highly esteemed in this and other countries for the past ten or fifteen years as one of the few laboratories in which a serious attempt has been made to apply accurate physico-chemical methods to the study of proteins. Whilst the output of work from the Department has been large in amount and of high quality it is in my opinion subject to the criticism of a certain lack of inspiration. That is to say it is thoroughly sound but pedestrian in conception and unexciting in results . . . This general criticism . . . is subject to certain reservations, since major contributions have appeared . . . It happens, however, that during the past summer I had the opportunity of a long discussion with Dr. Cohn in which one of my physico-chemical colleagues took part. From this discussion I gathered that Dr. Cohn was planning an investigation of the question of distribution of charges on the protein molecule, and this I regard as a problem of more fundamental importance than those which he has hitherto attacked both in its chemical and biological relations.[11]

K. Linderstrom-Lang, who succeeded Sorensen at the Carlsberg Laboratory in Copenhagen, concluded a favorable letter by adding: "I don't think he (Cohn) is a very original thinker in theoretical respects but he has a clear sense of using the theories presented by others on problems of his own and his selection of these problems has been very fortunate."[12] Alexander von Muralt credited to Cohn the discovery of the important influence which proteins have on the dielectric constant and their strong effect on Coulomb forces. He predicted that this entirely new point of view will give rise to valuable insight into a great number of living processes, such as permeability, secretion, resorption and perhaps also excitation, and felt that these data will some day be the basis of a new science which will be called "theoretical biology."[13]

A.C. Chibnall of the Imperial College of Science and Technology in London described a visit to the Department of Physical Chemistry:

> Most of my time was spent with Cohn. I know of no other department of this size, devoted entirely to research, in which all the people seem so happy and get on so well together. A lot of the work they are doing is far beyond my comprehension—too physical by far—yet they certainly persuaded me that it was well worth while, and that it was part of a long programme that Cohn has mapped out for them during the next few years. The Ferrys, Edsall, Greenstein and the others are first rate team men and Cohn manages them extremely well, in spite of the fact that he appears to irritate a lot of people outside Harvard. I found the team so interesting that I called on them a second time, on my way back to New Haven, and I have carried away with me an impression of one of the most efficient University Research Laboratories that it has been my privilege to visit. I saw Hastings and went round his department, but most of his work is concerned with animal biochemistry.[14]

As the review by the Rockefeller Foundation proceeded, Warren Weaver recognized that the long-term support of Edwin Cohn's work by the Foundation could have implications for Harvard's central administration. On a visit to Cambridge late in 1937 he met with President Conant and discussed the scientific judgements of Edwin Cohn's work which were being received by the Foundation. He sketched the outlines of a continuing award he was preparing to recommend to the Trustees of the Rockefeller Foundation. President Conant pointed out that while the level of financial support being considered by the Foundation was less than what Cohn hoped for, it was also true that Edwin Cohn's aspirations tended to exceed realistic limits. From this discussion, a plan was devised to buttress the Harvard administration's continuing relations with Edwin Cohn. Warren Weaver agreed to write to Cohn outlining the funding plan being considered by the Foundation. In turn, President Conant agreed to call Cohn in to his office, and, as a condition of the approval by Harvard of a new Rockefeller Foundation grant, elicit from Cohn an assurance that he would be content to continue his studies within the level of funding contemplated by the Foundation together with the support Harvard was then contributing. In this way the Foundation could fulfill its desire to continue supporting

Cohn's research while strengthening the hand of the Harvard president in dealing with persistent pressures from Edwin Cohn for more money.[15]

New Scientific Achievements

In the spring of 1938 the Rockefeller Foundation announced the award of $100,000 to Harvard University "for researches in the Department of Physical Chemistry of Harvard Medical School under the direction of Dr. E.J. Cohn." Of this, $16,000 constituted a special reserve for equipment, while the balance was to be made available in annual amounts of $14,000 over seven years, until 1945.[16] Except for the special equipment reserve, the award was "given without defining the project in advance." In addition to assuring an important segment of the financial support that Edwin Cohn would need for the next seven years, this new award by the Rockefeller Foundation was a singular recognition of his stature in science. It placed him in an enviable position among his peers in the university. On the other hand, it tended to distance him and his department from colleagues on the medical school faculty. The original concept of a laboratory devoted entirely to research within the Harvard Medical School was a source of irritation to a growing number of basic medical scientists in medical school departments. Nor did Cohn's disdain for anything but pure scientific research generate much collegial support among the clinical faculty at the medical school. These feelings were reinforced by a perceived mounting affinity of Cohn and his laboratory to the university rather than to the medical school.

The new grant was generous in its span of committed years of support. It gave Cohn an opportunity to engage in long term planning. The timing of the new award was also almost perfect, occurring as it did just as the landmark work on the amino acids and peptides was reaching completion and he was turning his attention once again to the study of proteins. He had just passed his forty-fifth birthday, which meant that, with good fortune, he might have two more decades to work before retirement. He began to assemble the elements of a broad plan of research directed toward his primary goal: to learn about the structure of proteins and to elucidate how protein structure confers upon the proteins such a wide array of specific physiological functions. However, it was clear that the implementation of his plans would require more resources than would be forthcoming from Harvard. He therefore began to explore new approaches which would move him toward the goal of understanding the structure and function of proteins. New investigators should be brought in and new instruments for the study of proteins should be acquired. He visualized an exciting new period of exploration for which the newly accumulated knowledge and theories about the physical chemical behavior of the amino acids and peptides would provide a launching platform. Accompanying these plans, there was a crescendo of activities on a broader stage. He increasingly accepted invitations to speak about the work of his laboratory at national and international scientific gatherings.

In October 1938, Cohn was advised that the trustees of the Commonwealth Fund had approved a grant of $20,000 to the Harvard Medical School for the new ultracentrifuge. Grants of $3,000 and $5,000 were also received from the Genradco Trust and the Loomis Institute.[17] Important contributions to the project were made by the International Health Division of the Rockefeller Foundation in supplying blue prints and in consulting extensively with Oncley while the instrument was being built and installed. This included funds to support Charles Gordon, the instrument maker. The ultracentrifuge was installed in a new laboratory in the basement of Building E-2 at the Harvard Medical School. The installation included a reinforced concrete shield to protect personnel against injury should the dural rotor containing the protein solution fracture at its high operating speed—over 100,000 r.p.m. As installed, the instrument included a schlieren optical system which required a long optical bench. A camera located some distance from the spinning rotor recorded the data. Under the original plan, Gordon was to be kept on for an introductory period to operate the instrument. In the course of subsequent events, however, he spent the rest of his career performing ultracentrifugal analyses of proteins.

At the annual meeting of the American Society of Biological Chemistry in Atlantic City in April 1938, Cohn and his associates presented seven papers. Of these, five involved studies of proteins: globin, urinary proteins, two enzymes—asparaginase and pepsin—and proteins of smooth muscle. In 1938 and 1939, they published a total of forty-eight scientific papers describing their investigations of physical chemical characteristics of ten different proteins. Later that year, reports from the Department of Physical Chemistry occupied a prominent place in the annual Cold Spring Harbor Symposium on Quantitative Biology, with papers by Cohn, Oncley, John Ferry, John Edsall, John Mehl and Jacinto Steinhardt. Cohn's lecture bore the title "Number and Distribution of the Electrically Charged Groups on Proteins."[18]

His views about proteins at that time are revealed in his lecture, in which he told that proteins had very definite, if as yet incompletely apprehended structures. The specific groups of which they were composed was located in definite spatial relationships to each other within the vast molecules, with size and shape varying from protein to protein, as do the arrangements of the constituent groups. It was in the definite arrangements of aliphatic and aromatic, of various polar and non-polar groups that proteins differed from other colloids. Further, despite their size, proteins resembled the amino acids and peptides of which they were constructed. He pointed out that when there was an asymmetry in the arrangements of the charged groups of proteins, this gave rise to a dipole moment. As a consequence, such molecules would increase the dielectric constants of most solvents, including water. Further, Cohn emphasized that proteins often behave as though they were far more rigid than simple peptides. Electrochemical methods were thus capable of yielding information regarding the number and nature of the charged groups of protein molecules, as well as the symmetry of their arrangement, of their dipole moments, and of the dielectric constants of their solutions. Moreover, these parameters often determined interactions with other proteins.

A year later, in a Harvey Lecture at the New York Academy of Medicine, Cohn offered a confident projection of the advances which could be expected in the years ahead.

> In order to gain insight into the behavior of even the simplest proteins, we have thus been forced to explore the behavior of molecules of known structure such as the amino acids and peptides . . . The methods would appear to be at hand . . . for the study of all proteins. There is no theoretical obstacle to the isolation of all the protein

> constituents of any given tissue, or to their characterization as chemical substances, and to the study of their interactions as biological components. Proceeding thus, often with new techniques, but employing the classical methods of physical chemistry, we may hope in time to achieve an understanding of the morphology and physiology of biological systems in terms of the properties of their components.[19]

The advances in scientific knowledge about proteins being made at the Department of Physical Chemistry at that time represented the work of Cohn and about fifteen collaborators who, with their junior associates and students, worked in different institutions and departments, mostly in the Boston area. Cohn was the recognized leader. Although the epicenter of the efforts by the group was the weekly conference in Cohn's office, there was otherwise surprisingly little formal structure. If there was a common factor among the participants, it stemmed from a conviction on the part of each collaborator that the problems being addressed were interesting and important. They were bound together by little else.

Recommendations for faculty appointments and promotions in the Department of Physical Chemistry were made to Sidney Burwell, the new dean of the medical school. During the 1937–38 academic year Cohn recommended John Edsall for promotion to the rank of Associate Professor of Biochemistry. The matter came up for action in a meeting of the Committee of Professors that was chaired by President Conant in March 1938. A second case, of a "Dr R" who had been recommended by Baird Hastings for a similar promotion, was considered at the same time. Cohn had attended the meeting but was so concerned about its possible outcome that he mailed off a long letter to the President the next day. He feared that the discussion in the meeting had not focussed on the qualifications of the candidates but rather on the type of biochemistry a medical school should support. Pointing out that he had no reservations whatever regarding the desirability of promoting Edsall, he declared the "breadth of Edsall's learning impressive . . . not merely encyclopedic but of the kind that results in incisive experimental approaches to new problems."

However, he feared that because most of Edsall's recent papers were published in the *Journal of the American Chemical Society* and the *Journal of Chemical Physics*, his work might not be known or appraisable by members of the Medical School Faculty and wrote:

> This, I believe, was responsible for the different impressions you must have received regarding the two men, and not their intrinsic qualifications. Both are strong candidates. I know them both well, and though I rate Dr. Edsall higher, I believe that both should unquestionably be promoted . . . Had I appreciated in 1920 the strong bent toward supporting primarily such research with Medical School Funds as would be immediately appreciated by a faculty of medicine, I would of course not have gone to work at the Harvard Medical School . . . I must ask you . . . to promote John Edsall if he merits promotion on the basis of his scientific attainments and character. The proper solution of this case involves, I believe, the morale of many younger men.[20]

John Edsall was promoted to Associate Professor of Biochemistry in the spring of 1938. Oncley became a full time Harvard appointee in February 1939.

Attempts to Initiate an X-Ray Diffraction Study

Ever since his 1919-1920 postdoctoral studies in England and Scandinavia, Cohn had been aware of the potential of X-ray diffraction as a tool to study the molecular structure of proteins. The first evidence about the internal structure of proteins began to emerge from X-ray diffraction studies of dried proteins in the laboratory of W.T. Astbury at the University of Leeds in England in 1934.[21] That work had taken an important new tack when Bernal and Dorothy Crowfoot in England succeeded in devising a method for performing X-ray diffraction studies on wet protein crystals.[22] This yielded evidence showing that proteins had well defined three-dimensional structures. Svedberg had examined the behavior of a number of proteins in his ultracentrifuge and had found that, under certain conditions, some proteins dissociated into smaller subunits.[23] However, the amino acid composition of the first protein would not be known for another decade, and it would be a quarter of a century before the first amino acid sequence of a protein chain would be discovered. Meanwhile, Cohn had kept abreast of advances in this field, consulting with several experts including Bernal and Crowfoot in England as well as the mathematician von Neumann at Princeton about the best way to make an entree into this field. As the result of a conversation with Bernal at Woods Hole early in 1939, it was arranged that I. Fankuchen, an American, then spending a year with Bernal, would spend a year in Boston working with B.E. Warren at MIT who had a well equipped X-ray diffraction laboratory. Cohn succeeded in securing the necessary financial assistance to make this possible through a special grant from the Eli Lilly Company to the National Research Council to fund an NRC fellowship for Fankuchen. Over and beyond these arrangements, Edwin Cohn planned to keep in close touch with Fankuchen and furnish purified proteins for his study.

Unfortunately, darkening events in Europe overshadowed Cohn's plans. On September 1, 1939, Adolf Hitler's armies attacked Poland and Great Britain and France declared war on Germany. In October 1939, Bernal wrote to Cohn:

> Events seem to have decided things, and any cooperation in the protein field will have to count out most if not all Europe from now on. I am sorry because it seemed as if we might have contributed something to the solution of the main problems. It is particularly unfortunate with regard to X-ray work. I shall not be able to touch it for a long time, and I do not think that anyone else here will either. My only hope is that it may be possible to get it well started in the States through Fankuchen, who I am sending over with all my materials to carry on if he can find some means of doing so. I hope you will be able to help him in this along the lines that we discussed in Woods Hole. All he wants really is some apparatus and enough to live on, though it would be generally advantageous if he had a chance of showing the technique to some of the younger research workers.[24]

Fankuchen worked at MIT under this arrangement between 1939 and 1941. By then, however, Cohn had become involved in a much larger venture.

A Visit by Warren Weaver

At that time, Warren Weaver visited the laboratory for an afternoon and talked with Cohn, Edsall, Oncley, John Ferry, McMeekin and Hans Mueller of MIT who happened to be there. Weaver's notes provide an account of the visit and shed a rare beam of light on the relationship between a Foundation officer and its grantee.

> C. [Cohn] discusses at considerable length the arrangements he manipulated whereby Clowes of Eli Lilly will furnish to the NRC, for the committee on protein research, funds which will enable Fankuchen to spend this year with Bert Warren of MIT. Rose Mooney, on a Guggenheim fellowship from Texas came to C. and wanted to do X-ray work on proteins. She has also joined the Warren-Fankuchen team where her reputed excellence, the experience of Fankuchen, and the technical equipment and knowledge of Bert Warren should combine to make a splendid situation. C. will be in close touch with this work, furnishing them with purified proteins and getting from Northrop the ones which he [Cohn] cannot furnish. Fankuchen is the man who has done most of Bernal's actual experimental work the last few years. This program has had the benefit of direct advice from Bernal to Cohn. C. spoke of the fact that there were only four places in the country where Fankuchen could profitably go—Pauling at CIT, Lark-Horovitz at Purdue, Zachariasen at Chicago, and Bert Warren at MIT.
>
> C., who himself, apparently lives in a world of subtle and complicated intrigue, in which no one says just what he means, is forever reading deep and unintended significance in chance remarks of WW, later coming back with the proud report that he believes he correctly senses what WW meant when he earlier said something else. Thus C. now reports with considerable satisfaction that he believes he has figured out just what WW meant a year or so ago by certain wholly innocent and frank remarks concerning C.'s hopes to get a super-centrifuge. C. has obtained approximately $20,000 from the Commonwealth Fund and a splendid Beams-Bauer-Pickels instrument has been constructed, and funds are in hand to operate this laboratory for three years, Oncley now receiving all but $500 of his salary from Cohn and retaining a more or less formal connection with M.I.T. The work of the laboratory as a whole seems to be going forward splendidly, WW getting an excellent impression which fully

substantiates the complimentary remarks recently made by Tiselius, Bernal, and others.[25]

Cohn's Assessment of Significant European Science Threatened by War

In February 1940, Weaver wrote asking Cohn, in view of his special knowledge and competence, "to identify any minor or major fields of research in which the sole progress was being made by scientists in Europe, or to which European workers were contributing in such a large and significant measure that progress would be severely handicapped or even threatened if the efforts were to be terminated by the onset of hostilities."[26] Cohn responded in a remarkable letter in which he ranged over European science from Vienna to Uppsala and from Poland to Cambridge, England.[27] In Vienna, he opined that there had been ample time to disseminate and incorporate the work of the Viennese school in muscle chemistry. For Germany, he sketched the lineage of great organic chemists then led by Hans Fischer and Wieland continuing the tradition of Liebig, von Baeyer and Willstätter, and concluded that there had been enough time for the dissemination of that body of learning. He also noted that Bergmann, who had just come to the United States from Germany, was carrying on that tradition here. As for the physics and chemistry of isotopes under Hevesy in Copenhagen, Cohn considered that work in that field was well under way in the U.S. with Urey and Lawrence. He commented on the importance of the Otto Meyerhof school of biochemists whom he thought were "safe" in Paris. For tissue chemistry, he mentioned three scientists in central Europe: Weber in Munster, Parnas in Poland and Warburg in Berlin. As for Denmark, he believed that much of the new work there was now being developed in the United States.

In Cohn's view, the most original contemporary contributions were coming from Sweden, particularly from Uppsala. He commended the insistence of the Swedish school on a standard of accuracy that had not hitherto been achieved, as well as on their application of optical systems to their instruments. He concluded that "everything should be done to protect in every way possible the vital group now working in Sweden." As for English science, he expressed a view that, whatever the future held, a large number of good American students were being trained there.

Gaps between Scientific Aspirations and Fiscal Support

Although the new grant from the Rockefeller Foundation assured limited growth in the research activities of the Department of Physical Chemistry, Cohn began to realize that, unless additional research funds could be secured, there would be a gap between his own aggressive aspirations in science and his ability to continue the pattern of collaboration which had worked so well in the old "amino acid and peptide" days. Buoyed by the success in securing support from three different foundations for the new ultracentrifuge, he began to explore other sources of support from outside the university, including the possibility that developments in protein chemistry might find applications of industrial importance. After all, proteins were ubiquitous in natural products, from food stuffs to textiles, leather and plastics. Postulating that many of the problems of industry were closely related to the issues being addressed by protein chemists in his laboratory, he turned to his old friend, Wallace B. Donham, the Dean of the Harvard Business School for help and advice.

When they met for lunch in the spring of 1938, Cohn asked Donham for assistance in approaching firms in the corn products industry, suggesting that the corn products industry might well be interested in the work then being done on proteins at the Medical School since corn is rich in a protein called zein.[28] This contact with Donham resulted in a creative undertaking in which three firms—the American Maize Products Company, the Corn Products Refining Company and the A.E. Staley Manufacturing Company—each contributed $11,250 to a fund at the Harvard Medical School to be used for the training of protein chemists in the Cohn laboratory. This was followed later by gifts totaling $60,000 from the Corn Products Refining Company, made "with no strings attached, as has been the understanding in the past, but with the hope that, with it, you may be able to get such additional support as will maintain the splendid work that has been done in your laboratory."[29]

In the meantime, Cohn reminded President Conant of his commitment to explore the possibility of interesting "certain donors in capitalizing" the Cohn efforts. President Conant cautiously attempted to set up a conversation with a prominent individual associated with one of America's largest drug firms. But on January 22 1940, the President advised Cohn, "I am afraid my letter must have scared him," and suggested that Cohn visit an associate of the individual involved and "see if he can find out anything about why our friend has become so gun shy."[30] When this was done, Cohn responded, "I am reluctantly forced to the conclusion that his failure to write is evidence of the lack of interest in supporting the protein work."[31]

In the midst of this effort, President Conant sent a note to George H. Chase, the Dean of Harvard University, requesting that he

> get hold of Dr. E.J. Cohn and let him pour out his grief to you for several hours. (He is a very entertaining person, incidentally) . . . The question today is where does E.J.C. and his laboratory fit into the University? And more important, who is to finance it? E.J.C. feels it is my personal responsibility to go out and get funds for him because (1) I'm a chemist and (2) because my personal prejudices and actions have put him into the hole he now is in. From my point of view his work is good, but he can't cooperate with his equals or superiors—he is hopelessly "high hat" about all but "pure" or "basic" research. His budget is enormous . . . I'd like to have a committee between E.J.C. and myself. Why must a President of the University take on this man? . . . By listening and asking questions, you may steer him into a better mood and tell him the answer to a tough question, namely, how to keep him happy without making his support the first charge on my time and on the University's resources![32]

As for the attempted contact with the aforementioned prominent individual, Cohn advised President Conant that "We all agreed, I think, that people with the means to make the kind of gift that we have in mind are guarding their liquid assets very carefully in these uncertain days . . . My only regret, therefore, is that we did not start on this task three months ago. I feel certain we could have achieved the end you now have in mind, but I completely agree . . . that we are more likely to lose good prospects than to raise large funds at this moment." Cohn concluded: "As soon as the state of world affairs improves I shall be perfectly prepared to go vigorously to work."[33] President Conant responded with a single sentence: "The state of world affairs, I am afraid, may postpone consideration of this question of money for a long, long time."[34]

Illustration 1. The young Edwin Cohn (undated). Below: Marianne Brettauer as a bride in 1917.

Illustration 2. Above: Harvard investigators of proteins in wheat at the Wolcott Gibbs Laboratory in Cambridge during World War I. Theodore William Richards and Lawrence J. Henderson are seated in the foreground; Edwin J. Cohn is in the center of the back row. Photographed in 1918. Below, Alfred E. Cohn (left), as he appeared in June 1923; Walter B. Cannon (right), HMS Professor of Physiology and Chairman of the Committee on Transfusions of the National Research Council, who encouraged Cohn to undertake his pioneering studies of the proteins.

Illustration 3. Above: Marianne Brettauer Cohn with her sons, Alfred Brettauer Cohn (seated in her lap) and Edwin Joseph Cohn, Jr. in 1924; Below: The Cohn home at 183 Brattle Street, Cambridge, into which the family moved in the autumn of 1929.

Illustration 4. Scenes from Edwin Cohn's sabatical leave in Europe, 1931–1932: Above, The Cohn family about to depart from Naples on their 1931 Mediterranean tour; Fred and Ed Cohn in front, Marianne and Edwin Cohn in the rear. Below, left to right, Alice von Muralt, Edwin Cohn and Marianne Cohn taking lunch in San Martino di Castrozza (left); Edwin J. Cohn, Jr. and Alfred B. Cohn sitting in the Cohn automobile on the dock upon arrival back in Palermo (right).

Illustration 5. Above: At the Brettauer summer home in Lake Placid, New York in 1934. Fred Cohn (left); On the right, left to right, appear Mrs. Brettauer, Edwin J. Cohn, Jr., and Marianne Cohn. Below: George Scatchard (left) and Jeffries Wyman (right).

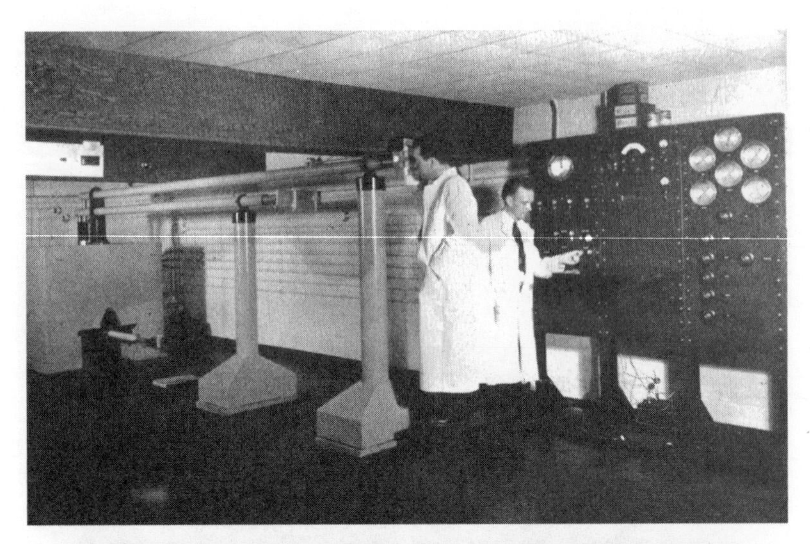

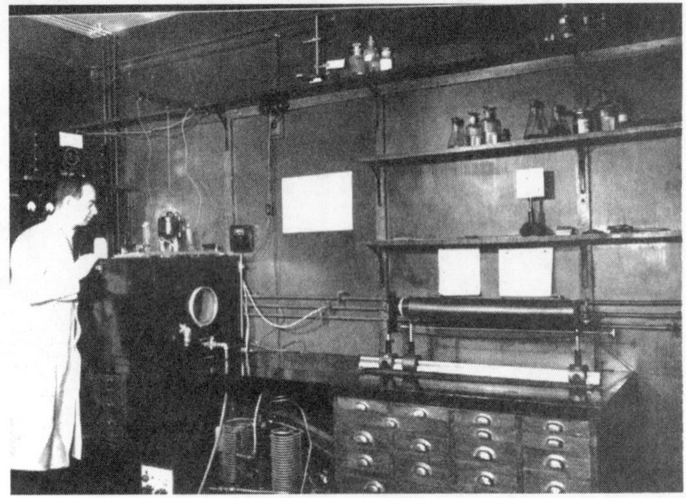

Illustration 6: Above: the Harvard ultracentrifuge installed in the basement of Building E of the Harvard Medical School in 1938. The machine itself was housed within a concrete structure near the back wall. The long tube coming forward carried the optical systems for viewing and photographing the sedimentation of proteins during the long measurement period. Viewing the status of the run are Dr. E. G. Pickels of the International Health Division of the Rockefeller Foundation and Mr. Charles Gordon, who built the ultracentrifuge under the direction of Drs. Pickels and J. L. Oncley. Gordon stands near the control panel. (Below): The Harvard electrophoresis apparatus being demonstrated by the machine's inventor, Dr. Arne Tiselius of the University of Uppsala, for attendees at the national meeting of the American Chemical Society in Boston on September 15, 1939.

6 War in Europe: An Abrupt Change in Direction

Hitler invaded Austria in 1938. Czechoslovakia and Poland fell into the Nazi orbit in 1939. Denmark and Norway were taken in April 1940. Belgium and the Netherlands were next. The Germans reached the English Channel in May 1940. France fell in June. England then came under fierce aerial assault by the Luftwaffe, but the Germans could not overcome the sustained valorous defense mounted by the British under the leadership of Winston Churchill.

At the time, the foreign policy of the United States was based on the Neutrality Acts enacted by Congress and signed into law by President Franklin D. Roosevelt between 1935 and 1939. These laws prohibited the sale or transport of arms and munitions to a belligerent, outlawed private loans to a belligerent, and forbade the entry of American ships into war zones. However, with Hitler in complete control of much of the European continent, with his armies separated from England only by the English Channel, and with England standing alone, the situation was far more grave than could have been foreseen during the earlier congressional debates on the Neutrality Acts. Would the U.S. come to the aid of the English people? That was the unanswered question.

The events on the continent had been closely followed in Washington. In the inner circles of government, the probability of the U.S. becoming involved in the conflict was viewed with mounting certainty. However, the U.S. was ill prepared for war. The armed forces lacked guns, ships, planes and men. Moreover, little was being done by way of research and development to prepare for a United States involvement in a new conflict. To address these inadequacies, a crescendo of activity was initiated. At first, it involved discussions aimed at assessing our position in critical areas related to military preparedness. Planning was begun. Experts were identified and consulted. In June 1940, President Roosevelt gave a clear signal in a commencement address at the University of Virginia: "We will extend to the opponents of force the material resources of this nation."[1]

The Blood Plasma for Great Britain Project

In the spring of 1940, Cohn learned from friends in New York of a unique effort called the "Blood Plasma for Great Britain Project." A community organization called the Blood Transfusion Betterment Association in New York City had received a request for assistance from Sir Edward Mellanby, Secretary of the British Medical Research Council. The MRC was seeking a supply of human blood plasma for use in treating Allied casualties while the local production of plasma from volunteer donors in Great Britain was being scaled up. A special meeting of the Blood Transfusion Betterment Association was held at the New York Academy of Medicine on June 12, 1940, just after the fall of France, to discuss supplying the Allies with plasma. Attending the meeting that was chaired by DeWitt Stetten were Alexis Carrel and others knowledgeable about the preservation of human blood and plasma. Captain Douglas B. Kendrick, Jr. represented the Army Medical Corps. Representatives of several U.S. pharmaceutical firms also attended. The outcome was a decision to undertake responsibility for the collection and shipment of plasma to England, assuming that the American Red Cross could be persuaded to participate.

The Trustees of the Association appropriated $20,000 to help finance the project. Subsequently, the Red Cross Board of Governors allocated Red Cross funds and conducted a publicity campaign to obtain the necessary volunteer blood donors. The "Blood Plasma for Great Britain Project" was formally launched in the late summer of 1940. The New York Chapter of the American Red Cross recruited blood donors and the Association arranged for the actual drawing of blood at donor stations in eight New York Hospitals. In August 1940, the first month of operation, blood was drawn from 1,044 volunteer blood donors. By October, the number of donors exceeded 4,600. By November, over 700 liters of plasma were in transit to England and another 800 liters were being readied for shipment. After collection, the blood was shipped to nearby pharmaceutical laboratories where it was centrifuged and the liquid plasma was drawn off into bottles. The American Red Cross took responsibility for shipping the liquid product to England. Unfortunately, what was originally conceived as a relatively straightforward humanitarian undertaking proved to be fraught with technical, logistic and even legal problems. Technical problems stemmed from the lack of a standardized protocol for collecting and handling the liquid plasma, which was shipped without an adequate preservative.[2]

The National Research Council

In the event that the U.S. entered the war, the military establishment would need to engage the scientific, technical and medical resources of the nation. The mechanism for doing so involved a quasi-governmental organization, the National Academy of Science.[3] Founded by President Lincoln in 1863 to provide advice to the government on "any subject of science and art," the Academy furnished important services during the U.S. Civil War. During World War I, the National Research Council had been established under the Academy to channel advice to the armed forces. In that conflict, the NRC had even served as the Department of Science and Research of the Council of National Defense and as the Science and Research Division of the U.S. Army Signal Corps. Near the end of that war, President Woodrow Wilson extended the responsibility of the NRC to include the ongoing survey and stimulation of research in the sciences. The National Research Council functioned through nine major divisions, each with a governing body consisting of representatives of leading professional societies.

In 1940, medical preparedness was the province of the NRC Division of Medical Sciences, whose chairman was Lewis H. Weed, Professor of Medicine at Johns Hopkins University. On April 26, 1940, when the Division held its annual meeting, Dr. Weed spoke briefly of the problems of defense and offered the cooperation of the Division to representatives of the armed services who were present. Within a few days, the Army Surgeon General, Major General James C. Magee, responded by identifying two subjects on which advice was urgently needed. These were the "use of chemotherapeutic agents in war casualties and the use of whole blood in individuals suffering from shock." In response, the NRC appointed two new Committees, a Committee on Chemotherapeutic and Other Agents, and a Committee on Transfusions.[4]

The NRC Committee on Transfusions

The first meeting of the new NRC Committee on Transfusions was held on May 31, 1940 at the National Academy of Sciences in Washington.[5] The attendees included Alfred Blalock, Professor of Surgery at Vanderbilt University; Walter B. Cannon, Professor of Physiology at the Harvard Medical School, Chairman of the Committee; Everett D. Plass, Professor of Obstetrics and Gynecology at the University of Iowa; Max Strumia, Director of Laboratories at the Bryn Mawr Hospital; and Cyrus C. Sturgis, Professor of Medicine at the University of Michigan. Also attending, without vote, were Major General James C. Magee, Surgeon General of the Army, Colonel George B. Callender, Colonel C.C. Hillman and Captain Douglas B. Kendrick, Jr. of the U.S. Army, and Commander C. S. Stephenson, of the Medical Corps, U.S. Navy. Weed opened the meeting on behalf of the National Research Council by explaining that the Surgeon General of the Army had asked the NRC to initiate work on the general subject of blood transfusion, surgical shock and blood banks. After briefly reviewing the role of the National Research Council, he turned the meeting over to the Chairman, Professor Cannon.

Cannon, a leading U.S. expert on shock, was a 1900 graduate of the Harvard Medical School. He had served as chairman of an NRC Committee on Physiology during World War I charged with the development of research programs to foster the health and safety of the U.S. armed forces, shortly after British surgeons became aware that more and more of the soldiers presenting to them for care mysteriously died of shock.[6] On accepting that assignment, Cannon put together a thorough study of what was known about shock, concluding that a failure in the circulation of the blood, i.e. circulatory collapse, was a cause of shock. He placed shock at the top of the committee's list of problems. In 1917, Cannon joined the Harvard Base Hospital No. 5 with the understanding that he would be sent to Europe and be allowed to investigate wounded soldiers in shock. Although initially frustrated at being stationed with a British Base Hospital well behind the lines in France, he succeeded in getting himself assigned to a British Casualty Clearing Station just behind the front lines. There he had seen shock at first hand when wounded soldiers streamed in during the battle of Lens. In the two years he was stationed in France, Cannon had learned a great deal and his findings contributed substantially to improved treatment of shock. In a monograph published in 1923, he had postulated that wound shock was produced by the absorption of toxic materials from the site of injury.[7]

Cannon opened the meeting by asking the representatives of the Army and Navy to identify the problems which confronted the Committee, referring specifically to blood transfusion, substitutes for blood, and the problem of wound shock. Stating that many trained investigators had voluntarily offered their services, Cannon suggested that if the representatives of the Army and Navy could formulate the problems that were pertinent and urgent, the Division of Medical Sciences would serve as an agency for diffusion of information regarding problems to be worked on and would disseminate information on the results that were obtained. The Surgeon General described the role of the new NRC Committee on Chemotherapeutic and other Agents, which had met for the first time on May 28, 1940. He described how that committee had already decided to form four Subcommittees, on Infectious Diseases, Venereal Diseases, Tropical Diseases, and Surgical Infections, respectively. He identified three broad topics on which he hoped the Committee on Transfusions would focus its efforts: blood transfusions; the use of dried plasma, its containers and intravenous sets for use in the field; and prevention of shock and of hemorrhage.

In amplifying on the Surgeon General's remarks, Colonel Hillman announced that the Army had little interest in the use of whole blood since it could not be stored for more than a few days. Even if combat in a future war fell outside the United States, mentioning Central America, South America and the West Indies, the Army would likely discourage the use of blood banks. Were war to be closer, the army might want to use blood that could be transported by air under refrigeration. In more distant places, where blood could not be collected locally, either liquid or dried plasma would have to be used. Speaking for the Navy, Commander Stephenson agreed, favoring dried plasma because it could be used in any form and without reactions. He emphasized the long-term stability of dried plasma, which meant that it could be kept for months without refrigeration and could be stockpiled in large quantities well in advance of military need. He pointed out that if plasma is dried immediately, it was unnecessary to add a preservative.

Stephenson noted that very few reactions had been experienced with plasma, whether used fresh, preserved by refrigeration, stored in the frozen state, or dried. Plasma offered the option of administration in concentrated form when there is embarrassment of the circulation. Since plasma, the liquid part of the blood, lacked the oxygen carrying capacity and other functions of whole blood, which reside in the blood cells, the postulated role for plasma in treating shock was based on the belief that infusion of plasma would have the effect of expanding the circulating blood volume. According to this logic, the supplying of plasma could even be viewed as more effective than transfusing whole blood. Nonetheless, the members of the Committee expressed considerable interest in improving existing methods for preserving whole blood. Everett Plass reported that he had succeeded in transfusing blood that had been preserved for more than thirty days. While this hinted at the possibility of preserving blood for long periods of time, the method used by Plass was too complex for general use at that time. The interest of the Armed Services was clearly focussed on moving ahead with plasma.

Cannon suggested that the problem presented to the Committee on Transfusions was one of recognizing and identifying differences in demands for transfusion support, at different stages in an organized line of care. At casualty stations in the field, products for transfusion should be easily carried by medical corpsmen without requiring refrigeration. On the other hand, more sophisticated products such as whole blood might be provided at hospitals in the rear. He suggested a tabulation of the various alternative products relating to transfusion such as whole blood, dried plasma, liquid plasma and blood substitutes—including their associated advantages and limitations in terms of military logistics. He also raised the possibility of developing a plasma from a lower animal that would be a good replica of human plasma.

At the time of the meeting, some members of the Committee knew of a study by Owen Wangensteen, Professor of Surgery at the University of Minnesota, suggesting that a plasma substitute might be obtained from animal blood. In very preliminary experiments, Wangensteen and a group of his colleagues had injected bovine plasma in fairly large volumes to a group of seriously ill, malnourished patients and had reported that it was well tolerated, was retained in the circulation and was apparently utilized. A brief report of these experiments appeared in a medical journal late in 1940.[8] This possibility of preparing a blood substitute from bovine plasma, coming as a serious suggestion from a prominent surgeon, aroused considerable interest on the part of representatives of the armed services and members of the committee. Such a product, if it could be successfully developed, would come close to being the ideal solution to the problem posed by the military planners. Not only was the raw material, plasma from bovine blood, already available in almost limitless supply in slaughterhouses across the country, it would free the armed services from depending on an untested voluntary human blood donor program to solve one of the most important medical problems then being faced.

In an afternoon session, the committee turned to a discussion of plans and organization. It requested that Colonel C. C. Hillman take up with the American Red Cross the interest of the armed forces in the collection of blood from the American public for use in treating casualties. It was further agreed that explorations should be begun with potential manufacturers of dried plasma in the hope of launching a program to evaluate its feasibility and costs. Representatives of the Navy agreed to convene appropriate experts to standardize the design of the containers and related devices for collecting and distributing blood and plasma.[9] Although it was recognized that extensive research might be required, the Committee on Transfusions urged that high priority be given to extending the period during which whole blood could be preserved for transfusion. A Subcommittee on Blood Substitutes was also established. With respect to shock, members of the Committee suggested that the Army organize a separate group of officers—study teams—to examine cases of shock.

The same suggestion was made with respect to transfusions, noting that transfusions can be carried out more effectively if carried out by special teams. It is likely that members of the committee knew about the organization of a public blood transfusion service in England during the summer of 1939 for the purpose of providing blood for transfusion.[10] Just before the fall of France, Colonel L. E. H. Whitby, the director of this new service, had visited a program in France, primarily to weigh the relative merits of whole blood and plasma. As a result, he had returned to England strongly favoring the use of plasma, mainly because it could be preserved for longer periods of time. Despite minor technical difficulties in the operation of the program, large quantities of plasma in "field transfusion kits" were being prepared and distributed in England by 1940. It is worth noting that the British began the war with a firm policy, decided long before and adhered to during the war, that there should be a separate and distinct transfusion service. The Americans never adopted a similar plan.

Shock

Shock is not a single disease, but a group of symptoms that might be caused by several factors, including acute hemorrhage and wounds. The British physiologist, R. J. S. McDowall, noted in 1940 that hemorrhagic shock can be either obvious or concealed.[11] If the hemorrhage is obvious, treatment intended to replace lost fluid immediately suggests itself. On the other hand, if hemorrhage is concealed, the loss of blood may go unrecognized until it has reached a critical extent, placing the patient in a position of impending shock. Much of the shock during the war resulted from concealed hemorrhage. At about the same time, Alfred Blalock, a member of the new Committee on Transfusions, together with his associate, Minot, defined shock as peripheral circulatory failure resulting from a discrepancy in the size of the vascular bed (an intricate network of minute blood vessels that ramifies through the tissues of the body) and the volume of intravascular fluid. They observed that in hemorrhagic shock, the prevalent form of shock in wartime, a decrease in the volume of circulating fluid plays a primary role.[12]

Minot and Blalock also identified a range of difficulties with respect to replacing lost fluid, depending upon whether the replacement was attempted with plasma or with whole blood. "If the problem is merely to restore fluid and electrolytes to an acutely dehydrated patient with a normal amount of plasma protein and with uninjured capillaries, the treatment is relatively simple. On the other hand, the restitution and maintenance of an effective circulating volume of blood in a patient whose capillaries have allowed and are continuing to allow a loss of all the plasma elements including protein is far more difficult." When stated in these simplistic terms, the central problem in providing supportive treatment in shock was framed. This was the background on which the Surgeon General announced the Army's choice of dried plasma.

Much later, in the official account of the American blood program in World War II, Kendrick identified a set of four generally accepted concepts concerning shock as the United States entered the war. These reflected the general belief at the time that the infusion of plasma alone could compensate for the loss of blood in shock:[13]

1. The reduction in blood volume present in peripheral circulatory failure is the most important single factor, if not the initiating factor, in the production of the clinical picture seen in shock. It results from the loss of plasma, at first locally and then generally, into the extravascular tissue spaces. Most therapeutic efforts in shock must be directed toward overcoming this loss of blood volume.
2. The reduction in the rate of blood flow is associated with a diminished venous return to the heart, which results in a decreased cardiac output, though the heart and the nervous energies that control it are not incapacitated.
3. The vasomotor center remains active.
4. Hemoconcentration (increased concentration of cells in the blood, usually resulting from loss of fluid to the tissues) is usually present in shock not associated with hemorrhage and tends to reflect the amount of plasma lost.

Meeting of the NRC Committees on Transfusions and Surgery

On July 24, 1940, in measured acquiescence to the views expressed by the Surgeon General of the Army at its first meeting, the NRC Committee on Transfusions met with its sister Committee on Surgery in order to see if the academic surgical community would support the Surgeon General's plans for treating battlefield shock. At that meeting, two primary reasons favoring the use of plasma over whole blood were agreed upon. First, that plasma is easier to store and transport than whole blood; and second, that the need for typing and matching was eliminated when dried plasma was used. (Pooling of the plasma from a number of donors effectively suppressed the activity of isoagglutinins.) Other reasons, less well documented, were the belief that shock is the result of hemoconcentration, i.e. increased concentration of cells in the blood, and that plasma would be as effective as whole blood in treating shock. It was well known that whole blood could not withstand freezing and thawing without destruction of the formed elements of blood—red cells, white cells and platelets. Thus, were whole blood to be needed by the armed services, extensive scientific research and development would be required before whole blood reached the state of acceptance already afforded to dried plasma. Without the availability of dried plasma, the United States would have no agent, at least in the short term, for the treatment of large numbers of casualties.[14] This meeting assured the concurrence of the surgical community in the policy of the Surgeon General.

Communicating the Surgeon General's Plan to Military Surgeons

The responsibility for announcing the Surgeon General's decision to rely on dried plasma in the treatment of shock fell to Captain Kendrick, who spoke on "Prevention and Treatment of Shock in the Combat Zone" at the Forty-Eighth Annual Convention of the Association of Military Surgeons of the United States in Cleveland on October 10, 1940.[15] After briefly recounting the history of shock, Kendrick turned to the concept of shock accepted by the Army as the basis for understanding the shock syndrome. He began by stating that shock is not the result of inefficiency of the heart or of interference with the nervous mechanism controlling the function of the heart. Instead, shock results in a marked decrease in the volume of circulating blood in the vascular system, explaining that the lost volume can be accounted for by stasis of red cells in the capillaries or by diffusion through membranes into the tissues. Whether produced by trauma or by hemorrhage, this fact is an outstanding feature of shock.

He then turned to the most important form of therapy for shock—replacement of the lost fluid. After ruling out the use of saline and glucose solutions and gum acacia, a water soluble gum from the acacia tree that had been used in World War I, he turned to blood banks and ruled them out as well. Since blood could then only be used within five to eight days after collection, it was impracticable for use by the armed forces in the forward areas. Citing the work of Strumia, Hill and Elliott, working separately in the treatment of shock, with good results, he averred that a satisfactory substitute for blood had been found. Then, asking the rhetorical question, "Are we providing adequate replacement of the lost fluid in shock, hemorrhage and burns when plasma is used," he responded,

> when the alterations in physiology are considered, it becomes clear that we are. It appears from the encouraging reports on the use of plasma that we now have a very excellent blood substitute. For our purposes in warfare, plasma is infinitely superior to blood . . . It can be safely stored . . . it can be used without typing, it can be transported without alteration . . . it is ready for instant use and can be injected fairly rapidly, and it does not increase the concentration of red corpuscles when injected. We have an agent of tremendous military surgical importance.

Red Cross Collection of Blood for the Production of Dried Plasma

In 1936, a few Red Cross Chapters had helped recruit blood donors for patients in need.[16] The first hospital blood bank in the United States had been established at the Cook County Hospital in Chicago in 1937.[17] Blood transfusions were being performed in some hospitals, often as a last resort in seriously ill patients. Of the transfusions being carried out, some institutions were cautiously shifting over to the transfusion of blood which had been stored for five to eight days in a refrigerator, although many transfusions were still being carried out directly between a donor and the recipient lying side-by-side. The inability to preserve blood safely for more than a few days was a serious obstacle. The scientific grounds for preserving red blood cells in viable form had not yet been mapped out.

At the Headquarters of the Red Cross in Washington, a special committee had been appointed to study whether or not this donor recruitment service should be authorized as a regular program of the Red Cross, but chapters had been directed to await completion of the study before initiating the service. In April 1939, a newly reconstituted Red Cross Medical and Health Advisory Committee, of which David Edsall was a member, had voted in favor of encouraging Red Cross chapters to establish blood donor services. The blanket prohibition against providing blood donor recruiting services was rescinded in March 1940, and the Red Cross began encouraging its chapters to establish blood donor recruiting services under appropriate conditions and where authorized by National Headquarters. Organized volunteer blood donor programs did not exist, the single exception being the Blood Plasma for Great Britain Project. Individuals experienced in donating blood were relatively few. Those who donated blood were often paid for their services.

A method for drying plasma had been developed and patented during the 1930s, thus making it possible to preserve plasma in the dried state for months or even years.[18] However, blood and plasma shared a common weakness as candidates for treating shock by the armed forces: to provide either or both in the volume that would be required to support the armed forces in a global war would be possible only if the American public could be persuaded to donate large volumes of blood. However, the willingness of the public to give blood had never been tested. As a result, there was considerable skepticism in 1940 that the armed services could rely on this source.

Cannon Enlists Edwin Cohn's Collaboration

Upon his return to Boston after the May 31 meeting of the Committee on Transfusions Walter Cannon went to see Edwin Cohn, bringing with him David Edsall, who was by then Chairman of the Red Cross Medical Advisory Committee. Cannon reviewed the discussion at the meeting of the NRC Committee on Transfusions, stressing that the Surgeon General of the Army had attached great importance to acquiring an agent for use in the treatment of battlefield shock. He told Cohn about Wangensteen's experimental infusions of bovine plasma into patients, and reported the favorable impression this report had made on the representatives of the Army Medical Corps who attended the Washington meeting. They also briefly discussed the difficulties that would have to be faced in meeting this need by providing human plasma or an agent prepared from human plasma. Cannon then turned to Cohn and came to the purpose of his visit. Could Cohn "undertake an investigation to determine whether the plasma of animals could be made safe for human transfusion?"[19]

What Cannon had in mind was an agent that would simulate the normal property of the plasma proteins in controlling fluid balance. It should be a colloidal substance of appropriate molecular size such that, when infused by vein, would be retained for a reasonable period in the blood stream, thus restoring and maintaining the fluidity of the blood. Such a plasma substitute should carry no risk of adverse long-term effects to the recipient. It should be issued as a concentrated solution that could withstand the range of temperatures in the field, from extreme cold to very hot. Military logistics also required that it have a long shelf life, thus making it possible to stockpile huge reserves for use anywhere and at any time.

Given the ready availability of bovine blood from slaughterhouses, Cohn could appreciate the logistic advantages of a possible new product from bovine plasma. At the same time, he foresaw possibly serious problems resulting from side effects of the administration of animal proteins in man. He reminded his visitors that there was a considerable body of knowledge dating back to the late nineteenth century concerning the clinical use of antitoxins prepared from horse and rabbit sera. Diphtheria antitoxin prepared from the sera of immunized horses had been particularly effective in reducing mortality from diphtheria infection. However, its use frequently gave rise to serum sickness, an allergic reaction to the injection of foreign protein manifested by itching, skin eruptions, swelling, arthritis and fever. He noted that while horse antitoxins were usually administered intramuscularly and in quite small amounts, large volumes of the planned bovine agent would be indicated to treat shock and would have to be given intravenously. Viewed overall, he saw that attaining the desired objective of a safe bovine agent for the treatment of shock would require the solution of two interrelated scientific problems, one physical chemical and the other immunological. The challenge for the physical chemists would center on isolating purified protein fractions for study as possible products. The immunological challenge would center on the conduct of clinical tests of the bovine protein fractions to assure the safety of any new product.

There were almost no precedents at the time for an undertaking of this magnitude. The only protein products listed in the AMA's *New and Nonofficial Remedies* in 1940 were insulin and a group of hyperimmune sera and vaccines.[20] Insulin isolated in highly purified form from animal pancreas glands had been in use for treatment of diabetes for a number of years. Cohn assured Walter Cannon that he would turn his efforts to the exploration of bovine plasma as a source of a protein which could be used to treat shock, the only proviso being that adequate clinical testing facilities also be made available.

The Subcommittee on Blood Substitutes

The new NRC Subcommittee on Blood Substitutes met for the first time on November 30, 1940.[21] Its members included C.C. Sturgis, M.M. Strumia, E.D.Plass, O.H. Robertson, O.H. Wangenstein, J. Scudder and E.L. DeGowin. Also attending were Walter Cannon, William DeKleine of the American Red Cross, Cornelius P. Rhoads of New York's Memorial Hospital, Commander Stephenson, Lt. Commander Newhouser, Lt. Colonel Munly, Captain Kendrick, and L.H. Weed. The agenda included reports by Strumia on the drying of plasma, by Plass on the preservation and transportation of whole blood, and by Scudder on the Blood Transfusion Betterment Association's project in New York City. At that meeting, Plass and DeGowin presented some new data from a study on the preservation of whole blood at the University of Iowa. Their procedure involved adding an equal volume of three percent glucose solution to a unit of citrated blood (the volume of blood in a single donation). In one experiment, twenty units of blood were shipped by air from Iowa City to Oakland, California and back. The blood was carried in covered milk cans filled with ice and re-iced every twenty-four hours. On return to Iowa after ten days the blood units had been transfused to patients with no untoward effects. The large volume of "diluent" glucose solution added to the blood was viewed as an advantage since it would be transfused to the recipient. This experiment contributed substantially to the development of present day preservation of blood in blood banks.

Abrupt Change

When Cannon and Edsall called on Cohn after the meeting at the National Research Council, the manuscripts of two scientific papers in a new series from the Department of Physical Chemistry on the "Preparation and Properties of Serum and Plasma Proteins" were ready for submission to the *Journal of the American Chemical Society.* The first manuscript bore the title, "Size and Charge of Proteins Separating upon Equilibration across Membranes with Ammonium Sulfate Solutions of Controlled pH, Ionic Strength and Temperature."[22] In that study, Cohn, McMeekin, Oncley, Newell and Hughes had separated normal horse serum into successive fractions precipitated by equilibration across cellophane membranes with ammonium sulfate solutions of known pH, concentration, volume and temperature. The amount of protein that precipitated into each fraction had been measured. Electrophoretic and ultracentrifugal analyses were included on specific fractions. This fractionation procedure incorporated several innovations. It was the first study in which pH and temperature were controlled as variables in the procedure. It incorporated a novel technique for adding ammonium sulfate by dialysis across a cellophane membrane into the protein solution over a period of forty-eight hours. By controlling the rate of addition of the precipitating agent over time, the proteins were precipitated, i.e. were "salted out" in a granular form that was easily removed by filtration rather than by centrifugation. On the other hand, the continued reliance on the archaic salting out method to precipitate the proteins presented some disadvantages. Ammonium sulfate lacked sensitivity as a protein precipitant, i.e. it did not discriminate well between proteins, and it was toxic. Moreover, before products so treated could be used in man, it would be necessary to remove all traces of ammonium sulfate. Nevertheless, the studies reported in the new paper represented a substantial technological advance over the current state of the art in protein fractionation. The second paper, by McMeekin, described the crystallization of a carbohydrate-containing albumin from horse serum.[23]

In assuring Walter Cannon that he would investigate bovine plasma as a source of an agent to treat shock, Cohn was confident that this new ammonium sulfate procedure could easily be adapted to prepare bovine albumin instead of horse albumin. He set McMeekin, Hughes and Armstrong to work on that task immediately. There was an added element of urgency in that McMeekin was about to take up a new post at the U.S. Department of Agriculture, while Hughes, an organic chemist who had recently received his Ph.D. degree from MIT, had just joined Cohn's staff. Cohn wanted Hughes to learn as much as possible from McMeekin before the latter departed. Starting with fresh bovine blood obtained from a slaughterhouse in nearby Somerville, and employing the ammonium sulfate procedure, McMeekin and Hughes were soon producing bovine albumin. Although the amounts that could be prepared at first were quite small, both the quantity and purity of bovine albumin could be increased over time. Samples of some of the first bovine fractions were sent out to Wangensteen at the University of Minnesota for examination.

At the same time, Cohn launched a separate investigation with Luetscher, Davis and Weare aimed at preparing bovine serum albumin by using ethanol instead of ammonium sulfate as the protein precipitant. Five years before, while pursuing their studies on the amino acids and peptides, Cohn, McMeekin, Greenstein and Weare had observed that the effects of electrostatic forces on the solubility of peptides were greater when the dielectric constant of the medium was decreased by the addition of ethanol.[24] In a related set of experiments, with Ronald Ferry and Ethyl Newman, Cohn had investigated the effects of sodium chloride on the solubility of egg albumin in 25% ethanol at −5° C.[25] While the observed effects of the salt on the solubility of proteins were qualitatively similar to those previously observed on peptides, Cohn had been surprised to find that egg albumin, a protein quite sensitive to denaturation, not only tolerated the exposure to ethanol at low temperatures, but actually crystallized—a good sign. This gave him the idea that ethanol-water mixtures at controlled low temperatures might open the way to a completely new system for protein fractionation.

Ethyl alcohol had been used in England for studies on proteins by Mellanby in 1908[26] and by Hardy and Gardiner in 1910.[27] But it never found much favor. Cohn's rediscovery of the utility of alcohol in protein chemistry was driven by the new set of circumstances he faced. Alcohol possessed important advantages over ammonium sulfate. In addition to being non-toxic, ethanol is volatile and could be easily removed from proteins by drying from the frozen state. A minor risk associated with the use of alcohol in working with proteins stemmed from the fact that some heat is generated when pure alcohol is added to the aqueous solution of the protein. Cohn insisted that special precautions be taken to control the temperature during the fractionation process, particularly while alcohol was being added to the protein solution. In a few weeks, the Luetscher team had devised the framework of a method that obtained reasonably pure bovine albumin in good yields. Once satisfied that the new method yielded good bovine albumin, Cohn arranged for it be tried out on a small volume of human plasma that was collected from volunteer blood donors by Carl W. Walter at the Peter Bent Brigham Hospital.

Cohn was so confident that this new procedure using ethanol could be developed that he delayed submission of the two manuscripts already completed for publication so that a third paper describing the new fractionation system could be included. He counted on the favorable effect that a published scientific article describing the new method might have on the members of Cannon's NRC Committee on Transfusions. In the meantime, Cannon had arranged with Soma Weiss, the newly appointed Hersey Professor of the Theory and Practice of Physic at the Harvard Medical School and Physician-in-Chief at the Peter Bent Brigham Hospital, to evaluate Cohn's protein fractions from the point of view of safety in man and their effectiveness in the treatment of shock. In turn, Weiss designated Charles A. Janeway, then working in the bacteriology and immunology laboratories at the Brigham, to conduct the needed clinical testing. On the advice of Walter Cannon, Cohn also notified Dean Burwell that

> at the request of Professor W.B. Cannon, Chairman of the Committee on Transfusions of the National Research Council, the Department of Physical Chemistry at the Harvard Medical School, in collaboration with the Department of Medicine at the Peter Bent Brigham Hospital, has undertaken a study of the separation and preparation in powder form of fractions of the plasma of species other than man for purposes of transfusion.[28]

Cohn identified three tasks being undertaken in this investigation. These were: development of a new ammonium sulfate fractionation process for bovine plasma using McMeekin's procedure; development of a low temperature ethanol fractionation method that would be more readily adaptable to large scale preparation; and conduct of clinical and immunological studies of the various fractions so developed. He also informed Dean Burwell that a new method for removal of ethanol and water by drying the fractions in the frozen state was being explored with the help of experts at the Research Laboratory of Physical Chemistry at MIT and stated that, in the interest of time, the studies of fractions prepared from horse plasma had already been initiated by McMeekin. The latter were to be replaced as soon as comparable fractions from bovine plasma could be prepared. All fractions were being carefully characterized by measurements by Luetscher and Armstrong of electrophoretic mobility in the Tiselius apparatus and of sedimentation velocity in the new ultracentrifuge by Oncley.

Shortly thereafter, Wangensteen's associates in Minneapolis published their preliminary findings of skin tests carried out with bovine albumin and globulin prepared at Harvard by McMeekin and Hughes.[29] As expected, these revealed large differences among the fractions as seen by skin sensitivity tests. Bovine albumin caused fewer reactions than bovine plasma while bovine globulin was intermediate when so tested. The crude bovine albumin resulted in moderate skin reactions in forty-seven of the subjects. This report served to confirm Edwin Cohn's belief that efforts to develop an agent from bovine blood to treat shock would rest heavily on clinical testing in man.

Publication of the New Plasma Fractionation Procedure

On August 17, 1940, a manuscript describing the new method was sent off to the *Journal of the American Chemical Society* along with the two earlier papers whose submission had been delayed. The new third paper, which bore the title, "Preparation and Properties of Serum and Plasma Proteins. III. Size and Charge of Proteins Separating upon Equilibration across Membranes with Ethanol-Water Mixtures of Controlled pH, Ionic Strength, and Temperature," was written by Cohn, Luetscher, Oncley, Armstrong and Davis.[30] Its opening paragraph included the prescient statement: "It has recently seemed of importance to standardize a method, capable of being employed for large scale preparations, for the separation of plasma into as many as possible of its component proteins." The fractionation procedure involved the stepwise adjustment of the pH, ionic strength, ethanol concentration and temperature of bovine plasma in four stages, a protein precipitate, or fraction, being formed at each stage. Successive fractions designated I, II, III and IV were removed by centrifugation at each stage, leaving a final clear supernatant solution from which the fifth protein fraction, fraction V, was obtained by drying. When the procedure was refined in later investigations, fractions II and III were precipitated together as Fraction II + III; Fraction IV was removed as before; and Fraction V was precipitated by a final pH adjustment rather than by drying. Fraction V included most of the bovine albumin originally present in plasma.

The paper provided the directions for fractionating two liters—about four pints—of fresh bovine plasma, yielding about fifty grams—less than two ounces—of albumin. At a time when a single gram of bovine albumin was almost impossible to obtain, this was a satisfying result. From this relatively crude prototype the elegant fractionation system that emerged during the coming months came to be known as the ethanol-water fractionation system. In its technically most advanced version, this alcohol-water system invoked the adjustment and precise control of even more variables to fractionate plasma. The concluding paragraph of the paper stated:

> Examples of procedures that have thus far been found useful in the preparation of human, horse and bovine gamma globulin and albumin are given in detail as illustrations of the general method which can, of course, be modified to some extent with the species of serum or plasma, the fraction sought in the condition of greatest purity, and perhaps also with the scale of preparation. The method of fractionation by equilibration through membranes with alcohol-water mix-

> tures or mixtures of related substances of controlled pH, ionic strength and temperature, appears to be applicable to the preparation of protein fractions for many purposes, some industrial and some clinical.

In order to obtain sufficient bovine albumin for initial clinical testing in man, the albumin fractions from successive runs, each starting with about two liters of plasma, were combined into larger batches. The development of this new fractionation system represented a major step forward. However, it also led to more challenges that lay ahead. For example, immediate consideration would have to be given to opening up a much larger source of bovine albumin for human clinical testing than could be produced in this limited manner in the laboratory at the Harvard Medical School. What Cohn needed was a pharmaceutical manufacturer with access to a source of bovine blood. When he raised this question with Dean Donham at the Harvard Business School, Donham advised him to contact Victor Conquest, an executive at Armour Laboratories in Chicago. After explaining the urgent mission he had undertaken for the National Research Council, Cohn asked Conquest if Armour could assemble a small pilot plant in Chicago to produce bovine albumin for clinical testing in man, using his new Harvard fractionation procedure. He assured Conquest that Armour technical staff would be trained at Harvard, and would receive all required scientific and technical support, including all biologic and clinical testing needed before release of each lot of bovine albumin for clinical use. Explaining that his work at Harvard was being supported by funds provided by Harvard and the Rockefeller Foundation, Cohn was careful to point out that no government funds were available, nor was he authorized to offer any assurance that Armour would be compensated for its efforts.

Conquest responded promptly by assembling a pilot plant and assigning a group of Armour scientists to work on the project. As it happened, the senior Armour scientist initially chosen to lead this effort had to be replaced a short time later because of difficulties in working with Edwin Cohn. Fortunately, the man assigned to take his place, Jules D. Porsche, proved to be a man of more phlegmatic bearing and a long fruitful relationship ensued between the Department of Physical

Chemistry and Armour Laboratories. Under these amicable circumstances, the Armour scientific endeavor became for all intents and purposes an arm of the Department of Physical Chemistry in pursuing the development of both bovine and human albumin.[31]

Human Albumin

As soon as the bovine albumin project was moving forward, Cohn could not help but recognize the potential of the corresponding human product—human albumin—as a completely safe alternative. Although doubts persisted at the National Research Council as to whether the American public could be counted on to donate blood, Cohn gave the human product increasingly detailed attention. By then, it was becoming evident that human albumin and bovine albumin were almost indistinguishable on physical chemical grounds. Both could be prepared by use of the new ethanol-water fractionation procedures. Nevertheless, the human product presented certain significant advantages over bovine albumin. Human albumin could be expected to be free of the immunological side effects of a bovine protein. It also presented no known obstacles to adoption for human use, although it would have to undergo thorough testing. Moreover, the byproducts of human albumin could be expected to yield additional protein components with potential uses in therapy. Nevertheless, Cohn never wavered from his commitment to devote top priority to the development of bovine albumin until it was either accepted or rejected by the Armed Services.

For Edwin Cohn, the significance of the Blood Plasma for Great Britain project lay in its demonstration that the public in New York City would respond generously by donating blood to help others. The implications were clear. If need be, given the proper organizational framework, human blood could be factored into providing a safe therapeutic agent for the treatment of military shock. From that time forward, he resolved that the development of bovine albumin and human albumin should be carried out in parallel at least until one of them was approved for the treatment of military shock. Since the collaboration with Armour Laboratories showed promise of providing adequate amounts of bovine albumin, Cohn began to consider how to assure a parallel supply of human albumin. This led to the decision to build and equip a small pilot plant to fractionate human plasma at the Harvard Medical School.

A pilot plant would serve at least three purposes. It would contribute substantially to the development and transfer of the plasma fractionation technology to the American pharmaceutical industry, a critical step if substantial amounts of either human or bovine albumin were to be provided to the armed services within the desired time frame. It would provide hands-on opportunities for training the managers of the pharmaceutical plants while the plants were being built. And, in the process, it would offer the Harvard scientists intimate experience with countless unforeseen problems that might develop. Moreover, by serving as a miniature pharmaceutical manufacturing facility, the pilot plant would generate a small but significant flow of human albumin and possibly other proteins during a critical clinical testing program.

A Question of Patent Policy

In September 1940, Cohn encountered President Conant in Detroit and they had lunch together. This provided an opportunity to discuss the implications of the new studies on the fractionation of the plasma proteins. In their conversation, Cohn raised the possibility of applying for a patent on the process, pointing out the many possible uses of the processes and products being developed, "some of therapeutic, some presumably of industrial significance, and some perhaps of value in connection with the national defense." On his return to Boston, he wrote a note to Conant stating,"I do not at this time see how we can separate the therapeutic from the industrial uses of the processes we have employed in our fractionations." He therefore requested that the newly discovered uses—therapeutic and industrial—be considered inseparable, and that the Harvard Corporation appoint counsel to advise on steps to take, including whether the processes are patentable, whether steps should be taken to prevent patenting by others, and whether to consider filing patent papers. Should patents be sought, he stated, "it would clearly be my understanding that in so far as therapeutic results arise from the work that we have done, neither the University nor I should in any way profit and that patents applied for should either be used for control, or if it seemed more desirable, dedicated to the public."[32]

An Important Meeting at Navy Headquarters

In November 1940, shortly after Franklin Delano Roosevelt was reelected to a third term as President of the United States, William DeKleine, the Red Cross Medical Director, came to Boston to meet with Red Cross volunteers and others for the purpose of stirring up interest in activating a Red Cross Blood Donor Center in Boston. While in town, DeKleine visited with Cohn at the Harvard Medical School. There, Cohn toured his visitor around the laboratory and showed him vials of human albumin and bovine albumin. In the course of their discussion, Cohn expressed his concerns that the risk of adverse side reactions could slow down or even ultimately halt the development of bovine albumin. Since human albumin would be perfectly safe, and other therapeutically valuable products could be obtained from human plasma, Cohn urged that the Red Cross move swiftly to collect blood on a wide scale to support the armed forces medical needs. DeKleine immediately recommended that the information which Cohn had disclosed to him be communicated to the U.S. Navy at the earliest possible time. As they parted, it was agreed that DeKleine would arrange a meeting.

On December 21, 1940, Cohn and DeKleine met with Commander C. S. Stephenson at the Navy Department in Washington. This was Edwin Cohn's first high level meeting with a representative of the Navy and he had prepared carefully for it. He had brought with him a copy of the December 1940 issue of the *Journal of the American Chemical Society* containing the scientific paper describing the new method for the preparation of bovine plasma. He also showed Stephenson a sample of bovine albumin, stressing how it had the potential of accelerating the development of a safe bovine product for the armed services. On the other hand, he cautioned Stephenson against being overly optimistic about a successful outcome of the bovine product, stressing how much depended on completely freeing bovine albumin from immunological reactivity, something that had never been done before. Under the circumstances, he advised Stephenson that it would be a mistake to delay keeping open the option of meeting the needs of the armed services with products from human blood.

As they departed, DeKleine reminded Stephenson that the Red Cross had still not been contacted by the armed services about initiating volunteer blood collections from the American public. He stressed the value of the New York program as a model for establishing blood donor groups in other parts of the country. The next day, Stephenson forwarded a memorandum to Admiral Ross T. McIntire, Surgeon General of the Navy, summarizing the meeting with Cohn and DeKleine, and reiterating Cohn's view that "it would be a mistake to delay much longer a set-up which would provide human blood from which plasma can be processed and he thinks the logical thing to do is to use the dried plasma. He urgently recommends that something be done at the earliest possible moment and await development of experience with bovine and equine plasma and such substitutes as may be developed synthetically".[33]

7 Preparation for War

In January 1941, President Roosevelt submitted to the new Congress legislation designed to circumvent the limits of the Neutrality Act. In March, Lend Lease became law. In April, all Axis shipping in American ports was seized, and the Navy began to patrol the sea lanes.[1] On February 1, in response to a request by the Surgeons General of the Army and Navy, the American Red Cross issued an appeal for volunteers to donate blood for an emergency blood bank for the national defense. According to its announcement, the purpose of the new blood bank will be "to form a readily available supply of transfusion blood in case of war or disaster."[2] The New York Chapter of Red Cross was chosen to launch this new program by virtue of its successful experience in the Blood Plasma for Great Britain Project. Shortly thereafter, regular weekly collections of blood to provide Cohn with plasma were begun by the Boston Chapter of the American Red Cross.

On February 11, 1941, Lewis H. Weed notified Edwin Cohn that the Administrative Committee of the National Research Council had approved the appropriation of $1,500 to Harvard University to support investigations recommended by the Executive Committee of the Committee's "Advisory to the Surgeons General." Weed added, "I am very happy that this sum becomes available for the support of your investigations in relation to the Defense Program. While the present appropriation does not extend beyond the end of the government's fiscal year, it is hoped that additional funds may be obtained for the next year. It is realized that the grant now made does not represent the optimum support of your research, but with the meager funds now in hand, no larger grant could be made."[3] Weed's communication provided the first indication that financial support for the scientific investigations that had been under way in Cohn's laboratory for over nine months might be forthcoming from Washington.

Edwin Cohn was appointed to the NRC Subcommittee on Blood Substitutes in the late winter of 1941. He attended his first meeting at the National Academy of Sciences in Washington on April 19, 1941. There he discovered that Robert F. Loeb, Professor of Medicine at the College of Physicians and Surgeons of Columbia University, had also been appointed to the Subcommittee and would serve as its chairman. Loeb was the son of Jacques Loeb who, more than twenty years earlier, had been instrumental in guiding Cohn toward a career in biological science. The continuing members of the Subcommittee were Elmer L. DeGowin, a pioneer in blood transfusion at the University of Iowa, John Scudder a surgeon at Presbyterian Hospital in New York, Max Strumia at the Bryn Mawr Hospital, O.H. Robertson, of the University of Chicago, Owen H. Wangensteen of the University of Minnesota and Cornelius P. Rhoads. Also attending the meeting were Captain C. S. Stephenson and Lt. Commander Lloyd Newhouser, U.S. Navy; Lt. Colonel W. C. Munly, and Captain Douglas B. Kendrick, U.S. Army; William DeKleine, the American Red Cross; Milton V. Veldee, Chief of the Division of Biologics Control of the U.S. Public Health Service; and Lewis Weed.

Loeb opened the meeting with the statement, "I take it that the consensus of the committee [arrived at in a previous meeting] is that either serum or plasma reduced to either a frozen or dried state is acceptable and that production should proceed at once with the understanding that in time other recommendations may be made." Hearing no dissent, he then summarized the responsibilities of the Subcommittee: "To conduct research work on the different blood substitutes and to establish their relative merits by clinical trial; to determine the best technique of preparing the different substitutes; to standardize the best method of collecting and dispensing both plasma and serum; and to select the type of container or containers for both preserving and dispensing the same."

In response to a question from the chairman, the representatives of the armed services stated that 200,000 units of an approved blood substitute had been set as the goal for the next fiscal year for the Army and Navy. However, no blood substitute had yet been investigated. Indeed, use of the term "blood substitute" led to some misunderstandings then, as it does today. After a brief discussion, the subcommittee defined a blood substitute as a substitute for a specific function of blood, not as a substitute for all the functions of blood. At that time, the only blood substitutes that could be considered were human plasma or human serum, stored in either the frozen or the dried state. Although this appeared to offer four alternatives, the question of which of the four to choose was quickly resolved when Veldee stated that dried human plasma had already been officially approved and minimum requirements had been adopted by the U.S. Public Health Service. Moreover, two U.S. firms were then licensed by the Public Health Service to produce dried plasma, using a patented process for drying the plasma from the frozen state. The realization that the process for drying plasma was already commercially patented raised some questions within the Subcommittee, but the issue was resolved subsequently when the holders of the plasma drying patent indicated that royalty free licenses to use the procedure would be granted to the United States Government for use in time of war.[4]

The Subcommittee then reviewed arrangements made for preparing dried plasma. The American Red Cross had agreed to collect blood from volunteer donors. Once collected, the blood was to be shipped to government contractors where the plasma would be separated and dried from the frozen state. DeKleine stated that the responsibility of the American Red Cross would be limited to the recruiting of blood donors in different cities and collecting the blood under governmental regulatory guidelines using approved needles, tubing and containers. The Red Cross would then turn the blood over to processing plants designated by the government for the production of dried plasma under government specifications. The Red Cross would furnish its services without cost to the government. Since some time would inevitably elapse before government contracts could be awarded for the preparation of dried plasma, the Subcommittee recommended that the blood then being collected by the Red Cross be centrifuged and the plasma be drawn off and be immediately frozen and held in the frozen state for later processing to dried plasma. The U.S. Division of Biologics Control would exert regulatory control over the product. It remained for the Subcommittee to appoint a working party to draw up specifications for the package of dried plasma, its accompanying container of sterile water, and the tubing and needles to be used in its reconstitution and use in the field.

A New Plasma Fractionation Method

In reporting on the plasma fractionation studies then under way in his laboratory at Harvard, Cohn described his new ethanol-water fractionation system for separating the protein components of plasma. This relied upon the strict control of several variables, including pH, ionic strength, ethanol concentration, and temperature. While he talked, he passed around a vial containing a few grams of white human albumin powder and a vial of 25% human albumin solution suitable for intravenous injection. He dwelt at length on the studies then in progress on bovine albumin, pointing out that while the purified bovine and human proteins were similar chemically, they might be quite different clinically. In that light he stressed that the fate of bovine albumin as a candidate blood substitute would rest on the outcomes of extensive clinical tests in man. He surprised those present by announcing that scientists from Armour Laboratories in Chicago had already been trained in Boston in the use of the new fractionation procedure and that Armour was assembling a pilot plant in Chicago for the production of bovine albumin according to Harvard specifications. Only in this way, Cohn stressed, could the needed amounts of bovine albumin be provided for testing in humans.

Cohn reported some initial clinical test results obtained by Weiss and Janeway with the first preparations of Armour bovine albumin. At first, tiny doses (0.4 grams) had been injected intravenously into eight individuals without adverse effects. In a second series, bovine albumin in doses ranging from four to sixteen grams was given intravenously to five individuals. One of these experienced serum sickness ten days after the infusion; the other four individuals tolerated the infusions without ill effects. Cohn also noted that the fractionation of human plasma to obtain human albumin had been underway for two months in his laboratory at Harvard using blood collected by the Red Cross in Boston. Approximately 100 grams of human albumin were being prepared weekly. Clinical tests of human albumin had not yet begun.

Support from the National Research Council and the American College of Physicians

Early in April 1941, Lewis Weed, the chairman of the NRC Division of Medical Sciences, notified President Conant of Harvard of the award of $5,000 to support Cohn's project for processing human plasma into its component protein fractions.[5] In his letter, Weed added that Harvard could expect a second grant of $5,000 from the American College of Physicians, which met in Boston a few days later. During that meeting, the *Boston Herald* carried a front-page story under the headline, "Use of Animal Blood for Human Transfusion Studied by Group of Chemists at Harvard," and announced the new award by the College of Physicians to Edwin Cohn.[6] Fortunately, Cohn had been contacted by the *Herald* reporter before running the story. As a result, he succeeded in persuading the Herald to soften its message. As published, the article stated that any tests on the use of animal blood in humans would not be completed for many months, and, further, "unless it can be demonstrated conclusively that animal blood materials can be used safely, continuing reliance must be placed on the use of human blood." The article also told that the same process was being used to break down human plasma into its protein components and that the funds granted by the American College of Physicians were being used for the construction of a small pilot plant in Building E at the Harvard Medical School in which larger scale production of human albumin would be undertaken.

Having opened the pathway for developing not one, but two possible new biological products for the treatment of battlefield shock—bovine albumin and human albumin—Cohn now had to steer the development of both products through to the stage of pharmaceutical production, or at least until it became evident that a bovine product completely safe for use in man had been produced. In that case, as Stephenson had pointed out at the meeting, the Navy would have no further interest in human albumin. On the other hand, should bovine albumin not prove safe for use as a human plasma substitute, then human albumin would become the agent of choice for the armed services. Moreover, time was of the essence. Whichever product was ultimately chosen, the technology for industrial scale production would have to be transferred to American pharmaceutical firms so that a biological product for treatment of battlefield shock could be stockpiled.

The Harvard Pilot Plant

Pilot plants are small production plants. They enable production of products on an intermediate scale between "bench top" experiments performed in the research laboratory and large-scale industrial production. In order safely and expeditiously to scale up chemical processes to the pharmaceutical production level, experience must be gained at the pilot plant stage. Pilot plants use tanks instead of beakers and work with other large processing equipment. Pilot plants often provide valuable insights into how increases in scale of operations can affect a chemical process. Increases in scale of processing often introduce unforeseen effects. For example, the time it takes to carry out certain steps, such as the addition of reagents, or the centrifugation of large volumes of suspensions of protein precipitates, may increase inordinately, due to the differing characteristics of the larger equipment that must be used. Furthermore, the dynamics of temperature control differ when the volume of liquid to be cooled is sharply increased. During the pilot plant stage, the Harvard scientists could identify and overcome the pitfalls that could be expected to arise during the transfer of new technology to the pharmaceutical industry.

Pilot scale operations offer other advantages. They facilitate the smooth transfer of a technical process to industry. Pharmaceutical scale requires that detailed written operating directions be made available, that supervisors and operators be trained, and that tight administrative control exist over every step in the process, from receipt of the starting material—plasma—to shipment of finished vials of product ready for human use. Under the circumstances then in effect, the pilot plant was an ideal setting for accomplishing these objectives. Of particular tactical importance, the pilot plant was capable of yielding a kilogram or more of albumin in a single run. The availability of human albumin in those amounts meant that the process for isolating human albumin could be adjusted to improve its stability and to develop and conduct requisite sterility and safety testing, as well as provide albumin for clinical testing. These tasks could all be accomplished while the basic fractionation process was undergoing fine tuning. Even more important, Cohn counted on the pilot plant to generate a steady flow of by-products, i.e. plasma protein fractions other than albumin. He was confident that these by-products from human plasma had the potential of providing additional therapeutically effective human proteins—the greater the number of byproducts produced, the more efficient the use of a precious plasma resource.

At this time, having received his Ph.D. in physical chemistry from Brown University in the fall of 1940, Laurence E. Strong came to the Department of Physical Chemistry expecting to work with Oncley on a study of the dielectric properties of proteins. On Strong's arrival, Cohn was designing a glass device for use in the equilibration of plasma with ammonium sulfate solutions. Cohn was not a glass blower, while Strong was a capable glass blower. He quietly stepped in and blew the glass device that Cohn sketched. With this "hands-on" encounter as a beginning, Strong, a quiet and thoughtful Quaker, was, before long, playing a key role in Cohn's developing organization. Although titles were seldom used in the laboratory, Cohn and Strong held titles as Director and Associate Director of the Harvard Pilot Plant, respectively, for the duration of the war.[7]

The fractionation of plasma by Cohn's new methods relied upon the use of certain operational steps commonly used in the chemical industry. In a precipitation phase, a new protein precipitate was caused to form from a solution by adjusting the physical chemical conditions of the solution to a desired pH, alcohol concentration, ionic strength, protein concentration and temperature by the addition of appropriate liquid reagents while stirring and controlling the temperature. In a centrifugation step, the precipitated protein fraction was collected by centrifuging in a refrigerated centrifuge. In this machine, the collecting bowl was a hollow stainless steel cylinder about five inches in diameter and thirty inches long. This bowl was spun at twelve thousand revolutions per minute while the protein suspension flowed in at its lower end. The precipitate collected inside the bowl as a thick paste with the consistency of butter. A clear effluent solution, termed the supernatant, emerged from the top of the bowl. After scraping the precipitate out of the bowl, it could be kept in a closed container within the cold room until ready for use in another stage. Another step involved drying the proteins from the frozen state. These unit operations were employed repeatedly throughout the fractionation process.

Although the blood coming from donors was drawn into sterile glass bottles, sterility was not maintained during the fractionation process. Nonetheless, from that point on, until the fractionation procedure was complete and the final biological products were again sterilized, strict standards of cleanliness were observed with precautions being taken to minimize bacterial contamination and growth throughout processing. A special water line in the Pilot Plant supplied hot pyrogen-free distilled water. (Pyrogens are fever producing substances thought to result from bacterial growth.) Prior to use, all tanks, hardware and the centrifuge bowl were rinsed with hot pyrogen-free water. The ethanol used in the fractionation process and the low temperatures maintained throughout the process contributed to maintenance of a bacteriostatic environment during processing. The volatility of ethanol assured its complete removal during the drying of the fractions so prepared.

The Harvard Pilot Plant was capable of fractionating forty liter batches of plasma at one time. This was the volume of plasma in approximately 160 blood donations. As the conditions were adjusted for each step in the process, the volume of liquid grew to about 200 liters at the stage when the albumin fraction, Fraction V, was removed by centrifuging. To handle batches of this size, the pilot plant was equipped with standard glass lined tanks of 120 and 240 liter capacities, mounted on wheels. Smaller stainless steel tanks accommodated volumes ranging from five to fifty liters. The Pilot Plant was housed in a twenty by six foot chamber that was divided into two spaces: a cold room maintained at a constant temperature of −5°C and a vestibule that was maintained at −10°C.

The heaviest piece of equipment was the refrigerated stainless steel Sharples centrifuge capable of centrifuging 200 liters of suspension in eight hours or less. Of the several physical chemical variables being manipulated during fractionation, the greatest attention was paid to maintenance of the temperature within prescribed limits at all times. The threat of a rise in temperature was greatest while concentrated ethanol was being added to increase the alcohol concentration of the supernatant protein solution and cause the precipitation of the next fraction. Primary control of temperature depended on continuous overhead circulation of cold air and use of a portable stainless steel cooling coil. The Sharples centrifuge bowl rotated 12,000 revolutions per minute during removal of protein precipitates. All fractionation operations, except for the drying of the products from the frozen state, were conducted in the cold room. A telephone mounted inside the cold room was intended to reduce unnecessary entry and egress. Immediately outside the cold room was a washroom where the tanks, centrifuge bowl and other small pieces of equipment could be cleaned. Space in a former corridor of the building was used to store the clean equipment. The drying of proteins from the frozen state was done in an adjacent room at ordinary room temperature.

The Massachusetts Antitoxin Laboratory

The protein fractions separated from plasma in the Pilot Plant had to be "finished" prior to their release for treating patients. Finishing involved a series of steps, starting with dissolution of the protein, sterile filtration, and filling aseptically into "final containers." A specified number of final containers from each lot of product were set aside and subjected to a prescribed panel of tests. (The term "lot" is regulatory agency parlance for "the quantity of uniform material identified by the manufacturer as having been thoroughly mixed in a single vessel.") Detailed accompanying documentation provided all pertinent data relating to the source and processing of the material in each lot.

Biological finishing services were performed for the Cohn laboratory by the Massachusetts Antitoxin Laboratory, a unique institution founded in 1894 to produce diphtheria antitoxin for use by residents of the Commonwealth.[8] This was the only laboratory in New England that was federally licensed to produce sera, antitoxins and other biological products for injection in man. Long associated with the Harvard Medical School and the Harvard School of Public Health, the Antitoxin Laboratory was located at the edge of the Arnold Arboretum in Forest Hills. In March 1941, the Massachusetts Attorney General issued a ruling permitting the Laboratory to cooperate with Cohn's NRC project. Since the Laboratory was already engaged in producing vaccines and sera, the incremental effort needed to finish the products produced in the Harvard Pilot Plant was provided without cost to the Federal government. Moreover, as the relationship flourished under the active collaboration of Elliott Robinson, its Director, it inevitably led to expanded interactions of public health importance.

Since no mechanism existed by which the Antitoxin Laboratory could be compensated directly for the services it rendered to the Harvard project, an informal barter system sprang up. Once during the war this unusual arrangement caused some raised eyebrows. George Scatchard received a call from the Treasurer of MIT stating that he was holding an invoice for two tons of hay. He wanted to know if the bill should be paid. Scatchard assured him that it should be paid. One of his research associates had immunized one of the Antitoxin Laboratory horses with human albumin in order to obtain a horse serum antibody that could be used in assaying for human albumin in experiments being conducted at MIT. The hay was simply a means to recompense the Antitoxin Laboratory for the valuable horse serum that was obtained.

Over and beyond the purely scientific implications of his new work, Cohn's departure into the study of human plasma inevitably drew his work into a larger and more complex set of issues concerning the supply and utilization of human blood. While the supply of bovine plasma was almost unlimited, only a trickle of human plasma was available. For human albumin to become a potential candidate in the race to develop a safe blood substitute meant that it would have to compete with dried plasma itself.

After the April 1941 meeting in Washington, Cohn returned to Boston with a new sense of the road that lay ahead. His appointment to the Subcommittee on Blood Substitutes signaled an important turning point in his career. It marked his emergence from two decades spent in pursuing theoretical investigations on proteins and amino acids in the relative obscurity of his department at the Harvard Medical School. His scientific work thus far had proved to be incomprehensible to most of his colleagues. Now he found himself facing an entirely new set of problems in a much larger universe. He was the only scientist on an important national committee charged with solving problems directly concerned with the national interest and welfare. The meeting in Washington made it clear that the fate of bovine or human albumin would hinge on the outcome of research he was directing.

Initiating Production of Dried Human Plasma

The pace of activities in Washington increased. The formal operating plan for the cooperation of the American Red Cross and the National Research Council was signed by Red Cross chairman Norman Davis and NRC chairman Lewis Weed in May 1941. With four blood donation centers then operating in New York, Boston, Buffalo and Rochester, the Red Cross announced a plan for establishing thirteen additional centers across the country, all to be located within seventy-two hours shipping time from a pharmaceutical contractor's plant.[9] On May 8, 1941, Cohn and Rhoads met with representatives of the pharmaceutical industry at the National Research Council in Washington to discuss the dried plasma program.[10] Among the issues raised by the firms was a concern about their ability to deliver dried plasma under government contract if the Red Cross failed to recruit the needed blood donors. In that eventuality, the procurement officers explained, the terms of the contracts could be extended. Another issue centered on specifying the volume of plasma to be dried in each package and whether the plasma from individual donors could be pooled prior to drying. The decision was made to dry 250 milliliters of pooled plasma in each standard Army/Navy container of dried plasma that had been designed by Captain Kendrick and Lt. Commander Newhouser for use in the field.

In order to conserve the plasma separated from the blood then being collected by the Red Cross, the specifications required that it be shell frozen and held in the frozen state while awaiting the installation of drying equipment. Even so, Kendrick pointed out that it would take at least a year before physicians in military hospitals on the North American continent could be trained in the use of dried plasma. Decisions on these points were made and approved by the full Subcommittee in July 1941.[11] In September 1941, contracts were drawn up and letters of intent were issued by the government to a group of firms that included Sharp & Dohme, Reichel, Lederle, Abbott and Eli Lilly. Others firms were added subsequently.

A Problem with Mercury in Dried Plasma

At a joint meeting of Walter Cannon's Committee on Transfusion and the Subcommittee on Blood Substitutes on September 19, Col. Hillman raised a question. He had learned that the dried plasma being produced for the Army contained a mercurial protective agent and questioned whether infusion of dried plasma so treated could lead to renal damage in wounded service men, especially since the infusion of large volumes of dried plasma was permitted under Army guidelines. Hillman's query surprised those present at the meeting, since the Subcommittee had specifically ruled out the addition of a mercurial antibacterial agent to plasma being dried for the armed services. On being investigated, it was revealed that the dried plasma contractors had complained to government contract officers about the absence of what they considered to be a standard requirement for biological products, the addition of a mercurial preservative. The ruling against use of mercury had been waived by the inspectors without notifying the Army Medical Corps. In the discussion that followed within the Subcommittee, it became evident that there was an underlying difference of opinion about the need to add the preservative. Weiss stated that the amount of mercury that would be administered with mercury-containing dried plasma exceeded the levels being used, for example, in the use of mercurial diuretics. On the other hand, W. G. Workman of the Division of Biologics Control held that adding a mercurial preservative was an added protection. In order to resolve the matter, Loeb requested that Veldee and Weiss review existing literature on the matter and report their findings to the next meeting of the Subcommittee.[12]

At the next meeting of the Subcommittee, on November 3, it emerged that Veldee and Weiss had been unable to agree on the interpretation of the published evidence concerning the effects of mercurial preservatives. Instead, each had submitted a report. Veldee's position was that the quantity of mercurial being added to the plasma was not sufficient to cause toxic symptoms in patients being treated for shock, burns or hemorrhage, while the mercurial being added had bacteriostatic and possibly bactericidal value. Weiss, who was unable to attend the meeting, had submitted a written opinion stating that mercurials would "at best . . . give only temporary growth-retarding action to a certain type of bacteria." He expressed the fear that, if very large volumes of plasma containing mercurials were administered to patients in shock, there could be harmful effects, particularly on the kidney. After some discussion, the Subcommittee approved the use of mercurial preservatives in dried plasma and set the maximum recommended dose of dried plasma at four liters in a twenty-four hour period.[13] Whether this action resulted in any adverse effects on wounded service men who received dried plasma is not known. However, the presence of the mercurial preservative in unused stockpiles of dried plasma remaining after the war presented an obstacle to the salvage of substantial amounts of clinically useful human plasma proteins.

Once the Subcommittee had formally approved the specifications for dried plasma, including the package designed by Capt. Kendrick and Lt. Commander Newhouser for its use in the field, Robert Loeb ruled that all future decisions concerning the production and release of dried plasma for use would be the responsibility of the Army and the Division of Biologics of the U.S. Public Health Service. At this stage, dried plasma was seen by members of the Subcommittee as a short-term issue that would ultimately be resolved in favor of bovine or human albumin.

Crystallized Bovine Albumin

The first preparations of bovine albumin were of variable purity, being comprised of about 95% albumin and 5% globulins. When tested in dogs, they seemed to be effective in restoring blood pressure and maintaining blood volume; when tested during the summer of 1941 at the Peter Bent Brigham Hospital in sixteen human subjects receiving doses of bovine albumin between 0.4 grams and 29 grams, there were no immediate reactions. However, four of the sixteen subjects later developed skin sensitivity to some component of the injected material. Further testing of bovine albumin was halted until the product could be further purified.[14]

In June, crystals of bovine albumin were discovered in a solution of crude bovine albumin that had been left in a cold room at the Medical School. Since crystallization offers a means to purify chemical substances, Cohn and Hughes immediately turned their efforts to identifying conditions under which crystallized bovine albumin (CBA) could be harvested in large quantity —a problem they had solved by late August 1941. Fortunately, Porsche and Lesh at Armour were by then providing Cohn with large quantities of amorphous bovine albumin. Hughes immediately set out to crystallize this Armour product in his laboratory at Harvard. By November 1941, following transfer of the crystallization technology to Armour, CBA became available in quantity and clinical testing was resumed. In the first phase of renewed testing, a large group of Harvard medical students volunteered to be skin tested with CBA—all without untoward effects. Planning for more definitive clinical testing was then initiated.[15]

Human Albumin

During the first half of 1941, the only human albumin in the United States was being prepared on laboratory scale at the Harvard Medical School. The first human albumin solution to reach the clinical testing stage was prepared by combining the albumin fraction from several laboratory scale preparations. Janeway began clinical trials with this albumin in April 1941. Cohn reported the first result to the Subcommittee on May 8, 1941. Thirty-two grams of human albumin as a 20% solution had been infused to a man from whom a pint of blood had been removed. The albumin produced a "rapid and satisfactory hemodilution."[16] Kendrick reported his first clinical experience with human albumin at the same time, describing his patient as a groggy, irrational twenty-year-old man who was admitted to the hospital sixteen hours after bilateral compound fracture of the tibia and fractures of five ribs. His blood pressure was 76/30. Two bottles of 25% albumin were infused intravenously, resulting in a rise in blood pressure to 130/80. During the next twelve hours the patient received nothing intravenously, but was given 1,250 milliliters of fluid by mouth and one liter of physiological saline subcutaneously, and the blood pressure remained above 130 throughout that period. Twelve hours later the patient was clear mentally and appeared to be feeling better.[17]

In commenting about these events, Cohn drew attention to some of the advantages of albumin as a biologic product in comparison to frozen or dried plasma. Albumin could be sterilized by filtration at the end of the process, it being unnecessary to go to the pains of conducting the fractionation procedure in a completely closed system. All that was required during fractionation was the application of ordinary precautions against gross contamination. No new or expensive equipment or techniques were required. He also hinted at what was to become human albumin's most important property, an unusual long-term stability of the solution. In one experiment, the sterile solution of human albumin had been incubated for two weeks at body temperature, 37°C, and had developed only a slight haziness—an encouraging sign. Walter Cannon stated, "I have high hopes for this product. Concentrated albumin might be given by vein, and water and salt by mouth. This would eliminate the necessity of supplying distilled water along with the blood substitute."[18]

The first lot of human albumin prepared in the new Harvard Pilot Plant, designated HA-4, passed its final safety tests at the Massachusetts Antitoxin Laboratory and was released for clinical use on July 9, 1941. Cohn estimated that between 500 and 800 grams of human albumin—one to two pounds—were then being produced weekly in the pilot plant.[19] By July 18, human albumin had been tested in eighteen cases at the Peter Bent Brigham Hospital and other hospitals. These included three cases in which the transfusions followed intentional blood withdrawal and eight patients with shock. There had been only one mild reaction.[20]

Heartened by these results, the Subcommittee reiterated its interest and enthusiasm for the work underway in Boston, urging that Cohn continue to produce human albumin at Harvard so as to support expanded clinical trials. At the same time, the Subcommittee formally recommended that the National Research Council take steps to provide ongoing funding for the human albumin project; that, at the appropriate time, commercial scale production of albumin from human plasma should be under the direct supervision of Edwin Cohn; and that the American Red Cross should continue to recruit the blood donors for the albumin program. The Subcommittee also recommended the expansion of clinical testing of human albumin beyond the Peter Bent Brigham Hospital. The expanded group of hospitals included the Presbyterian Hospital in New York, U.S. Army and Navy hospitals in Washington, Bryn Mawr Hospital outside Philadelphia, Strong Memorial Hospital in Rochester, Johns Hopkins Hospital in Baltimore, Cincinnati General Hospital, Grady Memorial Hospital in Atlanta, University of Chicago Hospitals, and Iowa State University Hospitals. On the other hand, the Subcommittee directed that testing of bovine albumin should continue to be restricted to the Peter Bent Brigham Hospital until further notice.

In a lecture in Cleveland on the "Properties and Functions of the Purified Proteins of Animal and Human Plasmas" as part of a symposium on "Blood Substitutes and Blood Transfusion" sponsored by the American Human Serum Association in June 1941, Cohn pointed out that plasma is a complex solution of many constituents. Neglecting for the moment the electrolytes, carbohydrates and phospholipids in plasma, he pointed out that there are many proteins that differ from each other in size, shape, electrical charge and other properties. Stressing the specialization of these proteins, he concluded that no one plasma protein could be expected to be capable of substituting for another. Thus, no amount of albumin could possibly substitute for the proteins that are involved in bringing about the clotting of blood. By the same token, fibrinogen and prothrombin that play important roles in blood clotting would hardly serve as substitutes for the treatment of shock. In response to a question regarding the possibility of using heterologous plasma, i.e. plasma from an animal species other than man for treating military casualties, Cohn declared that even if that proved possible, he would still favor the use of highly purified agents specifically aimed at correcting the impaired functions.[21]

Janeway reported in an article on "War Medicine with Special Emphasis on the Use of Blood Substitutes," published in the *New England Journal of Medicine* in September 1941, that "one outstanding feature of the war has been the heavy aerial attacks to which the cities of Great Britain have been subjected. This type of warfare, aimed at the civilian population, is new," he stated. "Observers returning from England, as well as reports in the British literature, stress the severity of the casualties." In alluding to an existing dispute about many details of the shock syndrome, Janeway credited Cohn for developing an entirely new approach to the therapy of shock in which he proposed to use fractions of the plasma for specific purposes. Noting that the important aim in treating shock was to restore the blood volume, Cohn's use of albumin in shock took advantage of albumin's greater net charge and smaller molecular size. Albumin was not only present in plasma in greater amounts than other plasma proteins, but was also more stable and more soluble, properties which made albumin a theoretically ideal substitute for whole plasma in treating shock.[22]

By mid-September 1941, over three kilograms, about eight pounds, of human albumin had been prepared in the Harvard Pilot Plant. Of this, 500 grams were infused to eighteen subjects in amounts varying between seven and sixty-one grams, with no untoward affects. Four of these cases involved tests in the treatment of shock, while the balance of the tests assessed the safety and lack of toxicity of albumin. Given this modest record of safety, members of the Subcommittee urged Cohn to accelerate clinical testing so as to expedite the ultimate approval of albumin as a new blood substitute. In particular, emphasis was to be placed on validating the safety of albumin in man, and gathering evidence on its efficacy in the treatment of shock. In an effort to expedite the clinical testing of human albumin, Soma Weiss secured the assignment of two young physicians, Lieutenant L. T. Woodruff (j.g.) MC-V(G) U.S.N.R. and Lieutenant S. T. Gibson, (j.g.) MC-V(G) U.S.N.R. to Cohn's laboratory. Woodruff and Gibson promptly established liaison with the emergency rooms at the Peter Bent Brigham, Massachusetts General, and Boston City Hospitals, with orders to be notified of cases admitted for trauma and burns. When a suitable patient was identified, they arranged to have the patient receive a 100 milliliter bottle of 25% human albumin, with changes in blood pressure and other appropriate laboratory measures to be monitored.

In order substantially to increase the supply of human albumin that could be made available for clinical testing, Cohn recommended that OSRD contract with Armour Laboratories to fractionate human plasma in Chicago. Armour was a natural choice for this purpose since it was already experienced in using the Cohn plasma fractionation procedures. In this way, it was projected that the first lot of Armour human albumin could be ready for clinical use by the end of January 1942. The human plasma for this new contract was to be separated from blood collected at the Michael Reese Hospital in Chicago.

Stabilizing Human Albumin

From the outset, the ability to dispense albumin in solution, rather than as a dry powder that had to be dissolved before use, was an important feature to the Navy. Cohn therefore pressed his associates to devise a plan of research on factors influencing the long-term stability of albumin solutions. At first, these studies consisted simply of observing changes in the appearance of sterile albumin solutions during prolonged storage at room temperature. One of the first observations from this effort was the discovery that albumin dissolved in dilute sodium chloride solution remained clear for weeks, while albumin dissolved in distilled water quickly became cloudy under the same conditions. At the next Subcommittee meeting in Washington, Cohn reported that the best samples of albumin appeared to be stable at temperatures between –40 and +45°C. In response, Captain Stephenson urged Cohn to press on to extend the range, preferring that albumin withstand temperatures between –50 and +50°C.[23]

The Globulin Components of Plasma

The 1940 issue of *New and Nonofficial Remedies (NNR)* offers a perspective on the state of the art of U.S. biologics production and use at that time.[24] The few protein-containing products listed consisted mostly of horse and cow sera and crude extracts of animal organs, as well as antitoxins to diphtheria, bovine tetanus, meningococci, pneumococci and scarlet fever. When administered in adequate doses, antitoxins could alleviate the course of infectious diseases, although their use was frequently accompanied by serum sickness. *NNR* listed a crude human placental globulin called Immune Globulin (Human) that was available commercially. This product was obtained by extracting freshly ground human placentas with cold saline solution, after which the extract was concentrated, yielding a turbid reddish brown solution that contained a mixture of antibodies and other soluble proteins. As listed in *NNR*, this product was said to contain "immune factor or factors against measles." However, since there was no way of assaying the efficacy of such a product against measles, its potency could not be stated, although it frequently produced reactions that were "not always mild."

Late in 1940, Elliott Robinson of the Massachusetts Antitoxin Laboratory had alerted Cohn to the potential public health importance of the human antibodies that might be isolated during the plasma fractionation project. In the summer of 1941, Joseph Stokes Jr., the Professor of Pediatrics at the Philadelphia Children's Hospital, came to Boston to see Cohn, having heard about the separation of the plasma globulins in the main fractionation procedure. He told Cohn that at its November meeting the Board for the Investigation and Control of Epidemic Diseases in the Army would be discussing "the possible use of human globulin (Cohn) in the passive immunization against measles." Cohn foresaw the potential of the proteins other than albumin, including this globulin fraction, in the plasma being collected by the Red Cross. Assuming that the antibodies could be isolated in a highly purified state, which he never doubted, and provided that the levels of specific antibodies in the pooled plasma from hundreds or even thousands of blood donors were sufficiently high, he could foresee important public health uses.

By mid 1941, the ethanol-water fractionation procedure had been modified somewhat. Some fine-tuning had been applied to the separation of Fraction I, which represented 5% of the plasma proteins and was rich in fibrinogen, the precursor of the fibrin blood clot. Fractions II and III had been combined into a single fraction, II+III, referred to as Fraction two plus three. Fraction IV had been split into two fractions, called IV-1 and IV-4. Of the fractions from plasma, Fraction II+III was the second largest fraction, representing about 25% of the proteins originally present in plasma. The only previously known proteins thought to be present in Fraction II+III were prothrombin and the antibodies. When analyzed by electrophoresis, the major components of Fraction II+III were identified as alpha, beta and gamma globulins. Fraction IV-1 and Fraction IV-4 included 15% of the plasma proteins. Once the pharmaceutical firms began to produce human albumin, large amounts of these fractions would begin to accumulate.

In the short term, they could be stored in the frozen state, but if permitted to accumulate, the possibility of having to discard kilogram quantities of frozen human plasma fractions other than Fraction V, the albumin fraction, was a distinct possibility. That would have been abhorrent to Cohn. Accordingly, he pressed his colleagues to turn their attention to the development of subfractionation procedures for the major fractions other than albumin.

In 1940, while Cohn and his associates were in the throes of developing the new plasma fractionation method, John Edsall had left to spend a sabbatical year with Linus Pauling at the California Institute of Technology. Before Edsall departed, Cohn, who by then was totally immersed in the new plasma fractionation work, persuaded Edsall to take over the task of reading the final proofs of the new Cohn book, *Proteins, Amino Acids and Peptides.* On Edsall's return to Boston in September 1941, Cohn asked Edsall, Ronald Ferry and John Ferry to open a new attack on the proteins involved in blood coagulation. Edsall, with his encyclopedic knowledge and long experience in the laboratory, took the responsibility for preparing human thrombin by activating the prothrombin in Fraction II+III. John Ferry, an expert on the mechanical properties of substances of high molecular weight, undertook an important series of studies on fibrinogen plastics. With his strong clinical experience, Howard Armstrong played an important role in linking the scientists in the Department of Physical Chemistry to colleagues in the clinical disciplines, particularly to a new group of surgeons who collaborated in studying products made from fibrin. Armstrong also directed the electrophoresis laboratory during this period.

In September 1941, the English pathologist Howard W. Florey attended a joint meeting of Walter Cannon's Committee on Transfusions and the Subcommittee on Blood Substitutes. Florey reported that the British Medical Research Council was providing pooled liquid human serum that had been sterilized by filtration to treat air raid casualties. Blood was being collected locally; if blood was unavailable, dried plasma or serum was being used in its place. A dried human serum was also being prepared for use of British forces fighting abroad. In the ensuing discussion, Col. Hillman confirmed that the use of preserved blood was not seen by the U.S. Army as feasible for meeting the needs of U.S. armed forces for blood.

Recommendation for Government Funding

In August 1941, the NRC Committee on Medical Research recommended government funding for Cohn's continuing studies. The initial mechanism was an Office of Scientific Research and Development contract (OEMcmr-22) with Harvard University entitled "Chemical clinical and, immunological investigations on human and bovine serum albumin." The subject of the Harvard contract was described as the "production of sufficient amounts of plasma fractionation products to permit evaluation by clinical trial of the usefulness and proper methods of employment of these agents." At a meeting of the Subcommittee on Blood Substitutes on November 3, 1941, a motion was unanimously passed, recommending that the armed forces should use whole blood transfusions in the treatment of hemorrhagic shock whenever possible. However, this action by the Subcommittee was not recorded in the minutes of the meeting, and it was not until September 24, 1943, when the matter was raised again, that all members of the Subcommittee confirmed the original action.

A Meeting of the American College of Surgeons

When the American College of Surgeons met in Boston in November 1941 for its annual meeting, its members were greeted by a headline in the *Boston Herald*, "Navy utterly lacking blood plasma for war." [25] The *Herald* article cited Admiral Ross T. McIntire, Surgeon General of the Navy and Maj. Gen. James C. Magee, Surgeon General of the Army, for urging a nation-wide campaign to collect and store huge supplies of blood plasma, claiming that "this country does not have on hand even a fraction of what would be required in a shooting war." At that meeting, University of Pennsylvania surgeon Isadore S. Ravdin stressed the importance of plasma in treating shock, pointing out that a severe case of shock often requires multiple transfusions of plasma. The surgeons attending the meeting also heard the medical officer of the British aircraft carrier H.M.S. *Furious* describe a plane crash in which forty men were burned, of whom twenty died. Many of these injured men might have survived, but there had not been enough plasma aboard to treat even one man adequately. In a presentation on current research at the meeting, Ravdin cited Cohn's work as "the most promising lead at present."

A Day of Infamy

On Sunday December 7, 1941, the "day that shall live in infamy," six Japanese carriers launched an air attack on the Naval Base at Pearl Harbor resulting in more than 2,000 dead and almost 1,200 wounded. An estimated 150 American planes were destroyed on the ground. Six battleships were knocked out, as were a number of other Navy vessels. On December 8, the U.S. Congress unanimously declared war on Japan. That same day, A.N. Richards, Chairman of the NRC Committee on Medical Research, telephoned Edwin Cohn to request that all available vials of human serum albumin in Boston be dispatched immediately to Pearl Harbor. On the evening of December 8, 1941, twenty-nine vials of human albumin that had been prepared in the Harvard Pilot Plant were on the way to Pearl Harbor with Isadore Ravdin, who had been assigned to consult with the staff of the Pearl Harbor Naval Hospital concerning the treatment of the casualties of the Japanese attack.[26]

8 The Aftermath of Pearl Harbor

At Pearl Harbor, Admiral McIntire reported, "it was burns, burns and more burns." Between 8 a.m. and midnight on December 7, 1941 Navy physicians treated 960 casualties, 60% of whom suffered from scorching wounds. Dried plasma—750 units were used during the first day alone—was an important factor in the remarkable efforts to save lives.[1] Ravdin reported on the aftermath of the Pearl Harbor attack at a Conference on Albumin arranged by the NRC Committee on Transfusions on January 5, 1942.[2] Among the attendees at the meeting, which was chaired by Soma Weiss, were A. Baird Hastings, who was a member of the Office of Research and Development (OSRD) Committee on Medical Research, Blalock, Cohn, Janeway and Weed, with Stephenson and Newhouser representing the Navy and Kendrick the Army.

Ravdin, a man of short stature but with the imposing bearing often seen in surgeons (he retired with the rank of brigadier general at the end of the war), delivered his report with stark clarity and directness. When he arrived at Pearl Harbor ten days after the Japanese attack, seven of the most severe burn cases had been selected to receive human serum albumin. Each had been treated with varying amounts of human plasma, glucose-saline solution and whole blood prior to being seen by Ravdin. As first seen, the patients had very low hemoglobin readings, low red cell counts and hypoproteinemia. Each was edematous and losing plasma from burned sites. Harvard albumin was administered to each patient. After the second injection of albumin, all showed general clinical improvement, with definite evidence of hemodilution and decreased concentration of blood cells.

One patient was unconscious and in critical condition, leading an attending physician to question the advisability of giving him albumin. He was given a quantity of albumin equivalent to more than a liter of plasma in the morning and by afternoon he was delirious but talking. The following morning he was given more albumin. On the third morning the edema had disappeared and the patient was able to eat breakfast. A second infusion of albumin to this patient was well tolerated. There had been one reaction, a chill, in the group of patients treated. The staff of the Naval Hospital in Hawaii had been impressed by the ease of administration of the albumin and the small storage space required. The volume of albumin equivalent to six liters of plasma could be stored in the space occupied by a quarter liter of dried plasma and its accompanying bottle of distilled water.

In the ensuing discussion, Blalock, the only other surgeon on the panel, termed the results "encouraging." He expressed the opinion that most severely burned patients die within forty-eight hours following burns, and wondered if albumin would be as effective if used early in these cases. Ravdin summarized that albumin did everything osmotically that plasma could be expected to do, produced hemodilution, and resulted in an increase in the albumin content in the bloodstream. Among patients who had not lost too much blood, he concluded that albumin served a satisfactory function. Encouraged by Ravdin's report and sensing the inclination of the Subcommittee on Blood Substitutes to take action, Weiss summarized the clinical trials on human albumin up to that point. "Harvard albumin had been used in eighty-seven cases, with only four minor reactions. The results in thirty-four cases confirmed the safety of albumin in man. Of the remaining fifty-two cases, forty-six showed either an improvement in clinical condition, a blood pressure rise or a decrease in hematocrit."

Cohn then summarized the factors that were retarding the production of albumin. These were the limited number of volunteer blood donors in Boston, and the reorganization of the Boston Chapter of the American Red Cross. Both or these issues were being addressed. He also reported an unexpected problem that arose while the first lot of Armour human albumin was undergoing final animal safety testing.

During a routine animal safety test involving the injection of samples of the albumin into guinea pigs, all the animals had died. This problem had never been encountered in testing the human albumin lots prepared at Harvard that had been filled into final containers at the Massachusetts Antitoxin Laboratory. Indeed, Armour's own bovine albumin had routinely passed this safety test without incident in the past. A toxic substance was suspected. All the chemical reagents that had been used by Armour had been analyzed, without uncovering any anomaly. Continued investigation revealed that the new Armour human albumin solution contained mercury. The source of the mercury had been traced to the blood bank at the Michael Reese Hospital in Chicago, where a well-meaning staff member, on his own initiative, had added merthiolate, a preservative, to the plasma before it was shipped to the Armour plant. This single episode delayed the production of Armour human albumin for six weeks. The person responsible for this episode felt the full fury of an angry Edwin Cohn.

Human Albumin is Officially Approved

On January 5 1942, the members of the NRC Conference on Albumin unanimously adopted the following resolution:

> Based on recently accumulated clinical evidence, normal serum albumin (human) concentrated has been proven to be of value in the treatment of shock including burns. The Conference on Albumin empowered by the Committee on Transfusions of the National Research Council, recommends to the Surgeons General in addition to the continuance of the program of securing plasma, the immediate adoption for clinical use of serum albumin for the following reasons:

1. It can be packaged and stored in less than one-tenth the space required for dried plasma (Standard Army-Navy package).
2. It is ready for immediate emergency without reconstitution.
3. It has been proven to be stable in solution at temperatures as high as 45°C (113°F) for protracted periods of time.
4. The adoption of human albumin will accelerate and supplement the procurement of satisfactory blood substitutes for war use.

The Subcommittee on Blood Substitutes recommended that the Surgeons General ask the American National Red Cross to recruit voluntary blood donors through its chapters and that this project be made a part of the overall national blood substitutes program serving the Armed Services and the civilian population. It also recommended that Edwin J. Cohn be asked to assume general supervisory direction of the processing of plasma in the commercial laboratories.[3]

In this way, only a few days after the declaration of war, and less than twenty months since Cohn and his colleagues had developed the new ethanol-water fractionation process, a critical step was taken toward making human serum albumin available to the Armed Forces of the United States. The first lot of Armour Human Serum Albumin was received at Harvard for clinical testing on January 26, 1942, and when Soma Weiss died suddenly on January 31, at the age of forty-three, Charles Janeway was appointed to replace him as Chairman of the Conference for the Study of Albumin.

Transfer of the Plasma Fractionation Technology to the Pharmaceutical Industry

Admiral McIntire authorized the drawing up of Navy contracts for the production of albumin. The Navy thus assumed responsibility for procuring human albumin for both armed services very much as had been done for dried plasma by the Army. Further clinical testing of albumin was immediately discontinued. All units of albumin produced at Harvard were ordered to be shipped to the Navy for military use, "since its value and uses were considered to have been demonstrated and the small compact, emergency unit was in demand by the Navy." Pending the start of albumin production under the Navy contracts, Cohn placed the Harvard Pilot Plant on an around-the-clock basis to maximize albumin production. Admiral McIntire also ordered "that the technicians who are to be in charge of the fractionation of albumin in the contracting laboratories shall be sent to the Harvard Fractionation Laboratory for such instruction as will be specified by Dr. Cohn or his designated representatives."[4]

Within the Navy, the task of drawing up the formal documents that would be used in preparing contracts with the pharmaceutical firms was long and drawn out, involving a series of meetings spanning the period from January to May 1942.[5] Since all the technical knowledge needed to draw up and manage the contracts was located at the Harvard Medical School, the burden fell on Cohn and his senior associates. Comdr. Cushing initiated the process on January 6, 1942 by advising Cohn that the Navy wanted 50,000 units of albumin (a unit consisted of 100 milliliters of 25% solution) as rapidly as possible. The next day, Comdr. Newhouser advised Cohn that the Surgeon General of the Navy had "authorized me to take whatever steps . . . to produce the newly designed albumin package, including the double ended glass bottle, tubing and needles which could be sealed into a metal can." A week later, Cushing wrote Cohn requesting that he prepare written specifications for use by the Navy in drawing up contracts, including the chemical, biological and safety requirements for acceptance of the final product by the Navy and for the training of personnel to operate the plants. Late in January 1942, Newhouser and Kendrick came to Harvard to assist in this task. The first draft was completed on January 31. On February 8, Newhouser forwarded a draft with more detailed specifications to Cohn, commenting that "in our dried plasma specifications we left too much up to the individual firms. With your assistance I am sure we will not repeat the error." On March 14, 1942, Admiral McIntire authorized Cohn "to undertake instruction of technicians in the fractionation of human serum albumin."

On March 27, Cohn received the first of several communications from Captain Melhorn in Navy procurement:

> This office has been directed to initiate procurement of fifty thousand packages (forty thousand for the Navy and ten thousand for the Army) of human serum albumin prepared in accordance with the process developed under your supervision . . . In order to proceed, however, it is desirable that information at hand be verified, the proposed program outlined, and a clear understanding of the data to be incorporated into purchase contracts be established, before negotiating contracts.

On April 9, Cohn added further suggestions, pointing out the advantages of engaging only pharmaceutical plants with large production capacity and experience, and recommending that Melhorn visit the

Harvard Pilot Plant so as to see at first hand the kinds of equipment involved in fractionating plasma.

Shortly thereafter representatives of six pharmaceutical firms arrived at the pilot plant to begin training. In due course, "letters of intent" to execute contracts were sent to six pharmaceutical firms: Lederle Laboratories, Upjohn Co., Eli Lilly Laboratories, E.R. Squibb, Cutter Laboratories and Sharp and Dohme. Originally, the Navy intended to contract for 8,500 vials of albumin, each containing 100 milliliters of 25% solution from each of the six firms. Overall, this would require the Red Cross to collect a total of 180,000 additional blood donations. A few days later, a revised target for albumin production was received. Melhorn advised Cohn, "according to word just received from Capt. Stephenson, that the Army is going to request two hundred thousand units and the Surgeon General of the Navy wants two hundred and fifty thousand units, making a total of four hundred and fifty thousand units. It is understood that arrangements have been made with the Red Cross to furnish donors."

The Red Cross response was hardly one of acquiescence. Caught in a bind, its new director G. Canby Robinson hastened to the Army Surgeon General's Office to plead for a lower plasma quota. In a letter to Cohn he reported that he had advised the Army to wait until the pharmaceutical plants were in production before expanding the albumin project. Cohn angrily fought back, pointing out the unwisdom of basing Red Cross goals on present Red Cross blood collection capabilities rather than on future national needs for albumin. He argued that while a pharmaceutical plant capable of large-scale production could fractionate relatively small volumes of plasma quite efficiently at the outset, the opposite was not the case. It would be impossible to increase the capacity of a small plant should the need arise at a later date.

Meanwhile, faced by some criticism of albumin from proponents of dried plasma, and increasingly impatient over delays in startup of albumin production by the pharmaceutical firms, Loeb convened a new Conference on Albumin on May 26, 1942. The criticisms of albumin centered on three points. First, albumin was seen to be less effective than plasma. Second, there were concerns that albumin might not prove to be stable when dispensed in solution. And third, concerns were raised about the possibility that use of albumin could result in dehydration in some patients. The albumin program was clearly in trouble. The Conference issued three statements:

> Human serum albumin in 25% solution has been recommended to the Army and Navy to fulfill . . . specific requirements for a blood substitute, which can be transported and administered in small volumes with great facility and with safety. As has been etched on the containers, and as is indicated in the directives to the Navy . . . albumin should not be administered in severe dehydration without the simultaneous administration of fluids.
>
> Human serum albumin has been established as a blood substitute of proven value in that it causes hemodilution and raises blood pressure in a manner similar to blood plasma or serum. On the basis of clinical tests, human albumin produces no more reactions than does plasma.
>
> The stability of human serum albumin solution as packaged has been established over a period of one year at room temperature in temperate climates. Such material has been carried at sea in the sick bay of a cruiser for three months at a temperature reported to average 37°C without deterioration.[6]

The Conference then turned to a consideration of the frustrating delays at the chemical and technical level in starting large-scale plasma fractionation. Loeb opened this part of the meeting by asking Cohn if he was certain that his methods were adequate to produce the units of albumin required by the Navy and Army. Cohn affirmed his complete confidence in the ability of the new fractionation procedure to produce acceptable albumin. In his view, the problem lay entirely in the inability of the pharmaceutical firms to obtain the needed processing equipment. A key element in the delay was caused by difficulties in procuring the needed Sharples centrifuges. He read a letter from the manufacturer of the centrifuges revealing that several of the firms had not yet ordered the centrifuges they would need. Stephenson, citing correspondence between the Surgeons General of the Army and Navy, reminded the Conference that the 360,000 units of albumin—a new production target to be acquired in the fiscal year ending June 30, 1943—would require the collection of 1,250,000 units of blood by the Red Cross. Those present decided to convene a meeting with representatives of the pharmaceutical firms, the Army, Navy and other government agencies.

An extraordinary daylong Conference on the Preparation of Normal Human Serum Albumin was held three weeks later, on Saturday June 6, 1942, in the Faculty Room of the Harvard Medical School. Among more than forty individuals attending were the chairmen of four NRC Committees and Subcommittees, W.J. McManus, a member of the War Production Board, representatives of six pharmaceutical firms, the Pfaudler Co., manufacturers of tanks, the Sharples Corp., members of the U.S. Public Health Service, the American Red Cross, the Navy, the Army, the Massachusetts Antitoxin Laboratory, and the Department of Physical Chemistry. Edwin Cohn chaired the meeting.[7]

Loeb opened the meeting by citing the desperate need of the Armed Services for a safe, efficacious blood substitute. He stressed the premium being placed by the Armed Services on providing a therapeutic agent which could be dispensed as a stable concentrated solution for use in the field. Human serum albumin had been found to meet these requirements, had been subjected to careful clinical testing and had been approved for use by the Armed Services in January 1942. He emphasized that the approval of human serum albumin was not intended to affect the existing dried plasma program. He then stated that it was up to those attending the Conference to arrive at a plan by which to collect the needed blood and produce albumin and distribute it to the Armed Services in the needed amounts as rapidly as possible.

Stephenson, the senior spokesman for the Armed Services, surprised most of those present by announcing that the Army and Navy were increasing the human serum albumin production target to 360,000 units for the fiscal year ending June 30, 1943. This represented a seven-fold increase over the 50,000 units originally communicated to the American Red Cross and the pharmaceutical industry. Stephenson's announcement elicited a frustrated response from Canby Robinson of the Red Cross. Pointing out that he had heard about the albumin target only the day before, Robinson cited the magnitude of the task that the Red Cross had assumed, painting a picture of an inexperienced organization overextended by present commitments and lacking the organizational skills to expand into new areas across the country. Physicians were proving difficult to retain. While there was no shortage of willing blood donors in most localities, it was the plethora of issues and bureaucratic requirements that overwhelmed him. Shortages of equipment such as needles, rubber tubing and bottles had to be dealt with on a daily basis. After only a year of experience in collecting blood, Robinson was staggered by the new expectations being thrust on Red Cross.

Stephenson gave no quarter, criticizing the considerable pessimism expressed about the entire albumin program. "If the material cannot be obtained, if the work cannot be done, then we must ask ourselves, shall we give up this entire albumin program or shall we go ahead and try to get on with the work? When I return to Washington I must advise the Surgeon General concerning the albumin program. I must know, therefore, whether or not you fellows are going through with this program or are dropping it." In the ensuing discussion, Newhouser revealed that, due to the unavailability of chemical equipment for the new albumin plants, particularly the large stainless steel centrifuges needed to remove the protein precipitates in the Cohn fractionation process, it would be months before large quantities of albumin could be produced. Moreover, he reported that some of the pharmaceutical firms, not having completed contract negotiations with the government, had still not ordered the centrifuges they would need.

The mood of the participants calmed down when McManus, the War Production Board representative, observed "The whole trouble is that no one has ever worked out what is actually needed for this program." In response, Cohn identified the two major problems facing the conference. The first was the allocation of blood to the albumin program. The second was the procurement of the needed equipment; he asked McManus and two others to deal with it. Cohn then directed that each pharmaceutical firm submit a list of needed equipment, and when it was needed. He also asked each firm to indicate their planned albumin production capacity. As the meeting came to an end, Stephenson, undoubtedly aware that Admiral Yamamoto had just sustained a crushing defeat in the Battle of Midway a few days before, brought the meeting to an end by asking that everyone get started at the earliest possible moment and work as hard as possible.

Later that year, three medical reports appeared in the *U.S. Naval*

Medical Bulletin in an effort to acquaint military physicians with the use of human albumin. Noting that the onset of hostilities had resulted in urgent demands for treatment of clinical shock, the first paper provided a preliminary evaluation of the physiologic effects of albumin in man under conditions in which the blood volume had been depleted.[8] Its principal conclusion was that in nine subjects, after removal of approximately 20% of their blood volume, infusion of albumin resulted in a restoration of the volume, each gram of albumin being capable of increasing the blood volume by an average of 18 ml. This result was in good agreement with the value predicted by Scatchard, Batchelder and Brown from osmotic pressure measurements in a laboratory at MIT.[9] A second paper, on "The Clinical Evaluation of Human Albumin,"[10] reported evidence demonstrating the safety of albumin in 200 patients. In patients with clinical shock, albumin brought about prompt recovery and improvement that was attributable to the action of albumin in bringing water into the bloodstream. Later in the fall of 1942, Heyl, Gibson and Janeway published the results of a much larger series of clinical cases showing that each gram of human or crystallized bovine albumin, when injected into the bloodstream, increased the circulating blood volume by approximately seventeen milliliters.[11] This corresponded to a blood volume increase of just under one pint for each unit of 25% albumin.

Shortly after the Boston conference, it was revealed that contract negotiations with one pharmaceutical firm had hit a snag over costs. This firm proposed to charge the Navy more than twice the cost per unit of albumin estimated by the Harvard pilot plant. On investigation, it was found that the proposed costs were technically in error; Loeb ruled that resolution of the problem rested with the Navy rather than with the National Research Council.[12]

Navy Specifications for Human Serum Albumin

Control by Cohn over the quality and safety of human serum albumin being produced for the Navy was exerted primarily through two documents: *Directions for the Preparation of Normal Human Serum Albumin*[13] and *Specifications for the Preparation of Human Albumin from Human Blood Plasma*[14]. The *Directions*, periodically updated, spelled out in detail the ethanol water fractionation procedure, beginning in 1941 with Method 2, a modification of the original procedure developed in 1940. In addition to describing the precise conditions of pH, ionic strength, alcohol concentration, temperature and protein concentration for each step in the process, the *Directions* provided comments and advice on important details that bordered more on art than science. Although the firms all used refrigerated Sharples centrifuges for separation of the protein fractions, other pieces of processing equipment used were not always identical. This occasionally led to minor difficulties, in which case a brief report would be passed on by Harvard to all producers. Gradually, these accumulated annotations became a valuable resource. For example, in a note issued in November 1942, it was suggested that a great deal of trouble and valuable human plasma could be saved by periodically testing the water and other reagents used in the processing of plasma for contamination with pyrogens. In another instance, the contractors were notified of a positive pyrogenic reaction in albumin from one firm that had been traced to newly purchased filter pads.

The Navy specifications covered the starting plasma, the fractionation procedure to be used, the packaging of the final product, and an elaborate series of tests and standards that every lot of product manufactured by the pharmaceutical firms was required to meet before acceptance by the Navy. The starting plasma had to meet the Minimum Requirements of the Division of Biologics Standards of the National Institutes of Health. The specifications incorporated the detailed Cohn procedure for fractionating plasma. Human serum albumin should be at least 98% pure by electrophoretic analysis. It had to be dispensed in the standard Army-Navy package in a volume of 100 ml. of 25% solution. The solution had to be "without significant turbidity or precipitate and should not undergo significant visible change when heated at 50°C for twelve days." As new scientific advances were made, these specifications were periodically revised and tightened.

The manufacturer, on completion of all manufacturing and testing steps for a lot of product, was required to submit a protocol, i.e. complete written documentation establishing the history of the lot, including the results of tests carried out in-house. The required tests covered the purity, stability, sterility, animal safety, and freedom from toxicity and pyrogens. The protocol was then sent to the Plasma Fractionation

Laboratory at Harvard along with a specified number of randomly chosen final containers from the lot. The final stage of testing, conducted at Harvard, was intended to confirm the results of tests reported by the manufacturer, and occasionally went beyond to document the results of special tests, such as ultracentrifuge analyses. A human safety test was performed on every lot of albumin produced during the war by a Clinical Testing Committee under the chairmanship of Charles Janeway. The Navy specifications identified Cohn "as the consultant in all matters related to the processing method. No variation in the manufacturing process shall be employed without his approval and each laboratory shall be open to inspection by Dr. Cohn or his representative at all times during working hours."[15] The overall effect of these specifications was to place on Cohn final responsibility for the safety and efficacy of every lot of human serum albumin released during the war. Between June 1942 and July 1943, while the new pharmaceutical plants were being readied to produce human serum albumin, the Harvard Pilot Plant operated with three shifts on a full time basis.

Each pharmaceutical firm holding a Navy contract set to work to plan and construct a new plant. Each firm designated two senior staff members to be trained at Harvard on a rotating basis while the plants were being constructed. While one man was being trained in Boston, the other oversaw the construction and equipping of the new plant back at the firm. While at Harvard, the trainees worked eight hour shifts with pilot plant staff members. The trainees were taught all special techniques, including the conduct and interpretation of electrophoresis and ultracentrifugal analyses. They attended Cohn's regular weekly luncheon meetings and participated in other meetings, formal and informal, with scientists in Cohn's department. A visitor to the pilot plant in the late evening might encounter a senior pharmaceutical firm manager clad in arctic clothing emerging from the cold room pushing an empty glass-lined Pfaudler tank out the door to be washed. Between 1942 and 1943, about 2,500 × 100 ml vials of 25% human serum albumin were produced in the Harvard Pilot Plant.

Cohn was strict about adherence to specified fractionation procedures at all times. He was particularly sensitive about temperature changes resulting from too frequent opening of the doors to the cold room. Recognizing the importance of the trainees for the successful transfer of the fractionation process to the industry, he frequently stopped by the Pilot Plant to monitor the work being done. He particularly liked to see with his own eyes the appearance of a precipitate, or examine a sample being studied. Although he never donned arctic gear before entering the cold room, the Pilot Plant personnel were expected to do so and avoid unnecessary opening of the cold room doors. As a safety precaution, there was a telephone in the cold room with a direct outside line. Returning home late in the evening from a dinner party, Cohn sometimes called the men in the Pilot Plant to describe an experiment that he wanted to have carried out, expecting to hear a report when he came in the next morning. If these telephone conversations were lengthy, the trainees shivered in the cold room, while Cohn was warm and comfortable at home.

In June 1942, the Subcommittee on Blood Substitutes approved the Standard Army and Navy Package for Albumin that was designed by Commander Newhouser. This was a special double ended glass container of "Normal Serum Albumin (Human), Concentrated," the formal name assigned by the Division of Biologics Control of the U. S. Public Health Service. When filled with albumin solution, the container was sealed with a rubber stopper at each end. It was contained inside a protective easy-to-open tin can, together with an intravenous set consisting of rubber tubing and needles needed to infuse the contents into a patient and a set of instructions for its use. In October 1942, Newhouser demonstrated the package to the Subcommittee on Blood Substitutes, showing how it would be used in the field. The can even floated if immersed in water.[16] A short paper describing the standard package and its contents was published in the *U.S. Naval Medical Bulletin*.[17]

Identification of Some Antibodies in Fraction II + III

Once the work on albumin was under way, Cohn launched two efforts aimed at opening up knowledge about Fraction II+III, which was the next largest fraction after albumin. In the first approach, he drew on the expertise of John Enders and William Boyd, both experienced immunologists. In the other, Oncley took a physical chemical approach to the isolation of the protein components of this fraction. Enders and Boyd began by identifying the antibodies that could be found in the major fractions of plasma, concentrating at first on known antibodies that could easily be tested for. Antibodies to typhoid were quickly identified in Fraction II+III, as were the antibodies responsible for blood groups, termed isoagglutinins. In contrast, only traces of these antibodies were found in the Fractions I or IV. These preliminary results were so encouraging that, with the assistance of the Massachusetts Antitoxin Laboratory, Cohn arranged to prepare 2,000 tiny vials of sterile Fraction II+III solution for distribution to collaborating investigators in other laboratories testing for new antibodies.

A special NRC survey committee consisting of A.R. Dochez, Boyd and Enders was appointed to oversee this effort. Dochez canvassed a large number of experts in infectious disease research, stating the desire of the Government to ascertain to what useful purpose the remaining fractions could be put. Vials of Fraction II+III solutions were sent to those investigators who indicated a willingness to collaborate. The first results emerged in April 1942, when Enders provided a table (Fig. 4) indicating the presence of antibodies to typhoid fever, influenza, pertussis (whooping cough)) and mumps in Fraction II+III. Later, polio antibodies were reported by S.D. Kramer at the Michigan Department of Health, and diphtheria antibodies by Geoffrey Edsall at the Massachusetts Antitoxin and Vaccine Laboratories. Several blood group antibodies were identified by Boyd. Cohn reported these findings at an NRC Conference on Albumin and Byproducts on May 11, 1942, at which new observations by Stokes identified antibodies to swine influenza, and by Enders, antibodies to mumps.[18] A laboratory test for measles antibody did not become available until many years later. However, during a measles epidemic in Philadelphia in the late spring of 1942, Joseph Stokes gave some of Cohn's Fraction II+III solution to infants and found that measles was prevented with doses as low as 0.25 ml of the II+III solution. This important finding provided the first indication of the potential importance of the antibodies in Fraction II+III.

Managing the Scientific Enterprise

The myriad scientific issues associated with transferring the technology for producing normal human serum albumin to the pharmaceutical firms held the attention of Cohn and his colleagues during much of

1942. Whenever possible, visiting scientists and physicians were scheduled on Mondays so that they could be shunted around to Cohn's associates before or after the weekly luncheon meeting. Cohn rarely used the library himself, relying instead on the superb reporting of John Edsall about new scientific discoveries that he ought to be aware of. The task of drafting reports was delegated to others. The writing of scientific papers ground almost to a halt during this period. Only twelve scientific papers were published during 1942; most of these were reviews or chapters for books. One was a memoir by Ronald Ferry on the life of Lawrence J. Henderson who died suddenly in February 1942.[19]

Once the transfer of the plasma fractionation process to the pharmaceutical industry was underway, Cohn directed the efforts of the scientists, wherever possible, to the isolation and development of additional plasma proteins of potential clinical importance. Anticipating these moves, the OSRD contract with Harvard was broadened in February 1942 to include "chemical, clinical and immunological investigations of bovine albumin and human and bovine fibrinogen, prothrombin and other serum globulins."[20]

George Scatchard once said that Cohn was the most meticulous man he had ever known. While the men and women Cohn brought together came from a broad range of disciplinary backgrounds and training, they learned to work under his leadership at high levels of proficiency on a singularly important mission in wartime. Yet despite the compelling circumstances under which they worked, the precision demanded of them, and the pressures of time, their differing personalities, ideals and dedication as individuals were never compromised. They flourished. Years later, while introducing Cohn for an award, Scatchard suggested that Cohn's greatest contribution to science lay in demonstrating that the whole can be greater than the sum of its parts; the "whole" being Edwin Cohn and his associates, and the "parts" being individual scientists and physicians. Trained to work on fundamental scientific problems, they quickly adapted, under Cohn's leadership, to conduct research in diverse fields of science. Scatchard was convinced that Cohn's style of leadership brought forth greater scientific achievements from his associates than could normally have been expected of them.[21]

The weekly luncheon meeting, usually held in Cohn's office, was the principal forum for directing the efforts of more than a hundred workers during the war. About forty of these were based in Medical School Buildings C-1 and E-2 on Longwood Avenue. Elsewhere in the Longwood area were Charles Janeway and George Thorn at the Peter Bent Brigham Hospital, John Enders and Franc Ingraham at Children's Hospital, Otto Krayer, in pharmacology, and Orville Bailey in pathology. Another group of investigators included Scatchard who was based at MIT, and Elliott Robinson, Geoffrey Edsall and F.D. Hager, experts in immunology as well as in the processing and testing of biologics at the Massachusetts Antitoxin Laboratory. Nine commissioned Army and Navy officers—Edgar A. Bering, Jr., J. Elliott, Sam T. Gibson, James T. Heyl, M.C. Hutchinson, E. J. Klein, L. Pillemer, H. L. Taylor and L. M. Woodruff—assigned to the Plasma Fractionation Laboratory were involved primarily with the clinical testing of serum albumin, immune globulin, fibrin films and foam. Some thirty technicians and special assistants worked primarily at the Medical School.[22]

Although Cohn was clearly the chief, there was surprisingly little evidence of formal organizational structure below his level. In today's terms, the laboratory functioned primarily in an informal matrix structure comprised of loose clusters of workers. The formation of these clusters occurred naturally, bringing together individuals with the special talents needed for the particular task. Each cluster had a natural leader, and its composition tended to change as the work progressed. At a given time, an individual scientist might be involved in two or three different clusters. Titles were seldom used, although Cohn referred to titles in official communications to Washington, e.g. Strong as Associate Director of the Pilot Plant, or Oncley as Associate Director of the Ultracentrifuge Laboratory. Certain support staff members were associated with only one task or one instrument. Charles Gordon, who built and maintained the ultracentrifuge, performed all the actual runs; Oncley and a technician analyzed and interpreted the results. All electrophoresis analyses were the province of Howard Armstrong, and, later, of Metchie Budka. A dedicated technical staff worked in the analytical laboratory under Marshall Melin; they performed at an extraordinarily high standard of proficiency.

Among the scientists, the two Larrys—Oncley and Strong—were based on the first floor of Building E-2, close to the ultracentrifuge laboratory and the Pilot Plant. The others worked on the fourth floor of Building C-1. During the day there was considerable traffic between the two parts of the laboratory. Since both Larrys normally had occasion to talk with colleagues upstairs in Building C-1, they frequently walked up together—Strong a taciturn Hoosier, Oncley a tall, gangly Kansan extrovert—usually with a bundle of folders full of data under one arm. On occasion, they pressed Cohn on some favorite topic, perhaps an experiment they thought should be tried. On those occasions, with a twinkle in his eye, Cohn would string them along just to tease them.

Subcommittee Meetings in Washington

In view of the importance of the work going on in Boston, preparation for the meetings of the NRC Subcommittee on Blood Substitutes in Washington represented a substantial added burden for Cohn. For each meeting, he had to be supplied with last-minute data, charts and other materials to be handed out at the meeting in Washington. The tasks of gathering everything together often fell to the two Larrys who sighed with relief after Cohn had boarded the Federal at the Back Bay station in the early evening. He always had a bedroom on the train and Marianne Cohn often traveled with him, especially when his schedule would keep him in Washington for more than one day. On arrival in Washington the next morning, he usually took a cab to the Carleton on 17th Street for a breakfast appointment before going on to the NRC meeting on Constitution Avenue. The return trip was invariably by train.

U.S. government contracts place on the contractor certain obligations and responsibilities. The government designates a project officer who represents the government in the oversight of the performance of the contractor. The government often sets performance milestones and reporting requirements. Much later, writing about government support of research, Cohn commented about research supported by OSRD. "The success of many of the undertakings depended, in my estimation, upon the fact that both the members of the committees and the responsible investigators were scientists with a long tradition of the freedom necessary to ensure the exploitation of even a tactical advantage, and able to communicate with each other in a common language."[23] Although the work of the laboratory was classified during the war, the door to the Department of Physical Chemistry was always open to visiting physicians and scientists. However, administrators were not welcome. The fiscal affairs of the Department of Physical Chemistry were in the hands of Stephen Wheatland, a seasoned Boston businessman who dealt with the Harvard University Business Office and with government contract officers about contract related matters, effectively shielding Cohn. To meet the formal requirements of the contracts, brief unsigned "monthly progress reports" were submitted for the first year, and roughly bimonthly thereafter during the war. These constituted the only formal record of Cohn's progress in meeting contract requirements.

Early in 1942, Frank Blair Hanson, a staff member at the Rockefeller Foundation, aware that OSRD was then supporting Cohn's work, wrote to him suggesting that the Foundation would be willing to hold in escrow the unexpended balance in his Rockefeller Foundation grant for use after the war. In response, Cohn wrote that "no matter what proportion of our work is taken over by contracts with the OSRD, there is always the new angle of the research to be investigated and at every level we have found it invaluable to be able to use our grant . . . for such investigations until such time as their value as an applied research could be demonstrated." Citing new work on the globulins that might be important in immunity, he expressed confidence that the Foundation would agree that this "is a very proper use to which to put your grant."[24] Hanson concurred. Cohn liked to refer to the Rockefeller grant as his risk capital.

In March 1942, separate OSRD contracts were established to engage the expertise of four investigators at other institutions. John W. Williams, Professor of Chemistry at the University of Wisconsin, where there was an ultracentrifuge and Tiselius apparatus, began collaborating with the Cohn group, thus effectively expanding the ability to characterize the proteins which were being obtained from the byproduct fractions of albumin preparation. In addition, the Wisconsin group played an important role in evaluating the lots of albumin produced by the commercial laboratories. A similar contract was extended to J. Murray Luck, Professor of Chemistry at Stanford University. Luck made important scientific contributions relating to the stabilization of albumin to heat as well as by evaluating albumin lots produced by pharmaceutical firms in California. Hans Mueller, Professor of Chemistry at MIT, was brought into the program to design and build a nephelometer, a sensitive light scattering instrument, for use in evaluating the stability of albumin. A fourth contract was extended to Erwin Brand, Professor of Microbiology at Columbia University. Brand was then engaged in analyzing the amino acid composition of protein hydrolysates; his involvement would thus add important new information about the amino acid composition of the proteins of human plasma.[25]

In approving human serum albumin for clinical use by the Armed Services in January 1942, the NRC Subcommittee on Blood Substitutes carefully distinguished between the production of albumin through the plasma fractionation program, and the separate dried plasma program. Both programs were dependent on the blood collecting activities of the American Red Cross. Since the albumin program required a substantial expansion of blood collections, the Red Cross suggested that a public announcement concerning the albumin program would help in expanding the domestic blood donor base. This idea was rebuffed by the Navy for security reasons. At the same time, the Red Cross felt pressure from the Office of Civil Defense. Following the Pearl Harbor attack, and a later Japanese attack on Dutch Harbor in the Aleutian Islands, there came the realization that American cities on the mainland, particularly on the West Coast, could be vulnerable to attack. In response, a number of communities initiated attempts to collect blood for local use, thus running counter to the Red Cross National Blood Procurement Program. This led to discussions between the Red Cross and OCD, aimed at making some Red Cross frozen plasma available to OCD while at the same time resolving several related issues.

When a plan was presented to the Subcommittee on Blood Substitutes, the Subcommittee confirmed its original policy that the blood collected by the American Red Cross and the plasma prepared from it became the property of the armed forces "from the time it enters the processing plant."[26] Later, the Subcommittee amended its action by assigning to the Armed Forces the authority to process, prepare and *dis-*

pose of blood and plasma, after which the Subcommittee approved the release by the Surgeon General of the Army to OCD of 55,000 units of frozen plasma awaiting conversion to dried plasma. This solution to OCD's problem was possible since delays in starting up albumin production in the pharmaceutical firms, occasioned primarily by difficulties in acquiring critical processing equipment, substantially eased the pressure on the Red Cross and permitted the buildup of excess frozen plasma.

In this way, the Subcommittee on Blood Substitutes insulated the American Red Cross from dealing with pharmaceutical firms and other governmental agencies concerning future actions over which it had no control and which could potentially adversely affect the National Blood Procurement Program. Further, to relieve a serious problem within the Red Cross caused by difficulties in recruiting physicians to staff bleeding centers, the Subcommittee recommended that the Surgeons General give favorable consideration to the assignment of a small number of medical officers to temporary duty in Red Cross bleeding centers.

In late October 1942, the War Production Board and related agencies having expedited the procurement and delivery of the needed equipment, the pharmaceutical plants were nearing readiness for full operation. Practice runs had been completed by four out of the six pharmaceutical contractors. The purity of the albumin being obtained met specifications, that is, it contained less than 2% of globulin impurities. The final solutions contained 25% albumin in isotonic sodium chloride; thermal stability tests following storage at 50°C for twelve days proved to be satisfactory. As a result, processing of fresh plasma was begun in all but one of the firms. The start of processing in that firm was delayed by the abrupt departure of one of the trained men. Cohn sent Strong to supervise operations there while new personnel were being trained. Murray Luck at Stanford University and Jack Williams in Madison, Wisconsin aided in the startup of fractionation at Cutter Laboratories and Upjohn, respectively. H.B. Vickery, a seasoned investigator at the Connecticut Agricultural Experiment Station in New Haven, was appointed as Associate Director of the Plasma Fractionation Laboratory in Boston, with responsibility for appraising the quality of the normal human serum albumin produced by the contracting firms and for reviewing the records on each lot of product prior to their being forwarded to Cohn for final approval as the responsible officer for the Navy.

These requirements had been reviewed by the Subcommittee on Blood Substitutes prior to activation. At one point, a debate ensued over the number of vials of albumin from each lot that should pass final clinical tests. Loeb questioned whether three tests would be enough. Captain Stephenson commented that he feared trouble from the rubber tubing enclosed in the tin can with the albumin, rather than from the albumin itself. In this same discussion, the issue of the safety of administering the concentrated albumin solution to severely dehydrated patients was raised once again. Members of the Subcommittee affirmed their earlier belief that ample precautions had been taken; no further action was deemed necessary. Although Loeb asked DeGowin to prepare a statement about the problem of developing solutions for preserving red blood cells for transfusion late in 1942, this matter was not discussed further during that year.

A Report on the Mechanism of Shock

At the outset of National Research Council activities in 1940, the Committee on Transfusions had appointed a Subcommittee on Shock, with Alfred Blalock as its chairman. This subcommittee launched a series of studies in four general areas: mechanism of shock; effects of certain therapeutic agents, particularly blood substitutes; special inquiries into vasomotor behavior; and the specific problem of thermal burns. The team leader on the mechanism of shock was D.W. Richards, Jr. of Columbia University. Reporting on these studies after the war, Richards stated that, in shock following trauma, the decrease in the volume of blood was caused by the loss of whole blood at the site of injury, and that this had been discovered in 1942. This was the reverse of the conclusion reached earlier, i.e. that loss from the circulation in shock is the loss of plasma rather than the loss of red cells. Clearly, this finding, if confirmed, would have major implications for the treatment of shock in combat.[27]

Post-transfusion Jaundice Reported in England

Late in 1942, NRC received a cable stating that the British had reported some cases of jaundice, a symptom of liver disease, appearing between sixty and 120 days after transfusion of pooled serum. Similar reports of well-authenticated cases following administration of serum were received from the Russians. According to Veldee, no U.S. cases had been reported following injection with plasma. This report was discussed at a meeting of the Subcommittee on Blood Substitutes in October 1942.[28] In answer to a question, it was revealed that current practice in the preparation of dried plasma involved pooling of the plasma from twenty-five donors prior to filling into individual bottles for drying. Nothing was done to alter the process. However, concerns about the possibility of a new blood-borne viral agent prompted members of the Subcommittee to recommend that Red Cross reject blood donors with a history of jaundice within the previous six months. In addition, Cohn was asked to arrange for a study by John Enders of the viability of experimental viruses under conditions used to fractionate plasma. At the same meeting, American Red Cross reported that it had passed the 1,000,000 unit mark in total blood collections. The weekly collection rate was then about 33,000 units of blood. Seven new blood centers were being opened in Hartford, Schenectady, Harrisburg, Kansas City, Columbus, Minneapolis and Washington, D.C.

Edwin J. Cohn Jr. and Alfred B. Cohn

Ed Cohn graduated from Harvard College in 1942 and took a post with the Board of Economic Warfare in Washington. Fred Cohn entered Harvard in the fall of 1942, and after the first semester, he enlisted in the Army as a voluntary inductee. Before entering the service, he had a deep conversation with his father about how he approached his work. Fred began to appreciate that he had an extraordinary father and that his father's relationships to Boston's upper class were very important to him. In the army, Fred served in a military intelligence unit and wound up in Italy at the end of the war.[29]

9 The Norfolk Incident

The first preparations of bovine albumin that were described by Cohn and his associates in December 1940 proved to be of variable purity, being comprised of about 95% albumin and 5% globulins.[1] When tested in human volunteers, they led to immediate reactions in some subjects and to posttransfusion serum sickness in a few others.[2] Testing was halted until the product could be further purified.

Crystals of bovine albumin were discovered in June 1941 by Hughes in a solution of bovine albumin that had been left in a cold room at the Harvard Medical School.[3] Since crystallization offered a means for reducing and perhaps even eliminating globulin impurities present in the product, Cohn and Hughes immediately redirected their efforts toward identifying the conditions under which crystallized bovine albumin could be harvested in large quantities from the bovine albumin that was then being produced in large amounts by Armour. While this might result in reduction in the yield of the crystallized product, it should be purer. The challenge of producing a biological product in such high purity was daunting. No crystallized protein had ever been prepared on such large scale.

Crystallized bovine albumin (CBA) prepared in the Harvard Pilot plant became available in limited amounts for clinical study late in 1941.[4] As a first step, skin tests were run in a group of 153 medical students to determine the extent of natural sensitivity to CBA. This involved injecting a tiny volume of CBA solution into the skin of one arm, and a similar sample of amorphous, i.e. uncrystallized albumin, into the other arm. After forty-five minutes, the arms were inspected by a physician. Three of the students proved to be sensitive to CBA, suggesting that one out of every fifty to sixty persons might exhibit a natural sensitivity to CBA. The same three students were also sensitive to amorphous bovine albumin. All subjects were subjected to this skin test and those found to be sensitive were excluded from further testing. Cohn and Janeway reported their first results at a meeting of the Subcommittee on Blood Substitutes in Washington on January 2, 1942. Among the first group of eight subjects receiving intravenous injections of from 3 to 8 grams of CBA, no evidence of serum disease, food sensitivity or other delayed reactions was detected.

The preparation of additional CBA for continued clinical testing in man required that the Armour bovine albumin be crystallized in the Harvard Pilot Plant in Boston. After being redissolved, sterile filtered and filled into sterile vials, it was subjected to final safety testing at the Massachusetts Antitoxin Laboratory. Since this process was time consuming, further clinical testing of CBA was retarded until March 10, when fifteen injections and fourteen reinjections in the same subjects were conducted without incident. This led Cohn to state, "these results encourage us in the hope that purification of bovine albumin by crystallization has markedly diminished the incidence of serum reactions following its use."[5]

Later, the Subcommittee recommended that Armour set up a plant to produce crystallized bovine albumin in Chicago,[6] and steps were initiated to transfer the crystallization technology to Armour.[7] At the same time, the Subcommittee moved to expand the clinical testing of CBA by enlisting David Seegal at Welfare Island, New York, and Alfred Blalock at Johns Hopkins, in addition to Wangensteen in Minneapolis and Janeway in Boston. By May 1942 the number of infusions had grown to forty-six,[8] and by June, eighty-five persons had been infused, all without associated post-infusion reactions.

However, trouble appeared on June 10 with the eighty-sixth case. As noted in the Subcommittee minutes, a sixty-two year-old man was given 12.5 grams of crystallized bovine albumin, followed by an equal dose six days later. On the tenth day after the first injection, the subject developed an illness characterized by fever, anemia, arthralgia, edema, purpura due to capillary fragility, urticaria, hypoproteinemia and nitrogen retention. The fact that the bovine albumin disappeared more rapidly than usual from his blood and that he developed a strongly positive skin test as soon as it disappeared were interpreted as consistent with serum disease. He was given plasma and blood transfusions and improved in four days. On hearing about this case, arising as it did after such a long string of favorable results, Cohn immediately notified Loeb, who convened a special meeting of the NRC Conference on Bovine Albumin in Washington on July 16.[9] At the meeting, Loeb opened the discussion by stating that "whatever else the man may have had, he had serum disease which, fortunately for the problem, was severe, and therefore served three purposes: a) in orienting the general problem; b) in provoking a decision as to the amounts of albumin to inject; and c) in provoking consideration of the distribution of the responsibility for the reactions that may occur."

Then, weighing the meaning of this single test, Loeb favored accelerating clinical testing with larger doses of bovine albumin. "To do less," he argued, "would be to render the problem academic" from the viewpoint of the armed services. The senior representatives of the Armed Services, Colonel Prentiss and Captain Stephenson, concurred, pointing out that the issue was too important to be decided by one case of serum disease. Wangensteen reminded his colleagues that in approximately sixty cases he had personally encountered no reactions from use of Cohn's bovine albumin—both amorphous and crystallized. The Conference unanimously recommended that clinical studies of bovine albumin be continued with all possible haste. Encouraged by the growing volume of favorable clinical results, members of the Subcommittee on Blood Substitutes pressed to extend the scope of the trials so as to reach a larger cohort of subjects, particularly a cohort with demographic characteristics that would resemble the characteristics of men entering the armed services. The subjects should be volunteers, and should be "controllable," i.e. be available for observation over a protracted period of time so as to make certain that no delayed effects accompanied the use of this blood substitute.

At a Subcommittee meeting on June 23, Cohn had reported that, with the consent of the Massachusetts Commissioner of Prisons, sensitivity tests to CBA had been performed on the skin and conjunctivae of 2,600 prisoners, of whom only seven were sensitive to bovine serum albumin.[10] According to a report in *The Boston Herald*, "Prisoners Help Great Blood Test," Arthur T. Lyman, the Massachusetts Commissioner of Corrections, personally solicited volunteers among the inmates of the Norfolk Prison Colony, the Charlestown State Prison, Bridgewater State Farm and the Concord Reformatory. The response had been overwhelming—2,747 prisoners came forward.[11]

The events to that time, and some of those that followed, were summarized in a long letter from Vannevar Bush, Director of the Office of Scientific Research and Development, to Harvard President James B. Conant on August 29.[12] "As you are well aware," Bush's letter began, "Harvard University has contracted with OSRD to undertake investigations which are expected to yield information, highly important to the armed forces of the United States, concerning the possibility of utilizing proteins obtainable from beef blood plasma as a means of successful treatment of certain disabilities to which wounded soldiers are subject." Noting that in the preliminary phases of that investigation exceedingly refined methods had been developed for the separation and purification of bovine serum albumin, tests had shown that its physical and chemical properties, although strikingly similar, were not identical with human serum albumin that had proved efficacious in the treatment of such disabilities.

Furthermore, he went on, animal experiments had shown that relatively enormous quantities of bovine serum albumin had been repeatedly injected into the veins of dogs, rabbits and mice without producing demonstrable injury, and approximately 150 human volunteers had received injections without suffering serious injury, except in one instance in which serum sickness developed. The results of these trials were deemed so favorable by the Surgeons General of the army and navy that they now considered it imperative that a series of similar tests be organized in a controllable group of human subjects. Although the idea of conducting these tests on members of the armed forces themselves had been explored, Bush related, this was found to be impracticable since it was necessary that individuals tested be available for further study for a protracted period of time in order to make certain that no delayed injuries followed the use of this blood substitute. Bush added that Captain C. S. Stephenson of the Bureau of Medicine and Surgery had informed Dr. Cohn that he would be glad to share the responsibility of these tests and would come to Boston to be present when they were given.

"It is clearly necessary that a larger group of individuals be tested before one could be completely satisfied with the safety of this blood substitute," Bush pointed out. "One such group which would meet the conditions of a stable population is a prison colony." An approach had been made to the Commissioner of Correction of the Commonwealth of Massachusetts, Mr. Arthur T. Lyman, and through him to the Prison Council representing the prisoners at Norfolk, that resulted in the enthusiastic approval of a plan whereby the inmates of that institution could volunteer at their own request with no coercion of any sort for this service. It was made clear that risks were involved in these tests. Bush emphasized that responsible medical advisors of the OSRD and the highest officials of the Medical Corps of the Army and Navy agreed that the national interest required that this and attendant risks be accepted. "For these reasons," Bush ended his letter, "we request Harvard University to authorize those members of its staff who are in immediate charge of this project to proceed with these and other tests."

As originally planned the design of the trial called for each subject to receive an infusion of 25 grams of Armour crystallized bovine albumin, designated ACB, to be followed by a second 25 gram infusion within ten to twenty-one days. Then, after three months, all subjects were to be examined and given a third 25 gram infusion. Following a further three-month period, a skin test was to be performed and blood samples were to be collected for study. Five institutions were involved in the trial: the Norfolk Prison Colony, Peter Bent Brigham Hospital, University of Minnesota Hospital, Welfare Island Hospital and Johns Hopkins Hospital.

Early in September 1942, Commander Newhouser arrived at the Norfolk Prison Colony with orders from Admiral McIntire to get the clinical test of crystallized bovine albumin under way. Newhouser began by meeting with the men who had volunteered to participate in the test. He again explained the purpose of the trial and what would be involved, taking pains to describe the types of reactions and illnesses that had been encountered during testing of earlier subjects. About 200 men signed consent forms confirming their understanding of the risks involved and their willingness to participate in the trial.[13]

Infusions of crystallized bovine albumin were begun immediately and continued on subsequent days. Once infused, each man was examined daily thereafter. On September 14, about ten days into the trial, a prisoner developed symptoms characteristic of serum sickness. In the days that followed, other cases were detected. On September 18, faced with the realization that the incidence of serum disease was substantial among the men at Norfolk, Janeway halted further infusions there and advised the other participating institutions to discontinue their tests as well. On September 21, Cohn notified A.N. Richards at the National Research Council in Washington that he had ordered the termination of the trial. Cohn and Harvard Medical School Dean C. Sidney Burwell visited the sick prisoners at the Norfolk prison hospital with Janeway and Heyl on September 22. Two days later, Loeb visited the sick men and met with the Prisoner's Council.

Before the outbreak of serum disease was over, twenty-one men experienced symptoms attributable to serum sickness. They were cared for in the prison hospital or in the Peter Bent Brigham hospital, several for prolonged stays. On the morning of September 30, Arthur St. Germaine, a prisoner, was found to have died in his sleep in the prison hospital. Of twenty-one prisoners who experienced serum sickness, eleven, including St. Germane, were characterized as having had severe reactions. The others were characterized as mild. The magnitude of the adverse outcomes, totally unexpected, was devastating. On learning of the demise of St. Germaine, Cohn and Janeway drove out to Norfolk and met with the volunteers. One observer at that meeting remembered the effective manner in which Cohn dealt with what must have been one of the most difficult encounters in his career. Captain Stephenson flew up from Washington and visited the men who were still hospitalized, expressing the deep appreciation of Admiral McIntire on behalf of the Armed Forces for their patriotism. Some of the men expressed the hope that the Navy would continue the albumin studies without interruption if it would contribute to attaining Navy objectives.

An extensive investigation followed the aborted Norfolk clinical trial, culminating in reports at a Conference on Albumin Testing that was held at the National Research Council a month later. Those attending included C. A. Janeway (Chairman), R. F. Loeb, O. H. Wangensteen, E. J. Cohn, A. R. Dochez, H. B. Vickery, O. T. Bailey, D. Seegal, F. E. Kendall, J. T. Heyl, M. Ravitch (representing A. Blalock), Captain C. S. Stephenson (representing Admiral Sheldon), Commander L. R. Newhouser, U.S. Navy, Lt. Col. R. G. Prentiss Jr., Lt. Col. B. N. Carter, and Lt. Col. D. B. Kendrick, U.S. Army, M.V. Veldee, U.S. Public Health Service, W. M. Clark, Comm. E. H. Cushing, W. C. Davison, T. R. Forbes, O. Temkin and G. A Carden, Jr., National Research Council, G. C. Robinson, E. S. Taylor, American Red Cross, E. C. Andrus, Committee on Medical Research, and Professor Harold Burn of England.

As the first item on the agenda the Conference reviewed the status of production of normal human serum albumin. Edwin Cohn credited Mr. McManus of the War Production Board for supervising the procurement and delivery of the equipment for the fractionation plants. Human albumin was on the verge of production in four of the six contracting laboratories. At this meeting, Commander Newhouser exhibited the final container that had been adopted by the Navy for albumin. Satisfactory reports on the thermal stability of fifteen lots of albumin from three firms were presented. Results of clinical trials of albumin with eleven lots of albumin produced in the Harvard Pilot Plant, all favorable, were reported. The possible danger of using concentrated human albumin in severely dehydrated subjects was again discussed, ending in adoption of a resolution "that in the opinion of this group, ample precautions have been taken to safeguard the use of concentrated albumin solutions under conditions of dehydration."

When the conference turned to the major item on its agenda, the clinical trials of crystallized bovine serum albumin, it was revealed that there had been thirty instances of serum sickness among 277 individuals who had been infused with ACB in the five institutions, for an overall incidence of serum sickness of almost 11% among the participants. At the Norfolk Prison Colony, the incidence of serum sickness was 32% (twenty-one cases among sixty-six men infused with ACB). In contrast, the incidence of serum sickness at the other four hospitals was 4% (nine cases among 211 infused). In the autopsy report on

St. Germaine, the pathologist, Orville Bailey, stated that "if the autopsy had been performed without knowledge that a foreign protein had been injected, the latter could not have been determined from the autopsy findings."

During the investigation, a number of possible explanations had been suggested and explored. One line of investigation focussed on possible differences between the subjects involved in the clinical trials at Norfolk and those who were in the four hospitals. However, no credible explanation emerged. In the discussion, Stephenson, an experienced Navy physician, expressed the wish that the term serum sickness be dropped. In his view, "something totally unexpected had happened." Confessing that he had misjudged the significance of the reactions experienced by the men, he admitted to being greatly concerned at the severity of the disease that emerged. He concluded by emphasizing the needs of the military for a readily available blood substitute, and stressed the importance of preserving some of the evidence in a safe place. As the review ended, Stephenson recounted visiting the men in the hospital at Norfolk and paid tribute to their courage and patriotism, and particularly to their willingness to continue the study on behalf of men in the Armed Forces despite one of their number having lost his life while others remained sick and uncomfortable. He cited this "as one of the finest examples of heroism in the history of military medicine." Before adjourning the Conference, the Subcommittee extended to the Harvard group all the time they might need to deal with the problems posed by Armour crystallized bovine albumin, leaving it up to Cohn to decide when and if cautious clinical testing should be resumed. For security reasons, news of the Norfolk incident was withheld.

Arthur St. Germaine was posthumously pardoned by Massachusetts Governor Leverett Saltonstall in December 1942. The citation stated: "Arthur St. Germaine, with full knowledge and understanding of possible dangers and risks involved, voluntarily submitted to a vitally important research test that involved the possible saving of thousands of lives not only on the battlefront but among society itself."[14] Almost a year later, an Army bomber was christened "Spirit of St. Germaine."[15] Ralph D. Hamm, still critically ill, received a full pardon on May 27, 1943.[16]

In the midst of this turmoil, Edwin Cohn was awarded the Alvarengo Prize by the College of Physicians of Philadelphia. The Prize was established in 1883 under the will of Pedro Francisco DaCosta Alvarengo of Lisbon, Portugal, an Associate Fellow of the College, to honor "the author of the best memorial upon any branch of medicine, which may be deemed worthy of the prize." The award was presented at a joint meeting of the College and the Philadelphia County Medical Society that was held on the evening of October 14, 1942.

Cohn's accompanying lecture was entitled, "The Plasma Proteins: Their Properties and Functions."[17] After brief opening remarks in which he expressed his appreciation for the honor being bestowed by the College, he turned to the substance of his lecture. His penultimate paragraph suggests that his audience knew full well why he was being honored.

> In how far, therefore, will it ever be safe to substitute for therapy in man a protein of animal origin? This question can only be answered by clinical trial. Clearly from a physico-chemical point of view it is possible to substitute for a human protein an animal protein which will perform the same physical and physiological functions in nearly the same way. If an animal protein is to be reintroduced into the human body it should have the property of setting up as few as possible of the reactions to foreign proteins. The reactions to albumin containing as little as one hundredth or even one one thousandth percent of globulin are being carefully investigated, but we are not prepared to discuss the results at this time.

Several years later, while the writer was driving home to Cambridge with Cohn, he admitted to being puzzled by the differences in responses of the Harvard medical students and of Wangensteen's patients in Minnesota, on the one hand, and those of the Norfolk prisoners on the other. He wondered if the use of arsenical drugs to treat sexually transmitted diseases in the prison population might have been an important uncontrolled factor.

10 1943: The Critical Year

The British had entered the 1939 war with a firm transfusion policy that was learned during the 1938 Munich crisis. It was based on the concept that providing blood is a unique service and should be completely distinct and separate from other military services. The wisdom of this policy had been demonstrated in their combat experience in Europe before the fall of France. There, they had learned that large quantities of blood would be needed, and that while plasma was valuable, whole blood with its oxygen carrying and clotting capacity was essential during initial wound therapy and anesthesia. Based on their experiences, they had organized a completely separate transfusion service that was headed throughout the war by Brigadier Lionel E.H. Whitby.[1] By March 1943, utilization of plasma in British installations had been essentially abandoned in favor of whole blood.

In contrast, the Americans relied almost completely on the use of dried plasma. They believed that "plasma would be so effective in treating battle wounds that only a very small proportion of the wounded would require whole blood early in their treatment."[2] Accordingly, they carried an ample supply of dried plasma in sealed bottles, together with bottles of sterile water, and the necessary rubber tubing and needles with which to add the water to the dried plasma. The reconstituted plasma could then be infused into the vein of a wounded soldier. As for blood, were it to be needed, it would have to be collected from nearby military personnel, but the equipment for collecting blood—needles, bottles and tubing—had not been provided. Once

American ground forces were engaged in combat, it soon was realized that whole blood would be needed. The American medical authorities were forced to seek help from the British who provided them with a substantial volume of blood.[3]

In April, the Subcommittee on Blood Substitutes devoted a long meeting to an examination of the logistics of carrying out blood transfusions in the Army. In addition to the members of the Subcommittee present at that meeting, others included J.B Alsever of the U.S. Office of Civilian Defense, E.S. Taylor of the American Red Cross, O.F. Denstedt of the Canadian National Research Council, Major A.L. Chute of the Canadian Army, and Commander C.L. Best of the Canadian Navy. When Loeb asked that he describe the equipment for transfusing blood in U.S. Army medical installations, Lt. Col. Kendrick stated that blood transfusions could only be performed by use of an open beaker which, lacking a cover, could not be used for storing whole blood. Transfusions of plasma would be employed exclusively forward of evacuation hospitals. In evacuation hospitals, the Army proposed that only transfusions of fresh blood or reconstituted dried plasma would be administered. In general hospitals, a "closed" method could be employed, and blood so collected could be stored for a few days prior to use. However, lacking a standard blood collecting kit, it was expected that blood would have to be collected into reused one thousand milliliter Baxter flasks in which intravenous solutions had been shipped to the hospital. The original Baxter rubber stoppers would also have to be reused. As for blood donors, Kendrick explained that there was a "better chance" of securing donors at a general hospital, and that blood so collected could be shipped forward to the evacuation hospitals. In response, members of the Subcommittee, barely concealing their annoyance, asked their colleague DeGowin to prepare a summary statement reviewing pertinent experimental data and clinical information concerning solutions and equipment for whole blood preservation and shipment.

In the discussion that followed, Wangensteen questioned the advisability of preparing fluids for intravenous use in the combat zone. Major Chute of the Canadian Army stated that the British were then preparing fluids for transfusion in Cairo. In turn, this line of discussion raised questions about the wisdom of using untrained personnel for this purpose, to which Chute pointed out that it was definitely required in the Canadian Army that medical officers and laboratory assistants be specially trained for that purpose.[4] Members of the Subcommittee questioned the wisdom of setting up army laboratories to prepare intravenous fluids in the field, and asked how blood grouping would be handled. They also raised questions about the source of trained personnel to administer transfusions, keep records, control temperature for blood storage, and related issues. Vexed with the quality of the Army plans, the Subcommittee unanimously passed a resolution for transmission to the Army through Weed at the National Research Council:

> Whereas the Subcommittee on Blood Substitutes has now reached a point at which it can no longer function effectively without more precise information concerning field problems and conditions imposed by the military requirements of this war, and
>
> Whereas a considerable portion of our present fund of knowledge concerning the mechanism of shock and its treatment has come about through the physiological studies of specially qualified and completely integrated groups of investigators in the field [e.g. Cannon, Bayliss, Keith, Robertson et al.], it is the unanimous recommendation of the Subcommittee on Blood Substitutes that this policy be pursued by the appointment of a qualified fact-finding group without special function of command . . . The topics identified for study included: supply and distribution of substitutes and whole blood in various field units . . . dispensing equipment . . . facilities for procurement and storage of whole blood . . . organization, distribution and requisitioning of materials . . . appraisal of efficacy and the relative value of substitutes and whole blood, including identification of the criteria of clinical improvement, optimal dosage and the present rate of utilization per casualty, and untoward reactions . . . and table of organization.[5]

If this elicited any response, it was not recorded in later Subcommittee minutes. North Africa was cleared of the enemy in May 1943 with the surrender of 275,000 German soldiers. July 10 was D-Day for the invasion of Sicily. Axis resistance on that island gave way by July 17.[6]

The Possibility of Preserving and Airlifting Whole Blood to Europe

In Washington, the Subcommittee on Blood Substitutes initiated a series of conferences on the preservation of blood. The first such meeting, held at the Massachusetts General Hospital in May 1943, brought together a group of experts to review knowledge about the in vitro preservation of red blood cells. This conference began by discussing possible criteria for measuring the survival of red blood cells. Although a number of criteria were identified, it was quickly agreed that there was no single criterion that could be relied upon to quantify the successful preservation of human red blood cells. The most important development from this meeting was the presentation of a new method using radioactive iron isotopes to determine the life span of preserved human red cells following their infusion into the circulation of a recipient. Developed by Joseph C. Aub at the Massachusetts General Hospital, Robley D. Evans of the Massachusetts Institute of Technology and John G. Gibson of the Peter Bent Brigham Hospital, this revealed, for example, that 90% of the red cells in freshly transfused blood remained in the circulation for at least seven days. It offered a means for quantitatively evaluating the survival of preserved red blood cells once infused into the circulation of a patient. By using different radioactive iron isotopes, it was possible to assess the fates of red blood cells from two different and successive transfusions. With these methods, J.F. Ross and M. A. Chapin at the Evans Memorial Hospital were able to measure the effectiveness of a blood preservative diluent in prolonging the in-vivo survival of transfused red blood cells. One of their earliest observations was the revelation that prolonged storage of blood in citrate exerts a very deleterious effect on red blood cells.[7]

On August 10, 1943, Loeb summarized the consensus of the experts at the Boston meeting that red cells were preserved longer when they were refrigerated, and when dextrose was added at the time of collection from the donor. Cohn thereupon introduced a motion approving the addition of dextrose and specifying low temperatures for preservation of red cells. Strumia surprised his colleagues on the Subcommittee by demurring, explaining that he did not accept the evidence of superior survival of cells preserved in dextrose solution. He wanted to wait for results of the studies with radioactive iron. Moreover, he saw no advantage in preserving blood for over five days since this was the average time during which blood was kept in the bank.[8] Cohn withdrew his motion. The result was a delay of more than a month in reaching a decision on that important issue. When the Subcommittee met again on September 24, Loeb, undoubtedly having arranged in the meantime that Strumia be briefed by the Boston investigators, called the meeting to order by stating that "it had been generally accepted that citrated blood to which glucose had been added was safer to use for transfusion after being stored for three weeks than was citrated blood stored for the same length of time without the addition of the sugar." Cohn's original motion was adopted, with the additional statement, "In the light of present knowledge the (preservative) solutions of Alsever, of DeGowin and of Denstedt have proven effective."[9]

Important Input by a Senior U.S. Surgeon

Under a North Africa dateline on August 26,1943, the *New York Times* described an Army report by Colonel E.D. Churchill asserting the need for whole blood transfusions in the treatment of a significant proportion of wounded personnel in North Africa, and pointing out that plasma is not an adequate substitute for blood.[10] On leave from Harvard where he was Professor of Surgery at the Massachusetts General Hospital, Churchill served as a consultant to the Surgeon, North African Theater of Operations in Tunisia. He had been asked to make a study of the need for blood transfusion in Tunisia, specifically, to answer two questions. "With plasma readily available, was whole blood really needed? And, if whole blood was really needed, how best could it be provided?" In the first of several reports, Churchill emphasized that

> whole blood is the agent of choice in the resuscitation of the great majority of battle casualties, whole blood is the only therapeutic agent that would prepare seriously wounded casualties for the surgery necessary to save life and limb, that both the mortality rate and

> the incidence of wound infection are reduced by use of whole blood at the time of initial wound surgery, that plasma should be looked at as a first aid measure for dire surgical emergencies and as a supplement for whole blood, not as a substitute for it.[11]

In his visits with Army surgeons in Algeria, Churchill stressed certain aphorisms such as "the goal of resuscitation is not solely to save life but to prepare for necessary surgery which will, in turn, be accompanied by further blood loss;" "the wounded require replacement of the lost blood;" and "wound shock is blood volume loss; it is identical with hemorrhage." In stressing the importance of transfusing whole blood in North Africa to surgeons for whom blood was only rarely available, Churchill was politely received. His message contained an element of irony to any Army surgeons who had seen Kendrick's 1941 paper advocating dried plasma in *The Military Surgeon.* As Churchill's tour of duty neared its end, he was so disturbed about the lack of blood transfusion support for wounded soldiers in North Africa that he concluded "my only recourse was to talk to the New York Times and say you must break the story that plasma is not adequate for the treatment of wounded soldiers."

As published in the *Times*, the story included an indication of support from the Surgeon of the Northern Africa Theater, stating that there was a need for whole blood transfusions in treating a significant proportion of the wounded, and that plasma was not an adequate substitute for blood in these cases. The outgrowth of four months of research and questioning of medical officers in the combat zones, Churchill's reports proposed the establishment of a far flung system of blood banks in the rear echelons behind combat zones. These blood banks were visualized as functioning much like Red Cross blood drawing teams in the United States, collecting blood from volunteer donors among noncombatant troops, convalescent and mildly wounded patients, and soldiers from medical detachments. Churchill concluded by pointing out that "the single factor that stood out most prominently in the care of battle casualties in North Africa was the indispensability of whole blood before, during, and after wound surgery. Unless casualties were properly resuscitated and the resuscitation included whole blood, often in large quantities to replace what they lost, surgery would be attended with an excessive mortality rate. Plasma could not replace whole blood."[12]

U.S. troops from Sicily landed at Salerno on September 9, 1943, beginning what would be a long and difficult winter campaign in Italy.[13] Still lacking the capability to transfuse blood, the Americans again turned to the British for assistance. As a result, U.S. hospitals on the Anzio beachhead received about 4,000 units of blood, much of it donated by U.S. Army Air Forces personnel stationed in the area. The British supplied the collection equipment, collected the blood and performed the necessary processing of the blood.[14] On November 17, Kendrick informed the Subcommittee that reports had been received from various Theaters of Operations indicating that transfusions of whole blood were needed to supplement the use of dried plasma. A Blood Transfusion Service had been recommended to the Surgeon General's office, but had not been approved. However, the provision of refrigeration equipment for field hospitals, evacuation hospitals, and general hospitals had been approved. Furthermore, collecting bottles containing Denstedt's red cell preservative solution, microscopes and serological equipment for typing and crossmatching blood would be provided so that blood banks might be operated at those points.[15]

Human Serum Albumin

In the Harvard Pilot Plant, a new "run," starting with about forty liters of fresh human plasma, began every week. That meant that the albumin fraction would normally be "on the dryers" about seven days later. Most of the time required to produce a single lot of albumin, or any other product for that matter, was consumed during pauses not directly involved in the fractionation process. As a result, three or more months could elapse between the time a pool of plasma was collected and the ultimate release of final sterile vials of 25% human serum albumin solution for clinical use. For example, the large container of the sterile bulk solution of albumin was held up for more than a week while awaiting the results of lengthy tests to assure its sterility before being filled into final containers. Nevertheless, samples of human serum albumin from 113 commercial lots had been received and tested by March 1943. In every lot, the final product was found to pass the purity criterion: each contained less than 2% of globulin impurities. Even more encouraging, 100 of those lots contained less than 1% of globulin impurities. On this basis, Cohn concluded that none of the pharmaceutical firms were experiencing difficulties in using the Harvard procedure known as Method 2. Routine electrophoretic analyses of every lot submitted gave way instead to testing samples from a few selected lots. On the other hand, the yield of albumin, as measured in terms of the number of final units of albumin being obtained per liter of plasma, was on the low side. This was a source of concern to Cohn, for the pharmaceutical firms had contracted to produce 350,000 bottles of albumin.

At a review of the albumin program in Washington in March 1943, Cohn revealed that approximately 11,000 vials of albumin were available, including the 5,070 vials already released. The meeting reviewed the goals: 360,000 units already contracted for and an additional 350,000 units for the 1944 fiscal year. To conserve albumin, its use was limited by the Navy to advanced positions in the field. Although progress was being made, skeptics were occasionally heard from. One of these was Veldee, the senior government regulatory officer at the time, who once proposed that concentrated liquid human serum be produced instead of albumin. This drew a sharp response from Loeb who reminded him that there was no evidence that liquid serum could meet the viscosity and stability standards applied to albumin.

Cohn pointed out that the program would only be slowed in the long run by accepting substandard material. In his view, inferior stability and the occasional finding of haziness in certain lots after heating were the major problems. In the ensuing discussion, Veldee raised other questions about albumin, even suggesting that the heat treatment might be damaging the product. He also called attention to the fact that breakage of blood bottles received from the American Red Cross by the pharmaceutical laboratories was considerable and that residual plasma left in the bottles when the plasma was drawn off represented a sizable loss. Other sources of loss were reviewed. Another question came from Blalock who asked Loeb if he was not disturbed by criticism from studies by Myron Winternitz at Yale claiming to show that albumin was not effective in experimental shock. Loeb responded in the negative since the type of shock being studied by Winternitz was quite different than battlefield shock.[16]

In the early spring of 1943, the transfer of the plasma fractionation technology to the pharmaceutical firms being well underway, Cohn gave the go ahead to efforts by Strong in the Harvard pilot plant to introduce some modifications in the fractionation process. At the time, production of albumin amounted to about 10,000 vials of albumin each month despite the fact that the Navy contract goals called for more than 25,000 per month. Some technical problems had been encountered in the separation of Fraction IV, the fraction removed just prior to the precipitation of albumin into Fraction V. Fraction IV sedimented slowly, and difficulties had been encountered in clarification of the supernatant of Fraction IV prior to removal of the albumin fraction. This was a critical step because it directly affected the stability of the albumin that was precipitated in the final step of the fractionation process. In this instance an important modification discovered during the preparation of crystallized bovine albumin resulted in increased yield, improved stability, materially reduced processing time and reduced cost of human serum albumin. This formed the basis for Method 5 of plasma fractionation, a procedure which substantially increased the yield of albumin and reduced the time consumed in fractionating plasma.

One of the unexpected findings resulting from the transition from Method 2 to Method 5 lay in the discovery that certain ions other than sodium chloride significantly increased the stability of human serum albumin solutions to heat.[17] A new, more rigorous stability test involving heating for twelve days at 50°C was introduced, and a new optical instrument developed by Hans Mueller at the Massachusetts Institute of Technology came into use to measure changes in turbidity during heating. As a result, the stability specification for human serum albumin was increased so as to require that the final albumin solution remain free of turbidity after heating for 100 hours instead of forty hours at 57°C.

Although the Navy specified in its contracts that Edwin Cohn was responsible for approving all products made under Navy contracts prior to their release to the Navy, Veldee had the authority as the senior government regulator to review the manufacturer's protocols for each lot of albumin produced. In July, Veldee forwarded to Chairman Weed of the NRC Division of Medical Sciences some observations deemed by him to be "of urgent importance, not only to the Army and Navy but also to the Red Cross and the blood donor himself." Based on data from the first fifty-eight lots of albumin produced by six contractors, Veldee had noted that only about half the albumin contained in the starting plasma ended up in the units being shipped to the Navy. He therefore asked: "would it not be advisable to have this memorandum placed before the responsible authorities in the Army and Navy?"[18] Weed referred the matter to Admiral McIntire, who responded by confirming the importance to the Navy of having a very large stockpile of albumin on hand. Moreover, he added, albumin "has very definite advantages that cannot be disputed and that has to do entirely with packaging."[19] Veldee had been unaware that improvements in the fractionation process had meanwhile substantially increased the yields of albumin.

At a meeting with representatives of the pharmaceutical firms in Boston on July 19, 1943, Cohn read a letter from Admiral McIntire anticipating the invasion of Europe and urging that the production of albumin be accelerated since "it is evident that the need for blood substitutes will be tripled in the coming year."[20] As a result, production of albumin rose sharply during the summer of 1943, reaching levels of almost 30,000 units per month after September. Captain Newhouser estimated that the cost per vial of albumin was between seven and eight dollars and was approaching the six dollar cost originally predicted by Edwin Cohn. Canby Robinson of the Red Cross presented a preliminary analysis suggesting that a vial of albumin was being obtained from every three and a half donations of blood. This corresponded to a yield of one unit of albumin from eight tenths of a liter of plasma—about 60% higher than the yield estimated by Veldee.[21]

Substantial increases in the stability of albumin to heat were achieved. Scatchard noted that the thermal stability of albumin was increased approximately two-fold when the concentration of sodium chloride was increased from 0.15 to 0.30 molar. At about the same time, Murray Luck at Stanford discovered that the sodium salt of butyric acid was a far better stabilizer than sodium chloride. These were the first indications that the stability of albumin was influenced by the specific stabilizing agent as well as by the purity of the albumin being produced. They were a harbinger of progress still to come.

Clinical Evidence of Efficacy of Human Serum Albumin

Late in 1943, three clinical papers were submitted to the *Journal of Clinical Investigation* for publication under the general title, "Chemical, Clinical, and Immunological Studies on the Products of Human Plasma Fractionation," and were scheduled for publication in July 1944. The first of these papers, by Janeway, Gibson, Woodruff, Heyl, Bailey and Newhouser, was a collaborative study of the use of human serum albumin in the treatment of shock involving patients in the Grady Hospital in Atlanta, the Johns Hopkins Hospital in Baltimore, the Beth Israel Hospital, Boston City Hospital, Massachusetts General Hospital and Peter Bent Brigham Hospital in Boston, University Hospital in Iowa City, Memorial Hospital and Presbyterian Hospital in New York, University Hospital in Philadelphia, and U.S. Naval Hospital and Walter Reed Hospital in Washington.[22]

In the first part of this study, the theoretical and experimental basis for the use of albumin was examined in 200 cases, of whom seventy-five patients were suffering from shock and twenty-five from burns. In these cases, it was found that infusion of 100 milliliters of 25% albumin increased blood volume rapidly by drawing extravascular fluid into the circulation. This hemodilution was well sustained in patients with previously depleted blood volumes, but was transient in cases with normal blood volumes. When injected after hemorrhage, the standard Navy package of albumin (100 milliliters of 25% solution), was found to be equivalent in its osmotic effect to 500 milliliters of plasma, resulting in prompt hemodilution and clinical improvement. Albumin was not harmful in cases of shock with severe dehydration, but was more effective when water and salt were also administered. The Standard Army and Navy Package provided a stable blood derivative in compact form, instantly available for rapid administration. When properly prepared, in the absence of kidney disease, albumin was well tolerated and did not result in reactions. Albumin did not normally appear in the urine after injection. Very large daily doses of albumin were required to produce an appreciable rise in serum albumin concentrations in patients with chronic hypoproteinemia.

In a second study by A. Cournand et al. at the College of Physicians and Surgeons at Columbia University, the standard human albumin solution was administered to twelve clinical cases in varying degrees of shock. In those who were not actively bleeding or losing plasma into burned tissues, the albumin was retained in the blood, remaining for at least six hours. Albumin resulted in recovery from shock and an increase in blood pressure and cardiac output. In many cases, anemia persisted after therapy with albumin, suggesting that whole blood should be given subsequently when available.[23] In the third study, involving six patients in shock at the Grady Hospital in Atlanta, Warren et al. reported a distinct clinical improvement without any undesirable side effects when albumin was used.[24]

Although human serum albumin had been officially approved in January 1942, the only albumin delivered to the Navy by November 1942 was prepared in the Harvard pilot plant. Samples of albumin from the first production runs at three of the pharmaceutical firms had been received at the Harvard Plasma Fractionation Laboratory for testing and analysis. Driven at such a pace, Edwin Cohn occasionally displayed evidence of considerable strain. He became increasingly peremptory in the demands he made of his colleagues. The pace of the Monday luncheons changed. The agendas became more complex and detailed. In dealing with collaborators in his own and other academic disciplines, he became more demanding. With the government and the pharmaceutical companies, he found himself increasingly functioning as an administrator. At one point during this period, the representatives of the pharmaceutical firms protested the rigor of Cohn's specifications for human serum albumin. George Scatchard was often at his side in those situations. They were not above falling into a simple routine in which Cohn hinted that it was Scatchard who was influential in setting the specifications and suggested that they talk to Scatchard. Scatchard would then play his part. While he talked—he had the habit of interrupting himself and looking away in deep thought—they couldn't. Moreover, they found it hard to interrupt when they couldn't understand some arcane point that Scatchard raised, or when his comments seemed to be irrelevant. Scatchard remembered that "once this tactic failed, but my memory retains no facts of that confrontation. I do have the vivid picture of Horatius at the bridge with a loyal supporter at either side, Edwin Cohn in the center, Lloyd Newhouser on one side and George Scatchard on the other. We held the bridge."[25]

Within the laboratory, the pressure was manifested in various ways. Cohn took less time for casual conversation. Gentle suggestions, even by senior associates, would be brushed aside without comment. He became impatient in dealing with his colleagues. Once, in a luncheon meeting, he exploded in violent anger at John Enders for failing to finish a report that he expected to bring with him that evening to a meeting of the Subcommittee in Washington. Most of those present at the meeting knew that Enders' wife had died two days before. Enders said nothing. Despite such instances of boorish behavior, Cohn never acknowledged guilt or apologized.

During this trying period, Cohn was particularly fortunate to have the support of a group of key associates—Armstrong, Edsall, J.D. Ferry, Hughes, Oncley, Scatchard and Strong—who had been associated with him since before the wartime project was initiated, some for many years. They were familiar with his ways. Another group of colleagues in other disciplines—Robinson, Janeway, Boyd, Enders, Bailey, G. Edsall, Ingraham, Thorn and Krayer—contributed an enormous wealth of knowledge and experience in collaborating. They were accustomed to his imperious management style. Still others—Williams at Wisconsin, Luck at Stanford, Brand at Columbia and Mueller at MIT—provided important links to other universities and brought expertise in other fields to support the effort. Each was never more than a phone call away from Cohn. The staff of the Department of Physical Chemistry in Buildings C-1 and E-2 at the Harvard Medical School included more than forty men and women. In addition, a group of about ten Naval officers had been assigned to the Plasma Fractionation Laboratory to coordinate the clinical testing of the products produced in the Harvard Laboratory.

Gelatin and Pectin: Candidate Blood Substitutes

One of the initial charges to the Subcommittee on Blood Substitutes was the responsibility to survey and consider any blood substitute that might have potential value in treating shock. In order to stimulate interest in this topic, while at the same time shielding the Subcommittee from perfunctory proposals, Loeb prepared a statement for distribution to interested parties in the national community describing the criteria that were considered essential for any proposed blood substitute to be seriously considered for use in human beings:

> The colloid osmotic pressure should be equivalent to that of normal plasma.
> The product should be amenable to production with constant composition.
> The viscosity should permit easy administration.
> The product should withstand the range of temperatures encountered in a global war and remain stable.
> The product should be capable of sterilization.
> The method of preparation should exclude pyrogens.
> The product should be nontoxic, be readily metabolized or eliminated and should not be stored in organs.
> Repeated injection should not provoke sensitivity.[26]

By November 1942, the Subcommittee had received brief proposals from a group of investigators advocating the use of gelatin, a protein from animal tissue, as a blood substitute. In response, a Conference on Gelatin was convened in Washington. In addition to the members and staff of the Subcommittee, about twenty individuals representing academic and commercial institutions attended. Seventeen substantial reports were submitted by visitors at the meeting, and written summaries were left with the Subcommittee.[27] As an added inducement, Loeb offered groups working on gelatin the collaboration of the physical chemical laboratories of Cohn at Harvard, Williams at Wisconsin, and Luck at Stanford in physically characterizing and screening candidate gelatin products prior to conducting physiological and clinical studies. A similar meeting on pectin, a water-soluble polysaccharide found in fruits, was held on the following day, with ten experts attending. As with the gelatin meeting, Loeb invited the attendees to submit samples for physical chemical characterization by Oncley, Williams or Luck as a first step. Two of the studies submitted were by Frank Hartman from Detroit and by Glenn H. Joseph of the California Fruit Growers Association.[28]

A third conference on gelatin was held in Washington in September

1943. At that time, it was agreed that measurements of osmotic pressure and viscosity should be made by the manufacturers, but that measurements of ultracentrifugal behavior and of the modulus of rigidity would be provided by the university laboratories collaborating with the Harvard Laboratory where the needed instruments were available. At the next regular meeting of the Subcommittee, Loeb summarized the findings on gelatin. Several gelatin preparations had been found not to be antigenic, pyrogenic or toxic in man. The Subcommittee then recommended to the NRC Committee on Medical Research that further studies be made on gelatin solutions, particularly that gelatin preparations that were degraded as little as possible be prepared and compared with gelatin preparations that were extensively degraded. At the same time, the Subcommittee indicated that the prospect of preparing a gelatin blood substitute for use by the military in the field was remote.[29] Because no subsequent samples of pectin had been submitted to NRC since the previous meeting, the status of pectin was left open. In November 1943, Loeb commented that despite the Subcommittee offer of testing facilities for studies of pectin, no samples of pectin had been submitted.[30]

Gamma Globulin

In January 1943, the main plasma fractionation process having been transferred to the pharmaceutical industry, and the first commercial lots of albumin having been submitted to the Plasma Fractionation Laboratory in Boston for final testing and release, Cohn gave the first detailed report on the antibody studies at a meeting of the Albumin and By-Products group of the Subcommittee on Blood Substitutes in Loeb's office in New York. He prefaced his remarks by emphasizing the desirability of making the most possible use of the by-products resulting from the production of human serum albumin. Enders reported the scientific findings from a survey of antibodies found in Fraction II+III. These revealed that antibodies to diphtheria, dysentery, herpes simplex, influenza, lymphocytic choriomeningitis, measles, mumps, parapertussis, pertussis, perfringens, poliomyelitis, scarlatina, streptococcus, typhoid, and vaccinia were present in two to tenfold concentrations over plasma.[31] The evidence that such a diverse group of antibod-

ANTIBODIES IN GLOBULIN FRACTIONS FROM HUMAN BLOOD PLASMA

TABLE I

Antibodies in globulin Fraction II + III derived from human plasma[a]

Antibody	Type of antibody	Investigator	Institution	Concentration comp. to plasma
Anti-diphtheria	Antitoxin neutralizing	Edsall	Mass. Antitoxin and Vaccine Laboratory	10
Anti-dysentery	Agglutinins	Mudd	Univ. of Pennsylvania	2–10
Anti-herpes simplex	Neutralizing	Stokes	Children's Hosp., Phila.	*
Anti-influenza (human PR8)	Hirst inhibition	Hirst	Rockefeller Institute	4–8
Anti-influenza (human PR8)	Hirst inhibition	Eaton	Calif. Dept. Public Health	4
Anti-influenza (human PR8)	Hirst inhibition	Enders	Harvard Med. School	10–15
Anti-influenza (human PR8)	Complement fixation	Enders	Harvard Med. School	10–15
Anti-influenza (human PR8)	Neutralizing	Stokes	Children's Hosp., Phila.	10
Anti-influenza (human PR8)	Neutralizing	Enders	Harvard Med. School	9
Anti-influenza (swine)	Neutralizing	Stokes	Children's Hosp., Phila.	10
Anti-influenza (swine)	Neutralizing	Shaffer	Mass. Antitoxin and Vaccine Laboratory	4
Anti-lymphocytic choriomeningitis	Neutralizing	Stokes	Children's Hosp., Phila.	*
Anti-measles	Protective (human)	Stokes	Children's Hosp., Phila.	
Anti-mumps	Complement fixation	Enders	Harvard Med. School	2–10
Anti-parapertussis	Agglutinins	Mudd	Univ. of Pennsylvania	64
Anti-pertussis	Agglutinins	Mudd	Univ. of Pennsylvania	4–10
Anti-pertussis	Agglutination	Enders	Harvard Med. School	10
Anti-pertussis	Mouse protection	Bradford	Univ. of Rochester	4–10
Anti-perfringens	Protective	Hall	Univ. of Colorado	*
Anti-poliomyelitis	Neutralizing	Kramer	Mich. Dept. Health	10
Anti-poliomyelitis	Neutralizing	Stokes	Children's Hosp., Phila.	16
Anti-poliomyelitis	Rat and mice protection	Kramer	Mich. Dept. Health	10
Anti-poliomyelitis	Rat and mice protection	Stokes	Children's Hosp., Phila.	10
Anti-scarlatina	Neutralizing	Bradford	Univ. of Rochester	*
Anti-scarlatina	Neutralizing	Wadsworth	New York Dept. Health	5–10
Anti-streptococcus	Antitoxin	Wadsworth	New York Dept. Health	4–10
Anti-typhoid	H agglutinin	Enders	Harvard Med. School	8–10
Anti-typhoid	O agglutinin	Enders	Harvard Med. School	8–10
Anti-vaccinia	Neutralizing	Janeway	Children's Hosp., Boston	*
Isoagglutinins	Agglutinins	Boyd	Harvard Med. School	8–10

[a] These assays were undertaken at the request of Dr. A. R. Dochez of the Committee on Medical Research of the Office of Scientific Research and Development who wrote to the investigators listed in March, 1942, "In the process of preparing human plasma used for transfusion purposes in the armed forces of the United States, by Dr. Edwin Cohn of Harvard University, a number of fractions of the original plasma result. Only one of these, the albumin fraction, is used for transfusion. It is the desire of the Government to ascertain to what useful purpose the remaining fractions can be put. Among these fractions is one containing the α-, β-, and γ-globulins. As you doubtless know, this fraction contains whatever immune bodies may have been present in the original plasma. In the process of purification approximately ten times concentration of the immune body fraction is effected. It is hoped that these immune bodies may be used practically either for the prophylaxis or treatment of certain infectious diseases. In order to test the validity of such a procedure it is first necessary to titrate the globulin fraction for its content of specific antibodies. . . . The first titrations would be with mixtures of the α-, β-, and γ-globulins. Later fractionation of the different globulins will be performed and the specific immune body containing globulin will be furnished for a similar titration." We are greatly indebted to Dr. W. C. Boyd for compiling this table.

* Activity present but no quantitative data.

Fig. 3. Antibodies identified in globulin Fraction II + III from human plasma.

ies were concentrated in Fraction II+III, the second largest fraction, was impressive.

The only other proteins known to be present in Fraction II+III were the isoagglutinins—antibodies produced by individuals that cause the agglutination of red blood cells of other individuals of the same species, and prothrombin, an important protein involved in blood coagulation. When subjected to electrophoretic analysis, the major constituents of Fraction II+III's were alpha, beta and gamma globulins.[32] Late in 1942, the first attempts at subfractionation of Fraction II+III yielded several subfractions of which one, called Fraction II, contained a higher proportion of electrophoretic gamma globulin than the original Fraction II+III. However, the procedure for preparing this new fraction was cumbersome and particularly time consuming. Nevertheless, Cohn ordered that a small supply of this new fraction be prepared and filled into vials for clinical testing by Stokes and others in the next measles epidemic. The timing was important, since the only method for evaluating the effectiveness of Fraction II against measles involved clinical testing in children exposed to measles. The opportunity to test this material came later in 1942 in Philadelphia.

The first evidence that Fraction II contained measles antibodies was reported by Stokes to an NRC Conference on Immune Globulins on February 8 1943.[33] The clinical tests had been conducted in the Pennsylvania School for the Deaf during a measles epidemic. Among eighty-three exposed children between three-and-a-half and six years old given 2 ml of Fraction II solution, fifty had no measles, thirty had mild disease and three came down with unmodified measles. While noting that these results were subject to variables in age, weight, interval following exposure, intimacy of exposure and other factors, Stokes reported that these were the most satisfactory results of any agent he had used for the modification or prevention of measles, noting in particular "the striking value of Fraction II of the immune globulins for protection against measles." Doses of 0.5 ml of Fraction II solution, when injected within a suitable period in susceptibles exposed at home, afforded an approximately equal chance of preventing measles or of attenuating it, i.e. reducing the severity of infection.[34] Following the Stokes report, Colonel S. Bayne-Jones, the Army representative at the meeting, stated that Fraction II should be made available for use among Army personnel, particularly in protecting troops about to be sent oversees against measles and possibly also mumps. He requested that ten thousand vials of Fraction II be reserved for use by the Army.

Variations in the potency of concentrated measles antibody preparations could be expected to result in variations in responses in exposed individuals. The responses could vary from attenuation to complete prevention of measles infection, i.e. there would be a dose response effect. Treatment to achieve complete prevention would leave treated individuals susceptible to later infection (in such cases, the normal response in antibody formation in the exposed individual would not occur). On the other hand, modification of the course of infection—attenuation—would permit normal immune responses to proceed, thus causing the development of circulating antibodies that would provide permanent protection. Since measles is a childhood disease, modification rather than prevention would be the primary objective in most instances. For Fraction II to be useful in modifying the clinical course of measles infection, the potency of the antibodies in Fraction II would have to be standardized, but this could only be done during measles epidemics.

Accepting that reality, Cohn and Enders solved the problem by proposing to monitor in the laboratory three easily assayed viral antibodies *other than measles* in Fraction II as surrogates for the measles antibody. If this strategy was to be successful, it would still be necessary periodically to conduct clinical tests in exposed children. The decision to base this policy on other viral antibodies rather than on bacterial antibodies depended on the lack of evidence at the time concerning possible differences between bacterial and viral antibodies. This strategy was deemed particularly important during a period in which Oncley's team had not arrived at a final procedure for subfractionating Fraction II+III.[35]

A globulin control laboratory was organized in Boston along the lines already in use for albumin testing, with Enders responsible for immunological testing and Janeway for clinical testing. As the investigations continued, the results began to point clearly to the likelihood that the measles antibodies were also gamma globulins, i.e. had the electrophoretic mobility of gamma globulins. This became a certainty in March 1943 when the subfractionation procedure began to yield Fraction II preparations that were more than 97% gamma globulin by electrophoretic analysis while still retaining antibody potency against measles. It was on this basis that the name gamma globulin came into use in the Cohn laboratory and was adopted by the public press. At the time considerable interest had been expressed about the possibility of modifying Fraction II so that it could be used intravenously. However, when this was cautiously tried, some serious reactions were encountered. As a result, the use of gamma globulin was restricted to the intramuscular route. Many years were to elapse before a gamma globulin safe for intravenous use could be developed.

In the winter of 1943, severe measles epidemics in Philadelphia, Baltimore and Washington offered opportunities for further evaluation of Fraction II. In Philadelphia, Stokes confirmed the effectiveness of Fraction II+III and Fraction II, the latter being greater in potency. In a controlled family study of measles in Boston, Janeway and colleagues reported that gamma globulin resulted in protection against measles in 71%, modification of measles in 27%, and failure to prevent measles in 2% of individuals receiving it. Similar findings were reported from clinical trials in measles epidemics at Philips Academy in Andover and at Milton Academy in Massachusetts.[36] Normal human serum gamma globulin was recommended to the Armed Forces by the National Research Council on March 22, 1943.

A careful study of the effectiveness of gamma globulin was completed with the cooperation of the District Medical Service of the Boston Dispensary in families in which there was at least one case of measles in June 1943. In each family, the exposed susceptible children were divided into two groups, one of which was given gamma globulin while the other was left uninoculated. Among thirty-one children who received gamma globulin, twenty-seven did not come down with measles. In the uninoculated control group, only four out of twenty-four children had no measles or mild measles. The study confirmed that gamma globulin was effective in the prevention of measles. In a separate study aimed at modifying the severity of measles, a large dose of gamma globulin was given to a group of children on the ninth day after exposure to measles. Thirty-two out of thirty-three children in the inoculated group had no measles or mild measles. In the uninoculated control group, only four out of twenty-four children had no measles or mild measles; the others went on to have classical symptoms of measles. A separate study conducted in the private practice setting was aimed at testing the relative potency of a series of six different lots of gamma globulin. This revealed that all six lots had the same antibody activity against measles.[37]

A New Role for American Red Cross

Early in 1943, Lewis H. Weed became Chairman of the Medical Advisory Council of the American Red Cross. On April 6, Cohn wrote to inform him of an opportunity for the Red Cross to return to the civilian population some of the benefits that were being garnered in excess of military needs from the blood donated by the American Public. Cohn told Weed that from the point of view of the Red Cross, the by-products, such as the red cells which are separated from the plasma as well as the fibrinogen (in Fraction I), thrombin (in Fraction III-2), and the isohemagglutinins (in Fraction III-1), clearly cannot become a source of profit to commercial houses. According to the contract between the Navy and the commercial houses preparing normal human albumin, the by-products are the property of the Government and are being retained in the cold under conditions believed to be the best that can be specified at this time.

He pointed out that the opportunity confronting the Red Cross can be most clearly stated in connection with gamma globulin. Assuming that subsequent tests of gamma globulin bear out the encouraging results that have been obtained thus far, and assuming that all of the plasma being fractionated to obtain albumin under existing Navy contracts could be processed to yield this concentrate of human measles antibodies, something over two million doses could be made available.

> The necessity emerges of developing a policy so that the maximum amount of these valuable human proteins may be so used as to prove of the greatest military value, and, where the amounts available are in excess of military use, of the greatest social value. The human antibodies that can be made available and employed in the control of measles and, later, perhaps of other infectious diseases, should clearly not be permitted to deteriorate. The Red Cross, which has collected this blood for the armed forces might well, it seems to me, be put in the position of making a public restitution of those of the byproducts not needed by the armed forces—that is, insofar as so doing would not interfere with the war effort.[38]

At the next meeting of the Red Cross Advisory Council, Cohn and Loeb were asked to prepare a plan for the control and production of by-products. Their plan was drawn up and was approved by the NRC Subcommittee.

Fibrin Foam, Fibrin Film and Thrombin

Even before the development of the plasma fractionation program, there was a substantial body of knowledge about the proteins involved in the coagulation of blood, although much of it came from the study of animal rather than human samples. The initial focus of the studies of blood coagulation components in the Department of Physical Chemistry was on fibrinogen and thrombin. The unique position of Cohn and his colleagues was that they had a good supply of fractions prepared from fresh human plasma. In February 1942, John Edsall, Ronald Ferry and Howard Armstrong had learned how to prepare and isolate human thrombin from the prothrombin in Fraction II+III. Sterile preparations of thrombin had been made available to Ravdin and Wangensteen, who reported encouraging results in the topical use of thrombin in controlling bleeding from small, but not from larger vessels during surgery. This set in motion a broader program of research on thrombin with the objective of developing other clinically useful products. These began with efforts to assay and purify fibrinogen from Fraction I.[39]

At first, products that might be of value for the treatment of burns were investigated. A mixture of dry fibrinogen and thrombin powders was one of the first preparations to be tested. When the mixture was wetted by the burn exudate, a film of fibrin clot formed on the surface of the burned area. However, early films were too rigid and did not adhere to the site of the wound. Next, agents such as glycerol were added to soften the films. Powdered sulfa drugs were also incorporated in an effort to minimize infection. By the summer of 1942, tests with laboratory animals were begun by Orville Bailey and Clinton Hawn with encouraging results. By August, these preparations were being tested in human burn cases by E.D. Churchill at the Massachusetts General Hospital and C.C. Lund at the Boston City Hospital. For the latter studies, the fibrinogen, thrombin and sulfa powders were worked into a stiff paste by addition of glycerol, and then applied directly to the surface of burns. However, these encouraging studies were limited by the paucity of burn cases.[40]

In the spring of 1942, experimenting with so-called coarse fibrin clots, John Ferry and Peter Morrison had found that the clots could be compressed lightly to form strong, elastic, opaque sheets or films. A number of other film formulations were developed, including films backed with layers of related substances. These found their most important use in neurosurgery as a substitute for the dura mater, the fibrous membrane that covers the brain. The initial clinical studies with these films were carried out by Orville Bailey in the Harvard Department of Pathology and Wilder Penfield at the Montreal Neurological Institute in late spring 1943. When Penfield later left Montreal to serve in England, the collaboration continued with Franc Ingraham at Children's Hospital in Boston. Following studies in monkeys, one of the first patients in which fibrin film was used was a two-year-old child with lead encephalitis caused by gnawing on the railing of her crib. The child's spinal fluid pressure was elevated to nearly four times normal. Ingraham relieved the pressure by opening a window in the skull. Fibrin film was used to cover the swollen exposed surface of the brain. This was thought to be the first procedure in which fibrin film made from proteins derived from human blood plasma was used as a substitute for the dural membrane. Later, Ingraham and Bailey stated: "In our experience so far, fibrin film has proved more satisfactory than any other material tested as a dural substitute."[41]

Early in 1943, Tracy Putnam at Columbia University reported the use of oxidized cellulose sponge saturated with a thrombin solution to control bleeding.[42] This suggested the possibility of using a sponge-like form of fibrin instead of cellulose. Accordingly, Edgar Bering, a Navy officer assigned to the Harvard Fractionation Laboratory, began experimenting with fibrin that had been aerated during clotting. The resulting foamy mass was surprisingly rigid. It could be cut into pieces and dried from the frozen state. When immersed in a thrombin solution and pressed gently, it became sponge-like and absorbed the thrombin. Following experiments in animals, clinical trials were initiated by Ingraham at Children's Hospital in Boston with encouraging results. Confirming his impressions of the extraordinary value of fibrin foam as a hemostatic agent in neurosurgery, Bailey and Ingraham reported the use of fibrin foam in sixty patients.[43]

John Ferry's studies led to one unique use of fibrinogen and thrombin. John Dees, a Duke University urologist used sterile solutions of fibrinogen and thrombin to remove renal calculi (kidney stones). In a surgical procedure termed coagulation pyelolithotomy, he introduced the fibrinogen solution into the open renal pelvis. Thrombin was then added, causing a fibrin clot to form in the lumen of the pelvis. The clot was then gently removed with forceps, bringing with it the enmeshed kidney stones.[44]

Cohn had been intrigued by the possibility of preparing a plastic directly from fibrinogen that might be of use in tissue repair. To that end, John Ferry and Hughes developed a series of plastic formulations by heating fibrinogen with glycerol and other plasticizers under pressure. Plastics made in this way were then studied for their rates of absorption when embedded in the tissues of laboratory animals. In preliminary experiments, the most promising plastic was one made from equal parts of fibrinogen and ethylene glycol and heated for thirty minutes at 100°C at a pressure of 5,000 pounds per square inch. Plastics made in this way were also tested for use in repairs of tendon sheaths and in nerve suture.[45]

Blood Grouping Globulins

A Conference on Blood Grouping was held at the National Research Council on March 23, 1943 to discuss proposed changes in blood grouping techniques to be incorporated in the Army *Manual for Laboratory Technicians.* The key issue was presented by DeGowin who urged the Conference to recommend to the armed forces the exclusive transfusion of group O blood regardless of the blood type of the recipient, provided that an accurate check could be provided that the blood grouping of the donors had been accurately performed. All members of the conference voted to approve the recommendation.[46]

At Harvard, blood group A and B Isoagglutinins—antibodies that cause the agglutination of red blood cells of other individuals of the same species—had been prepared by W.C. Boyd from Fraction II+III in 1942. The possible use of these substances for blood typing of large groups of individuals had been recognized at the time, but interest was lacking in preparing large quantities of typing reagents from donor blood. Later that same year, the necessity arose at the Army Medical School in Washington of preparing large quantities of isoagglutinins for use in the typing of large numbers of individuals. Colonel G.B. Callender and Lt. Louis Pillemer provided a method for preparing anti-A blood grouping reagent from the plasma of group B blood donors, and of anti-B blood grouping reagent from the plasma of group A blood donors.[47] In due course, Pillemer was sent to Harvard to integrate the Army studies with work going on there in plasma fractionation. For this special purpose, blood donors of the same blood type, either A or B, were bled at the Red Cross Center in Boston and their plasmas were pooled. Following fractionation, type-specific Fraction II+III's were obtained, from which anti-A and anti-B reagents were isolated in good quantity. The products were tested and evaluated by a panel of consultants chosen by the Subcommittee on Blood Substitutes. As a result, the recommendation was made that these reagents be approved for use by the Armed Services for mass blood typing.[48]

A Conference on Shock

A second conference on shock was held in Washington on December 1, 1943. It called attention to the importance of determining the patient's blood volume as a guide to therapy. One of the chief problems in dealing with shock was the difficulty in supplying blood in the forward areas, a point that was made by H. K. Beecher who commented that somewhere along the planning line somebody seems to have forgotten that plasma lacks oxygen-carrying power.[49] This conference still fell short of identifying other key problems in dealing with shock in the field.

In its issue of November 29, 1943, *Time Magazine* stated that a generation's advance in knowledge of blood had been packed into the last three years, and that one of the great advances had been in the field of blood plasma. Plasma had been split into several useful components: albumins, which proved even better than whole plasma in treating shock; blood clotting factors (prothrombin and fibrinogen), which look very promising in the treatment of hemorrhage and burns; and antibodies, which have been tried as injections against viral diseases, have already worked well in preventing measles.[50]

11 The Invasion of Europe

In early 1944, the attempted invasion of the European continent by allied forces was clearly imminent. The Surgeon General, Major General Norman T. Kirk, was the chief medical officer of the U.S. Army. He was based in Washington. Under Kirk, Major General Paul R. Hawley served as Chief Surgeon for the European Theater of Operations (ETOUSA), with responsibility for oversight, support, and coordination of the medical programs of the five U.S. Armies assembled to mount the invasion of Europe. Hawley was based in England.

In November 1943, General Kirk had rejected a proposal from members of his staff stressing the need for stored blood in theaters of operations on the grounds that his observations in overseas theaters had convinced him that plasma was adequate for the resuscitation of wounded men; that, from a logistic standpoint, it was impractical to make locally collected blood available farther forward than general hospitals in the communication zone; and that shipping space was too scarce to warrant its use for sending disposable transfusion equipment overseas.[1] In response to a recommendation from the Subcommittee on Blood Substitutes in November 1943 that he give consideration to the transportation of whole blood by airplane to certain theaters of operations, General Kirk ruled that no whole blood could be expected in the theater from the Zone of the Interior, the latter being Army parlance for the forty-eight states.[2]

On January 2, 1944, General Hawley was informed by the Adjutant General that the provision of whole blood for combat casualties had been approved by ETOUSA at all levels down to division clearing stations. The effect of this policy was to assign to General Hawley the responsibility for establishing blood bank and transfusion services for the five Armies, including the recruitment of blood donors and collection of blood in ETOUSA hospitals. In response, Hawley, seeking a forceful executive with knowledge and experience in Army operations with professional experience in the use of whole blood, appointed Captain R. C. Hardin as ETOUSA transfusion officer. Hardin had received much of his training under DeGowin at the University of Iowa. One of Hardin's first acts was to establish contact with Brigadier Whitby, the head of the British Transfusion Service. That move proved to be one of considerable importance.[3]

In his final report as Consultant in Surgery to the North African Theater of Operations in 1943, Col. Churchill had recommended the establishment of an independent central blood bank to supply the Fifth U.S. Army with whole blood, plasma and intravenous solutions. However, before that recommendation could be implemented, Churchill had completed his mission in North Africa and had moved with the Fifth U.S. Army to the Mediterranean Theater of Operations, MTOUSA, in Sicily and Italy. In the interim, the British supplied blood to the Americans until an American blood bank was operational.

The first blood was collected by MTOUSA in Naples on February 23, 1944 by a staff consisting of an officer and two enlisted men. The blood was used in treating the wounded at the Anzio beachhead. In May, reorganized and expanded as the Naples Blood Bank, it functioned as a central blood drawing unit, collecting universal type O blood from soldiers and other personnel into a citrate preservative diluent for distribution to hospitals. The blood thus obtained had a useful dating period of about seven days. Subsequently, with a staff of five officers and thirty-eight enlisted men, the Naples Blood Bank collected and distributed about 300 pints of blood daily for transfusion in six nearby evacuation hospitals and three field hospitals. It was the only American blood bank to be patterned after the British model of a completely distinct and separate blood transfusion service. Had such a blood bank been established earlier in North Africa, it is likely that more American lives would have been saved. As it was, the blood collected in Naples was important in weaning the American forces away from their reliance upon dried plasma.[4] Almost 58,000 units of blood were collected in Naples before the end of the war in Europe.

In contrast to the campaigns in Africa, Sicily and Italy, the planned invasion of Europe would involve an entirely different set of circumstances. In the European Theater of Operations, five U.S. Armies would participate, each under its own General officer. Each had its own Chief Surgeon who would need an adequate blood supply at all times. The five Armies were being assembled in Great Britain in the spring of 1944. After crossing the English Channel on D-Day, each would fight its way across France toward specific objectives in Europe. Unlike the American experience in Southern Italy, weather would have an important effect on outcomes in Northern Europe. During the invasion, the blood needed for each Army would be transported across the English Channel. Once ashore in France, maintaining a supply of whole blood would not be easy.[5]

In March 1944, new and unequivocal advice from a National Research Council Conference on Shock in Washington reached General Hawley. Based on the results of studies of circulatory changes in clinical and experimental shock in dogs, this conference concluded that reduction in the circulating blood volume is the single most important factor in initiating shock. Whether the injury was mechanical or thermal, shock was seen as resulting from loss of blood or fluid at the site of injury. The conference had concluded by advising that whole blood be used without delay to treat shock, pointing out that "the first five hundred to one thousand cubic centimeters of blood should be given rapidly. Subsequent administration of blood should be at the rate of five hundred cubic centimeters per hour or less." This advice was consistent with the earlier position taken by Col. Churchill. However, it differed completely from the position advocated by Kendrick in 1941 when he addressed the military surgeons as the Surgeon General's spokesman.[6]

Red Blood Cell Preservative Solutions

Preservative solutions introduce the control of factors that extend the survival of red blood cells outside the blood stream. Commencing in May 1943, the Subcommittee on Blood Substitutes had provided the nidus for a concerted effort by a group of investigators that was aimed at two broad lines of study, first a study of physiological and biochemical factors in preserved red blood cells that might serve as indicators of their preservation status; and second, a study of factors that might serve as predictors of good red blood cell survival following transfusion into the circulation of recipients. One of the earliest preservative factors was sodium citrate, a nontoxic substance which forms a complex with the calcium in blood and inhibits blood clotting. Another factor was dextrose, which sustains red blood cells by providing a source of energy for cellular function. Preservatives containing citrate and glucose in specific proportions formed the basis for use of several preservative formulas, including those of DeGowin, Denstedt and Alsever.

A new red blood cell preservative, called the "Slough" mixture, was first mentioned at a March 1943 meeting of the Subcommittee. In a letter to the Subcommittee, Patrick Mollison in England suggested that this anticoagulant diluent for blood collection, incorporating sodium citrate, citric acid and dextrose, be used for the storage of whole blood. This diluent, also called Loutit-Mollison solution for the English investigators who published the definitive paper in England, quickly became known as acid citrate dextrose, or ACD. ACD was immediately subjected to intense study by the NRC preservation study group. It offered certain advantages as a preservative. By slightly increasing the acidity of the blood, by providing a source of dextrose for red cell metabolism and by regulation of the storage temperature between 4° and 10°C, a small volume of ACD extended the dating period of preserved blood to twenty-one days.[7] Meanwhile, the Army, pressed by the urgent need to provide a supply of blood, had adopted the Alsever preservative.

Planning to Meet the Needs for Blood

How much blood would be needed? Various approaches to estimating blood requirements were explored at meetings at the ETOUSA Blood Bank in Salisbury, England in the days before the invasion. On April 1, 1944, an estimate based on the experience of the British Blood Transfusion Service in North Africa suggested that one unit of blood would suffice for every ten casualties. An estimate based on American blood use in North Africa and Sicily was sought, but was not yet available. Planners for Operation Overlord—the code name for the invasion of the Continent—placed blood needs at 1,900 units of blood per day on D Day+90, but this was considered too remote to be of assistance to the invasion planners.

Other factors were weighed carefully. A few days after the April 1 estimate was made, a report revealed that the response of American troops in England to a donor recruitment drive had been "extremely disappointing." Since this pool of American blood donors would shortly be moving to the continent during the invasion, this report raised doubts about the likelihood of being able to recruit blood donations from soldiers. The proposed airlift of blood from the U.S. was frequently mentioned, despite the lack of any encouragement from the Surgeon General.[8] When, somewhat later, a complete report on blood use during the North African and Sicilian campaigns became available, it was revealed that one unit of blood had been used for every two casualties, a five-fold increase in blood use over earlier estimates. Since neither the Alsever preservative nor the ACD preservative was then being used, the lifespan of red blood cells for transfusion at the time was only about ten days. On April 6, Colonel Kimbrough submitted a daunting report to General Hawley. Based on plans for the invasion, slightly more than 4,000 units of blood would be available from Salisbury on D Day. During the next seven days after D Day, it was expected that 500 units of blood would be needed daily. However, only 200 units daily could reasonably be anticipated.[9]

June 6, 1944, the Invasion of Europe

During the first days after the Allied Forces were set ashore in Normandy, there was an adequate supply of blood. Col. Elliott Cutler wrote in his diary:

> The tremendous demand for blood completely justifies the establishment of the blood bank and from reports and observations it is clear we must have saved life by the establishment of an E.T.O. blood bank. Lieutenant Reardon of the blood bank is now on the far-shore.

> He has a large Navy-type refrigerator buried in the ground and (8) trucks (each taking 80 pints) are working well with the First Army delivering blood at this time. Almost all LST's and hospital carriers either gave up their blood to people on the far-shore or used it up on casualties on the trip back. Little was actually wasted. The major difficulty about blood has been the return of kits and sets and marmite jars.[10]

By 12 July, the ETOUSA blood bank was supplying 500 units of blood daily, although blood donors were becoming difficult to find. Pressure inevitably mounted to lay plans for obtaining blood from the U.S. On 24 July, there was a serious shortage of blood, causing the activation of a plan for rationing blood. In July, the demand for blood, reflecting increased military operations on the continent, far exceeded the supply. At the end of July, Colonel Kimbrough advised General Hawley that the shortage of blood was very serious. The need for blood was then approximately 1,000 pints per day.

In a tight spot, General Hawley ordered that immediate air transportation to the United States be arranged for Col. E. C. Cutler, the ETOUSA Chief Surgical Consultant, Col. E. F. MacFee, an experienced surgeon who was in command of an evacuation hospital in France, and Major Hardin, his transfusion officer. The stated objective of this mission was to arrange for an immediate supply of 1,000 units of airlifted whole blood per day from the United States. Once the travel arrangements had been secured, Hawley sent the following message by radio to Surgeon General Kirk: "Burden is being imposed that the ETO Blood Bank cannot meet in the demand for whole blood for the forces fighting in France. That blood is necessary and is saving lives, all are convinced. It is believed necessary that daily air shipment of one thousand pints of blood be sent. To coordinate this matter, returning to the United States are Colonel Cutler, Colonel MacFee and Major Hardin."[11]

On the third of August, General Rankin, the Chief Surgical Consultant in the Office of the Surgeon General in Washington, sent a memorandum to the Surgeon General advising him of the urgent need for blood in the European Theater and providing two alternative plans by which the blood might be obtained in the United States. The first plan called for blood needs to be met from blood collections by the Red

Cross in the United States; the alternative plan proposed salvaging the red blood cells from Red Cross blood collections being used in the preparation of dried plasma. This was essentially the same proposal that had been rejected by the Surgeon General in December 1943.[12]

Two days later, General Hawley followed up his radiogram with a letter to General Kirk that was intended to address some of the criticisms that the Surgeon General might make if he were leaning toward rejecting the proposal once again. Hawley began by reiterating the difficult position of the Army in France, pointing out that blood was hastening recovery and saving lives, and that blood was not being used extravagantly. He advised the Surgeon General that the ETOUSA blood bank, whose target was 300 pints per day, was then collecting almost 500 units daily and reported that the Air Transport Command was prepared to add one or two additional planes a day, if necessary, and could commit to delivering the blood in France within forty-eight hours of receipt.[13]

At the time, General Kirk was in Italy, where he had been shown a detailed report documenting the collection of 16,574 units of blood by the Naples Blood Bank between February and July. Back in Washington on August 10, 1944, he attended a conference that dealt with the details of initiating the airlift. At that meeting, Kirk reportedly stated, "If operating surgeons in the European Theater desire whole blood, they should certainly have it, and every effort will be made to provide what they requested." He sent the following radiogram to General Hawley on August 13:

> Whole blood is subject. This office prepared to ship 258 pints daily for first week commencing 21 August. This amount will increase to 500 as blood becomes available. Shipments will be made without refrigeration. Is sufficient refrigeration available in theater to accommodate shipments? Estimated weight first shipment 1200 pounds and 387 cubic feet. Request air priority and shipping instructions furnished this office. Request immediate reply.[14]

The first airlift, consisting of 258 units of type O whole blood in glass bottles that had been collected in Washington, New York and Boston, left Boston on August 21 for Prestwick, Scotland with a refueling stop in Gander, Newfoundland. From Prestwick, the blood was transported in a refrigerated truck to the ETOUSA Blood Bank in Salisbury, England where it was checked. It was then flown to France, arriving on August 27, six days after leaving Boston. Although the blood was chilled as usual before leaving the blood collection centers, its temperature was not controlled during the long transatlantic flight. Daily shipments of up to 1,000 units of blood followed. Beginning in October, the blood was flown directly to Paris by the Air Transport Command.[15] Each unit of blood (500 milliliters) had been collected into one liter glass bottles containing 500 milliliters of Alsever solution, yielding a liter of preserved blood plus preservative diluent.

On August 30, 1944, the Subcommittee in Washington reported substantial progress in defining the parameters that needed to be understood in order to evaluate proposed methods for blood preservation. The overall aim of this effort was the standardization of methods for the long-term preservation of red blood cells with a predictable ability to survive in the circulation of a recipient following transfusion. Following extensive discussion, the following resolutions were adopted.

> That this conference favors as a satisfactory criterion of the preservation of whole blood the survival in the recipient after transfusion of 90% of the transfused cells for 48 hours as optimal, and 70% for 48 hours as satisfactory.
>
> That this conference recommends to the armed forces that whole blood intended for transfusion be kept at temperatures between 4 and 10° C.
>
> That, in the light of information available from the medical literature and the results of experimental work presented at this meeting, the ACD solution is considered as the best available for the preservation of whole blood. There appears to be no evidence favoring the use of preserving solutions involving a considerable dilution factor. The optimum dilution factor is about 80% blood to 20% ACD.
>
> That it be recommended to the armed forces to provide transportation of whole blood at temperatures between 4 and 10°C from the time of collection to the point of final delivery.[16]

News that the Army was airlifting blood preserved with Alsever solution was not revealed to the Subcommittee in Washington until October 1944. At that meeting, J. C. Wearn, replacing Loeb as Chairman of the Subcommittee, stated that the majority of members of the Subcommittee favored the ACD preservative solution for whole blood. Wearn asked the members of the Subcommittee to begin testing Alsever's solution for presentation at the next meeting. He took pains to note that the studies on Alsever's solution had relied entirely on evidence of hemolysis and clinical reactions following the infusion of preserved blood as their criteria. Wearn stated that those criteria alone were not dependable when so important a decision depended on the outcome, and that further work must be done with Alsever solution. Nevertheless, Wearn believed that the evidence favored ACD as the better solution. In response, Capt. John Reichel, Jr. reminded the members of the Subcommittee that the Army had accepted the Alsever solution on the basis of a recommendation made by the Subcommittee back in September 1943. At the end of the discussion, the Subcommittee adopted two resolutions:

> That this subcommittee reaffirms its preference, expressed at its meeting of 30 August 1944, for the acid-citrate-dextrose solution as a blood preservative. This solution appears to be the most effective of any tested for the preservation of blood under both favorable and adverse conditions.
>
> That careful refrigeration of blood has been shown to be essential to prolonged preservation of red cells. This is especially important in the early period of preservation. It is recommended, therefore, that steps be taken to insure adequate cooling of the blood immediately after collection and to maintain its temperature at 4 to 10°C during transportation and until the time of its administration.[17]

In November 1944, the Navy, which, from the outset, had based its blood preservation studies on the use of ACD at 4° to 10°C, began to airlift blood to naval bases and Army troops fighting in the Pacific where ambient temperatures were very high. Furthermore, by using ACD rather than Alsever solution, the Navy substantially decreased the total volume of preserved blood to be carried across the Pacific from a total of 1,000 milliliters per unit (blood plus Alsever diluent) to 600 milliliters per unit (blood plus ACD diluent). That blood was collected by the American Red Cross from San Francisco, Oakland, Los Angeles, Portland Oregon, San Diego and Chicago.

Illustration 7. Edwin Cohn and a group of young colleagues on the front steps of Building A of the Harvard Medical School in the spring of 1942 at the beginning of the early experiments in the Harvard Pilot Plant directed at isolating human serum albumin from human plasma. Front row, (left to right), L. E. Strong, J. L. Oncley, E. J. Cohn, W. L. Hughes, Jr., J. N. Ashworth; back row, H. L. Taylor, L. Larkin, J. W. Cameron, D. L. Mulford, D. S. Richert, A. H. Sparrow

Illustration 8: Above: John F. Enders in his laboratory in the 1930s; Below, left to right: A. Baird Hastings, Louis K. Diamond, Carl W. Walter.

Illustration 9. The Harvard Pilot Plant on the first floor of HMS building E. It consisted of two connected refrigerated rooms. Beginning in 1942, the pilot plant was capable of fractionating forty liters of plasma at one time, using methods introduced by Cohn and his associates.

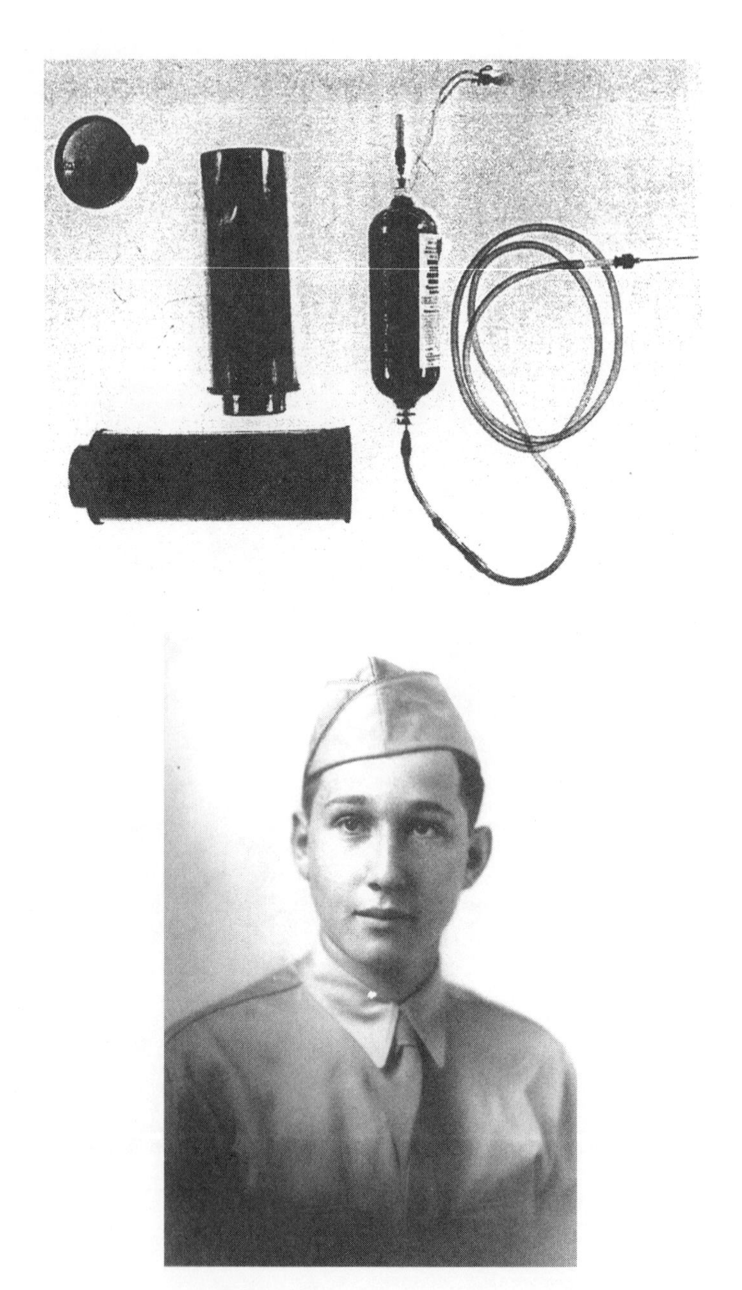

Illustration 10. Above: A single unit of 100 ml of concentrated human (25%) serum albumin assembled for clinical use. Designed by Dr. Lloyd Neuhouser, M.C., U.S. Navy, the whole unit was packaged in a cylindrical can; three cans occupied very little space. Below: Alfred Brettauer Cohn after he enlisted in the army in 1943.

Illustration 11. The Armour fractionation plant in Fort Worth, Texas enabled a much larger scale of fractionation in 1944.

Illustration 12. Edwin Cohn and his associates in the Department of Physical Chemistry at the peak of the work on plasma fractionation in 1944. Left to Right, Front Row: J. Edsall, Bailey, Oncley, Janeway, Cohn, Scatchard, Enders, Strong, Ingraham. Second Row: Woodruff, Brown, Sparrow, Hughes, Williams*, G. Edsall, Ferry, Richert, Mulford, Deutsch*, Bering. Third Row: Klein, Goodrum, Larson, Morrison, Gordon, Gross, Foster, Aldrich, Berenberg, Taylor. Fourth Row: French, O'Neil, Blanchard, Ashworth, Dale, Miller, Brooks, Barker, Gahm, Griffith. Fifth Row: Sanderson, Fleming, Cameron, Poole, Traill, Weeks, Smith, Beauchamp, Sullivan, Hasson. Sixth Row: Whitaker, Splaine, O'Reilly, Kottke, Budka. Not Present: Armstrong, J. Gibson, S. Gibson, Melin, Powell.

*Univeristy of Wisconsin

The Navy had used a disposable insulated container with a central separate chamber that was filled with ice to maintain the blood at the specified temperature during the long trans-Pacific flights. A special version of the shipping container enabled the Navy to parachute units of blood from helicopters to isolated outposts or ships.[18] The first shipment of 160 pints of blood left Oakland, California on November 16, 1944 and arrived at the Navy Distribution Center in Guam on November 19. The boxes were re-iced at Oahu, Johnston Island and Kwajalein during the flight to Guam. From there, the blood was shipped to Saipan, Iwo Jima, Okinawa, Ulithi and Leyte in the Philippines. The first units arrived at Leyte on November 22, 1944.

Human Serum Albumin

The production of albumin by the pharmaceutical firms attained its planned rate by the beginning of 1944. By March 2, a total of 224,000 units, hundred milliliter vials of 25 % albumin, had passed all required tests for acceptance and were released for delivery to the Navy. In February alone, almost 30,000 units had been finished. With Red Cross blood collections running at more than 100,000 units per week nationally, all plasma that had been contaminated and could not be used to make dried plasma was being processed into albumin, thus maximizing the utilization of the donated plasma collected by the Red Cross. When the albumin solutions from these runs were sterilized by filtration, they passed all tests. Notable progress had also been made in improving the stability of the albumin. Although dispensed initially in dilute sodium chloride solution, the use of 0.3 molar sodium chloride, introduced in 1943, resulted in a substantial improvement in stability, as evidenced by superior clarity of the solution after heating for prolonged periods of time.

The latest advance by Murray Luck revealed that substitution of the sodium salt of acetic acid for some of the sodium chloride resulted in a ten-fold increase in the stability of albumin. By early 1944, the specifications required that albumin withstand heating at 57°C for 100 hours without exceeding a specified increase in turbidity. Cohn was only waiting for the completion of clinical studies of albumin with this new stabilizer before introducing it into routine operations. By the end of

June 1944, 360,000 units of human serum albumin had been released to the Navy. A study revealed that slightly less than 3.5 units of donated blood were yielding one unit of human albumin. This agreed surprisingly well with the yield originally predicted by Cohn in February 1942.[19]

The ability to dispense albumin as a stable concentrated solution in a package requiring only a fraction of the space occupied by the equivalent amount of dried plasma had, from the outset, been seen as a distinct advantage by the Navy. Nevertheless, questions were raised by the Army once again in 1944 about the wisdom of administering 25% albumin to wounded men on the Italian front who might be severely dehydrated. In response, Dickinson Richards mounted a collaborative study of albumin use in shock patients in U.S. hospitals to try to answer the question once and for all. In Janeway's view, this new Richards study "justified the recommendation of 100 ml. of 25% albumin as the standard unit, which should result in an increase in the volume of plasma by approximately 500 ml. in the absence of continuing plasma loss or dehydration."[20] Every Landing Ship at the Normandy beachhead had albumin on board. Later, Lt. Comdr. Gibson received a letter from a physician on an LST citing three cases dying of shock while blood was running into their veins for whom rapid infusion of albumin directly into the intravenous tubing brought rapid responses and clinical improvement.[21]

Gamma Globulin

In 1944, Edwin Cohn led what proved to be the final effort within the Subcommittee to persuade Veldee and the U.S. Public Health Service to adopt a name for gamma globulin that would distinguish it from the quite different product, immune serum globulin, that was prepared from placental blood. Cohn favored returning to the name, gamma globulin antibodies.[22] A member of the Subcommittee suggested qualifying the source as *normal* to distinguish it from *convalescent*. Other members objected to the use of *serum* when the source was prepared from *plasma*. David Rutstein, the Associate Commissioner of the Health Department of the City of New York, citing a recently published report in the *Journal of the American Medical Association* of a death following the injection of placental extract that carried the same name, i.e. immune serum globulin, expressed a fear among some physicians in New York of using the placental product. The Subcommittee forwarded a recommendation that the Public Health Service adopt the name *normal human serum gamma globulin antibodies* for the Cohn product. Veldee was unyielding; he ruled that Cohn's product must carry the name immune serum globulin. However, gamma globulin was by then too well established to disappear from the scientific lexicon; it is used throughout this text.

The first commercial lots of gamma globulin for evaluation in measles prophylaxis were received early in 1944. Measurements of surrogate antibody titers by John Enders were proving satisfactory. Field tests during measles epidemics had been conducted by Janeway in Boston, by Stokes in Philadelphia, by Captain Newhouser in Washington, and by David Rutstein in New York City.[23] In Cohn's view, the experience with gamma globulin during this second epidemic year with measles would be very important. While all of the product used in the 1943 measles season had been produced in the Harvard Pilot Plant, 1944 would provide an important test of the scientific planning that had gone into transferring the process to American pharmaceutical firms. He was also confident that the base of clinical knowledge would be greatly strengthened as the result of another round of testing.

At the Harvard Medical School, Oncley and his group had introduced substantial improvements to the procedure for subfractionating Fraction II + III. Method 3, so-called, resulted in two important gains. The overall yield of gamma globulin was improved. Even more significant, the concentration of gamma globulin in the final product had increased from about 80%, the level in the 1943 trials, to 96%. This meant that the new trials could be based on a 16.5% solution rather than on the 20% solution that was used in the 1943 trials. In view of the spectacular improvements that the Harvard scientists were seeing following the introduction of stabilizing agents to human serum albumin solutions, the albumin stabilizers were tested for their effect on gamma globulin. Surprisingly, there was no effect. Neither sodium chloride nor the organic ions that had proved protective for solutions of serum albumin improved the stability of gamma globulin to heat. However, gamma globulin solutions could be stabilized by amino acids as well as by certain sugars. Glycine, the simplest amino acid, was adopted as the stabilizer of choice.

The 1944 winter trials of gamma globulin yielded a number of important findings. Typical of the new results were those reported by Rutstein in New York City when it was used in almost 800 contacts of confirmed source cases of measles: 77% of the contacts had no measles, 22% reporting mild measles, while only 1% had "average measles." Similar results were reported by Stokes when gamma globulin was tested among 107 Bryn Mawr College students who had been exposed to measles at the same time. Of these, sixty-five cases were treated while forty-two received no treatment.[24] The incidence of "average measles" in the treated group was 17% while it was 52% among the untreated students. Furthermore, when adjusted for differences in purity and in proportion of electrophoretically measured gamma globulin, the purest gamma globulin was no less effective against measles than earlier, less pure preparations. This strongly confirmed mounting evidence that antibodies against measles are gamma globulins by electrophoretic analysis. It also confirmed the previously established dosages of gamma globulin for prevention or modification of measles. On this basis, it was estimated that the anti-measles potency of the 1944 gamma globulin solution was twenty-five times greater than that of the average 1943 pool of plasma, confirming the value of relying on assays for certain known antibodies as surrogates of the measles antibody.[25]

Infectious hepatitis (IH), transmitted via the fecal or oral route, became a serious problem for American troops in the Mediterranean area. Stokes and Neefe found that gamma globulin was effective in modifying or preventing IH in exposed individuals.[26] Havens and Paul later confirmed its usefulness against IH during an epidemic among institutionalized children.[27] Gamma globulin was not effective in treating mumps or scarlet fever. The results of early studies in polio were encouraging when tested in monkeys, but had little effect on the course of the disease in humans who exhibited the symptoms of polio.

Fibrin Foam

In order to provide fibrin foam and thrombin for clinical trials, production was begun on a moderate scale in January 1944 under an OSRD

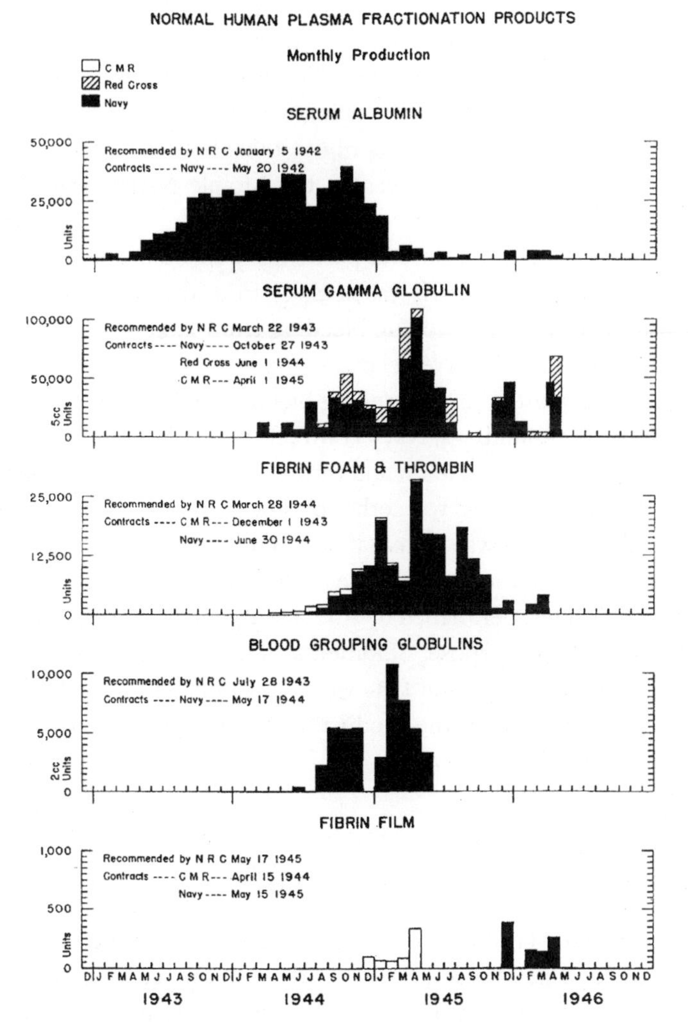

Fig. 4. Monthly production of normal human plasma fractionation products, 1943–1946.

subcontract between the Upjohn Laboratories and Harvard University. Vials of the product were distributed for testing and evaluation to a large group of neurosurgeons. This resulted in a series of reports so uniformly favorable that the NRC Subcommittee on Neurosurgery recommended fibrin foam and thrombin for use as a hemostatic agent on March 28, 1944. An important series of clinical studies was carried out at the Hospital for Head Injuries at Oxford. These resulted in a letter from Brigadier Cairns in Oxford to the British Medical Research Council on April 18, 1944, stating, "The Harvard fibrin foam is excellent as a haemostatic agent in stopping bleeding in neurosurgical operations . . . We could usefully distribute 1,000 bottles . . . at once. The highest priority should be given to the Mobile Neurosurgical Units with our armies in the field."[28]

Fibrin foam and thrombin was particularly useful in controlling oozing from small blood vessels, particularly from the beds of tumors. It was also effective in controlling bleeding from large venous channels and in dealing with blood vessel malformations of the cerebrum. However, it was not useful in controlling bleeding from large arteries. On the basis of these findings, Navy contracts were awarded in June 1944 and production of fibrin foam and thrombin began immediately at the pharmaceutical firms. As with other plasma products, each lot of fibrin foam and thrombin was tested at the Harvard Plasma Fractionation Control Laboratory for compliance with specifications that included a human clinical test performed by a neurosurgeon chosen from a group appointed for the purpose. Shipments of fibrin foam and thrombin reached England during the 1944 summer campaign. Over 4,000 clinical reports of the use of the product in the field were returned to the Harvard Control Laboratory.

Fibrin Film

The commercial production of fibrin films was delayed by difficulties encountered in sterilization of the product on the required large scale. The task was further complicated by the need to conduct careful physical chemical, biological and animal tests at each step in the process. With the close collaboration between the investigators in Boston and the men at Armour Laboratories, a satisfactory procedure based on steam sterilization was finally developed late in 1944.[29] These results prompted Warren Weaver at the Rockefeller Foundation to comment, "this laboratory and this long, patient effort thus constitute a really dramatic illustration of the truth of John Dewey's remark that it does not pay to tether one's thoughts to the post of usefulness with too short a rope."[30]

Serious Charges Before the Subcommittee on Blood Substitutes

In December 1943, Loeb received a strongly worded letter from Glenn Joseph of the California Fruit Growers Exchange leveling charges against the National Research Council, the Subcommittee on Blood Substitutes, and Edwin Cohn and George Scatchard concerning the Subcommittee's failure to give clearance to pectin as a blood substitute. Joseph stated,

> We cannot avoid concluding now that during the two and one-half years since the first publication appeared on the promising use of pectin in shock treatment, officially recognized progress has been practically zero. This is a surprising and deplorable condition because, as you know, our country is fighting the most gigantic war which ever clouded the surface of the planet, our young men and those of our allies are dying in appalling numbers. We have been led to believe that human blood and products derived from it are not available in quantities needed for the military. Civilian blood is being given to such an extent that medical authorities are becoming worried about possible resultant anemias. Our civilian centers are without adequate reserves of blood and plasma for possible disasters.

Joseph charged that Edwin Cohn had refused to read some clinical reports on pectin, indicating that he "would not waste time reading them." Joseph also cited a letter from Scatchard to a scientist in Detroit associated with the pectin project that stated "I am, therefore, holding your samples until I hear that lot PC-8 is an important one in physiological or clinical tests, and I suggest you send a sample of this same lot to Dr. Joseph." Joseph's letter concluded with the statement, "we see no reasons why your refusal to grant clearance for pectin to the military . . . should cause permission to use pectin for civilian emergencies to be denied."[31]

Loeb convened a special meeting of the Subcommittee on January 5, 1944 to deal with Joseph's charges. In addition to the members of the Subcommittee, those attending included Lewis Weed, George Guest of the NRC Committee on Medical Research, Col. R. G. Prentiss, Jr., Lt. Col. Kendrick, Capt. E. S. Taylor of the U.S. Army, Captain Newhouser, Captain E. H. Cushing, Lt. S. T. Gibson of the

U.S. Navy, M. V. Veldee, J. V. Alsever of the Office of Civilian Defense, W. Van Winkle, Jr. of the Food and Drug Administration, George Scatchard, and Mr. Joseph. At the outset, Weed reviewed the history and functions of the National Research Council, citing its role as a quasi governmental agency with responsibility to give professional and technical advice to government agencies when requested. He emphasized that NRC had no executive function. Loeb then reviewed the experience of the Subcommittee, gained over more than three years, which had taught that there were certain steps to be taken in the study of substances proposed as blood or plasma substitutes. Descriptions of these steps had been provided to all those attending previous conferences of the Subcommittee, including the initial conference on pectin held in February 1943.

Loeb cited the considerable progress that had been made in the development of another blood substitute, gelatin, following the same criteria. He emphasized the importance of characterizing proposed substances physically and chemically prior to evaluating osmotic pressure and conducting clinical studies. He pointed out that the workers on pectin had been repeatedly asked for samples on which such physical and chemical studies could be performed, but none had been received. He concluded by stating that the Subcommittee found the work on pectin to be "disorganized." On his part, Cohn confirmed the statement attributed to him, pointing out that he was not a clinician and had no competence in that field. Scatchard noted that Joseph had omitted the most important sentence in quoting his letter to Joseph, which was "I think it is quite important that the same lot which is being tested physiologically should be tested physical chemically and chemically."

The subcommittee then addressed the other charges made by Joseph. Col. Prentiss and Lt. Col. Kendrick, the representatives of the Army, stated that the supply of plasma for the Army was entirely adequate. This was confirmed by Capts. Newhouser and Cushing on behalf of the Navy. Joseph admitted that the source of his charge concerning the inadequacy of plasma supplies was a naval officer who had been in the South Pacific. Newhouser pointed out that there may be some officers, new to their duties, who do not know that medical supplies, including plasma, must be requisitioned. He stated that monthly reports from all naval bases came regularly to him in Washington and every one revealed that supplies of plasma were adequate. The discussion of pectin then came to an end with Joseph stating that he had misunderstood the function of the National Research Council.

When the Subcommittee reconvened, it unanimously approved a resolution proposed by Elmer DeGowin stating that the charges made by Dr. Joseph had not been substantiated by evidence brought forth by representatives of the Armed Services, the Office of Civil Defense, and the American Red Cross.[32] Final action on pectin as a blood substitute was taken in March 1944 when the Subcommittee concluded that, based on experimental data in animals and humans, "pectin does not appear to be a satisfactory blood substitute."[33]

The Subcommittee took a somewhat more favorable position in the case of gelatin. It issued a statement summarizing the chemical, physiological and clinical properties of gelatin solutions and identified a number of limitations and unanswered questions about gelatin. Gelatin was neither recommended to the Armed Forces nor rejected as a blood substitute. However the Subcommittee was firm on one point: the preparation and use of gelatin would in no way decrease the need for the procurement of blood by the American Red Cross and the preparation from it of blood substitutes for the Armed Forces.[34]

The Commission on Plasma Fractionation and Related Processes

On November 15, 1939, faced with increasingly complex issues with respect to University policy concerning discoveries and inventions bearing on health and therapeutics, the Harvard Corporation had adopted a statement of policy which included the following provision:

> No patents primarily concerned with therapeutics or public health may be taken out by any member of the University except for dedication to the public. In cases as to which it may be deemed necessary to take out a patent and dedicate it to the public in order in order to prevent others from obtaining a patent for their own benefit, members of the University are asked to report to the Dean of the appropriate Faculty any such discovery or invention made by them with a recommendation as to whether an application for patent should be

filed, in order that, if necessary, steps may be taken to obtain and dedicate the patent.[35]

While his investigations beginning in the summer of 1940 had led to the isolation of bovine albumin, then to human albumin, and other protein products, it became clear to Cohn that important new issues inherent in the nature of the work raised questions about patent policy. Of these, the central issue revolved around the intricacy of the procedures and the need for their meticulous control if the safety, quality and efficacy of the products were to be assured.

Acting under the 1939 Harvard policy, Cohn had met with President Conant in February 1942. This formed the basis for an agreement between Cohn and the President and Fellows of Harvard College which recognized the possible existence of "valid special reasons for departing from the usual policy of the University not to permit the patenting by members of the University of inventions having to do with public health." Under this agreement, Harvard consented to the filing of patent applications by Cohn covering the main fractionation procedures, and by John Ferry concerning fibrinogen plastics. At the same time, Cohn agreed not to dispose of any rights under the patents except upon the written consent of the President and Fellows. The patent applications were filed with the understanding that neither the inventor nor the University would profit in any way from such patents.

The next stage involved the development of a plan for effecting the controls over the safety, quality and efficacy of products that were made under the patents. During the wartime program, all production of plasma fractions was accomplished under contracts between specific pharmaceutical firms and the government in the form of Navy contracts. Acceptance of products for the Navy followed rigid chemical, immunological and clinical tests that were conducted by the Plasma Fractionation Laboratory at the Harvard Medical School. In addition, the National Institutes of Health released the products under a set of "minimum requirements."

By October 1944, requests to produce products under the Harvard patents had been received from other sources. One was from the Commissioner of Public Health on behalf of the Commonwealth of Massachusetts, which was planning to build a fractionation laboratory to serve the public in Massachusetts. Two commercial firms, Sharp and Dohme and Cutter Laboratories, submitted requests for licenses to produce the products of plasma fractionation on a commercial basis and discussions were under way with other firms. Moreover, the new plan for the distribution of excess gamma globulin over and above that needed by the Armed Services to the American Red Cross for distribution to the American public could also impact on the Harvard policy.

When Cohn conferred with Vlado A. Getting, the Massachusetts Commissioner of Health, a set of notes were recorded which initiated the process of formulating a clear policy on these issues. The resulting document began by stating the belief that dedication of the patents to the public at that time, the only alternative to a continuing control program, would defeat the public interest by permitting the release of inferior products that might fail to be of therapeutic value or even lead to injury to patients. It concluded that "such eventualities can be prevented by insistence on precautions such as are now exercised by the American Red Cross in the selection of donors for their Blood Procurement Program, and on continued rigid control such as is exercised under Navy contracts." Further, since the work of the Harvard Pilot Plant and its Plasma Fractionation Laboratory would presumably cease following expiration of Harvard's OSRD contract, it followed that prompt action would be needed to provide for continued control into the post war period. In developing a plan for effecting the controls under the patent, Cohn's discussion with Getting led to consideration of organizing a non-profit corporation to hold the patent rights and assume responsibility for all control measures. Informed about this, President Conant indicated that he preferred to explore the possibility of working through an existing non-profit corporation and mentioned the Research Corporation of New York.

The Research Corporation had been organized to administer patents on behalf of faculty members in American universities. In that capacity, Research functioned to pursue the commercialization of inventions, acting on behalf of the inventors, negotiating the terms of licensing and royalty agreements and disbursing the royalty income back to the inventors and their universities. Cohn saw the wisdom of using Research with its unique capabilities and experience. Although the pursuit of royalty income was not an objective of the plasma fractionation patents, Cohn harbored some aspirations to apply the same technology toward non-therapeutic ends. He visualized enlarging such applications by developing relations with the corn products industry. At the same time, his own inquiries revealed that the Research Corporation lacked the expertise that would be required to assure the rigid control of quality for therapeutic products. As a result, he readily acquiesced to President Conant's suggestion, but proposed that a self perpetuating Commission on Plasma Fractionation and Related Processes be established which would

> a) determine policy in the interest of the public health, b) serve without remuneration of any kind, c) authorize the distribution of new products after review of clinical appraisals and of specifications for control, d) designate control boards charged with arranging for and supervising chemical, clinical, immunological and tests deemed necessary, e) recommend to the corporation in which patents are vested, budgets for the costs of control, f) advise this corporation regarding the quality of products distributed by commercial firms and recommend either the granting or cancellation of contracts, g) make every effort, as a consequence of the policy of refusing royalties on products of therapeutic value, to assure the public of the safest products that can be produced at the lowest cost, h) in the case of commercial developments not of therapeutic value, authorize the corporation in which patents are vested to accept royalties and recommend the policy to be followed in their distribution.

In forwarding these ideas to President Conant, Cohn suggested that, whether established by the University or by Research Corporation, familiarity with the medical, public health, and technical problems would, he believed, be necessary to establish wise policy and to minimize admitted risks, other than legal risks, of establishing control in the public interest.

These suggestions were approved by the President and Fellows. The Research Corporation replied to Cohn on December 22, 1944, accepting the Harvard suggestions, and stating:

> Your underlying wish and ours is to provide for the management by us of the patents issuing on any of the above applications in the interest of the public health, subject to your technical and scientific ad-

> vice, without profit to you or us where the processes are practiced with human blood of the products produced are for human therapeutic public health purposes.
>
> You and your colleagues will create and provide for the management and continuity of a commission which will appoint a salaried executive secretary. When a commercial company shall apply to us for a license under these patents, we shall apply to the commission for advice as to the technical competence of the applicant. Subject to a veto by the commission on such grounds, our licensing policy shall be entirely within our discretion. The commission will establish procedure and standards for production and testing of products and prescribe them to the licensees; and will recommend to Research such institutions as may be prepared to test the licensees output against such standards, the results of such tests and release of the products to be evidenced by certificates issued as the commission may determine. Research will attempt to enter into a contract therefore with such testing institution or institutions, and will submit the terms asked to the commission that it may advise upon them.[36]

Cohn invited Roger I. Lee, John Enders, Vlado Getting, and George Scatchard to join him in the formation of the Commission on Plasma Fractionation and Related Processes. The Commission held its first meeting on December 29, 1944 with Edwin Cohn as its Chairman.

The Public Learns about the Plasma Fractionation Project

As written, the 1941 contract between the Office of Scientific Research and Development and Harvard University concerning Cohn's work prohibited the publication of scientific and clinical findings from the plasma fractionation project without government permission. In 1943, approval had been granted for the publication of a series of papers describing many of the results of the Harvard work. However, the relaxation of the restrictions against publication did not extend to publication of the chemical fractionation and purification methods. These were not published until after the war. Cohn followed a simple plan for the publication of the "results" papers. Under a special arrangement with the *Journal of Clinical Investigation*, twenty-three papers were published in the July 1944 issue of the journal. Each paper carried the general title "Chemical, Clinical and Immunological Studies on the Products of Human Plasma Fractionation," a phrase taken from the original contract between the U.S. Office of Research and Development and Harvard University under which the Harvard studies were conducted.[37] All of the papers focused on the characterization of the products of plasma fractionation and the appraisal of their uses. Cohn, Oncley, Strong, Hughes and Armstrong were the authors of the first paper, entitled "The Characterization of the Protein Fractions of Human Plasma."[38] This was the only paper in the long series that carried Cohn's name.

Use of the term "clinical" in the titles of these papers stirred surprising resentment on the part of two of Cohn's Harvard Medical School colleagues, Beecher and Altschule. In their 1977 book, *Medicine at Harvard,* they stated "the clinical mandate by Doctor of Philosophy Edwin J. Cohn, that brilliant, strong-willed, arrogant man, may have been at least a partial source of the disastrous overemphasis on plasma."[39] This referred to a polemic over the causes of American delays in providing whole blood for transfusion to the Armed Services during the war, in which, as we have seen, Cohn had no role.

Following publication of the papers in the *Journal of Clinical Investigation*, the public-at-large began to hear about the Harvard work. On February 11, 1944, Edwin Cohn was the featured speaker at a dinner meeting of the New York Section of the American Chemical Society at the Hotel Pennsylvania in New York. Introduced by an old friend, Vincent du Vigneaud of the Cornell University Medical School, he spoke on the topic, "Chemical Properties and Physiological Functions of the Plasma Proteins." The speech was given prominent press coverage the next day in the *New York Times* under a headline, "Seven Aids to Life Taken from Blood." The *Boston Globe* carried the story under the byline of Associated Press Science Editor Howard E. Blakeslee, which was picked up by AP affiliates across the country. Both stories followed the advance press release issued by the Chemical Society. Perhaps reflecting a resolve to avoid a replay of his unfortunate experience with the American Chemical Society news office and the press in Kansas City in 1936, Cohn did not make himself available to the press. A few days later, Cohn's work and that of H.W. Florey, one of the discoverers of penicillin, were cited by Raymond B. Fosdick in his annual review as

President of the Rockefeller Foundation, as reported in the *New York Times*. Fosdick called attention to the Foundation's support of Cohn's "fundamental research on the physical chemical properties of the proteins . . . that are now paying large dividends in human lives on our far flung battlefields."[40]

In April 1944, Cohn participated in a day-long symposium on Wartime Advances in Medicine at the American Philosophical Society in Philadelphia. The other participants included James L. Gamble, HMS Professor of Pediatrics; Allen O. Whipple, Professor of Surgery at the College of Physicians and Surgeons, Columbia University; Rollo E. Dyer of the National Institute of Health; Detlev W. Bronk, Professor of Biophysics at the University of Pennsylvania; Georges F. Doriot of the Army Quartermaster Corps; and Edward A. Strecker, Professor of Psychiatry at the University of Pennsylvania.

During May and June, Cohn's work was widely cited in articles in the *New York Times* and Scripps-Howard newspapers concerning the civilian distribution of gamma globulin by the American Red Cross for the control of measles. A *New York Times* story in September, "Harvard Study Finds New Plastics in Parts of Human Blood Plasma," stated that "one reason for the swift and dazzling variety of accomplishments in the blood bank was Dr. Cohn's assumption that in human blood everything is valuable and that the parts of man's blood are more useful than the whole blood when they are refined, concentrated and pointed for a specific purpose."

Cohn presented the Annual Sigma Xi Lecture at a meeting of the American Association for the Advancement of Science in Cleveland on September 12, 1944. William L. Laurence quoted Cohn in the lead sentence of his article in the *New York Times*, saying: "This war, which has caused the flowing of more rivers of blood than ever before, has also led to the greatest progress in our knowledge of this vast ocean of life, its multiplicity of constituents and the mechanism of its many systems."[41]

In the *New York Herald Tribune*, Cohn was quoted as declaring, "There will be a continuance of the collection of blood and the making available of various derivatives that have proved of definite clinical value . . . but whether it will be done as part of a public health program, by individual hospitals or by a commercial enterprise remains to be seen."[42] The Harvard stories, propelled by national news services, rippled across the country in successive weeks: "New 'Foam' Found in Blood Checks Bleeding in Surgery;" "New Miracles from Blood Plasma;" "Blood Plasma Fights Contagions;" "Brain Surgery Aided;" "The Story Behind Plasma Miracles;" "Army Conquers Measles Epidemics."

On June 5, the day before D-day, Edwin Cohn received an honorary Doctorate of Science from President Stanley King at the annual Commencement Ceremonies of Amherst College. The Cohn citation read:

> For three years a student at Amherst in the Class of 1914; for twenty-four years in charge of the laboratory of physical chemistry at the Harvard Medical School: Your abstruse and original research in blood plasma has produced results of extraordinary value in medicine and surgery, saving countless lives in every war theater and on the civilian front.

The issue of *Time Magazine* published that same day carried an article in its Medicine Department under the head: "Blood v. Measles." A picture of "Harvard's Cohn" bore the legend "He searches the devious ways of blood." According to *Time*, Cohn "refuses to talk about himself, talks about his laboratory only in scientific journals in abstruse articles on the devious ways of blood."[43] In December 1945, The American Pharmaceutical Manufacturers Association honored the National Research Council for its role in making available to Government Agencies the knowledge and advice of civilian experts in the medical sciences. Speaking at the award ceremony in New York, Cohn summarized the work on the plasma proteins with particular emphasis on its therapeutic value.

> Recognition of the role of the National Research Council in making available to Government Agencies the knowledge and advice of civilian experts in the medical sciences appears to be peculiarly appropriate at this time. These experts have met repeatedly with the designated representatives of the Surgeons General of the Army, the Navy, and the Public Health Service . . . inviting other experts to meet with them as the occasion demanded. Unburdened by an administrative hierarchy or intermediaries, they have transmitted their advice directly and supplemented and implemented their recom-

mendations with reports and memoranda on the value and safety of products and of chemical or biological processes.

Starting with the assumption, which must for the present remain an assumption, that every part of human blood performs an important natural function, Cohn told that "we must continue, as we have begun, to make available as many as possible of its diverse cellular, protein and lipid components, separated and concentrated as specific therapeutic agents, of value in different conditions, in the interests of the most effective and economical use by society of the blood which it contributes." [44]

12 Victory in Europe and the Far East

The winds of change began to blow. As Samuel Eliot Morison recounted the story, the Allied Armies were poised along the borders of Germany ready to plunge into its heart. But first there came a dramatic German counteroffensive that encircled a large Allied force in Bastogne. After a fierce battle, the original battle lines were reestablished and the final Allied campaign was launched into Germany from the west while the Russians launched a major offensive from the east.[1] While the climactic events of the fighting in Europe and the Pacific still lay ahead, there was a perceptible shift in priorities within the research and development phases of the war effort. Attention was increasingly focused on those projects which could still influence the outcomes of the war, particular those which could save the lives of the wounded.

On a visit to the Mediterranean Theater late in 1944, Col. Kendrick found that the European Theater of Operations had an estimated six months supply of dried plasma and that airlifted whole blood was being used at about the same rate as dried plasma in the forward areas. Moreover, there was a strong trend toward increasing use of blood.[2] On his return to Washington, Kendrick found that plans had already been made to cut back the production of dried plasma. Indeed, the Army began to receive requests from civilian sources for unused stores of plasma and its byproducts from community hospitals in the U.S., although none were granted. Production of human serum albumin continued with the product being shipped to the Navy for use in the Pacific.

At a National Research Council Conference on Blood Preservation in January 1945, the Subcommittee considered further technical improvements in the airlift program. A brief burst of interest in the use of corn syrup in place of ACD faded quickly following a thorough test by Ross. Similarly, evaluation by Gibson of the use of human serum albumin as a red blood cell preservative revealed no advantage on red cell survival. A brief interest in the possibility of salvaging red blood cells that were being discarded in the dried plasma program subsided rapidly when it was pointed out that such a proposal would require the revision of the entire Red Cross blood collection program. At the same meeting, the Army indicated that it was planning to provide refrigeration for the blood being airlifted to Europe. Over 101,000 units of blood had been airlifted by then, all reportedly in good condition. Attention turned briefly to the incidence of reactions following the transfusion of Type O blood to recipients of other blood types. A conference considered the addition of Group A and B blood group substances to blood being transfused in these instances, but it was decided that more laboratory work was needed before a decision could be reached on that question.

Brigadier General F.N. Rankin from the Army Surgeon General's Office attended another Conference on Blood in February 1944 to assure the Subcommittee that it was the desire of his office to deliver blood to overseas services in good condition, affirming his conviction that the use of blood, as then being delivered, was giving good results judged by the successful resuscitation of injured men. When Wearn asked if any tests other than spontaneous hemolysis in the red cell supernatant fluid had been used, General Rankin responded that the test of whether the blood resuscitated the patient was the chief test. In that same meeting, Col. B.N. Carter reported that Col. Churchill had sent a cable requesting that ACD blood be exclusively provided for blood being sent to the Italian theater.

On March 24, General Hawley cabled the Surgeon General with the urgent request that, effective immediately, all blood being shipped to the European Theater be collected in ACD, and that refrigeration be provided. Pointing out that Alsever solution had been satisfactory in respect to cell survival in recipients, he averred that the pressing need for refrigeration made it advisable to change to ACD, thus taking advantage of the smaller total volume of ACD blood in comparison to Alsever blood, and substantially reducing the total weight per unit of airlifted blood. Hawley added that he was willing to accept the NRC pronouncement that ACD blood was at least as satisfactory as Alsever blood.[3]

Renewal of the Rockefeller Foundation Grant

Since the 1937 Rockefeller grant was due to expire early in 1945, Frank Blair Hanson at the Rockefeller Foundation wrote to Dean Burwell early in 1945 suggesting that he and Cohn outline their hopes and plans for the continued operation of the Department of Physical Chemistry. In a response prepared in late February, Dean Burwell wrote that it may be said that a successful transition to effective work on war problems was possible because in 1940, with aid of the support provided by the Rockefeller Foundation, the laboratory and its staff was organized and integrated as a unit whose extremely able young men became the heads of the various subdivisions of the wartime laboratory. Dr. Cohn thus had the personnel and the specialized equipment to operate as an effective unit, and the responsibilities that have been assumed came to rest upon these men who were able to apply the principles and methods developed to new urgent and vital problems. When this highly successful application to the immediate urgencies of war can be discontinued, Burwell told Hansen, the need will be to go back again to fundamental investigation. The transition from applied to fundamental research should be at whatever rate is consistent with the Department's obligations, but the transition should begin immediately. He expressed the hope that Harvard could continue to count on financial support from the Rockefeller Foundation so long as Dr. Cohn was active.[4]

The Dean estimated that the Department of Physical Chemistry would require annual funding at the level of $70,000. Of this, Harvard expected to contribute $22,000, leaving $48,000 still to be found. While he avoided making a specific request of the Foundation, the implication was clear: he hoped that the Foundation would consider funding annual support amounting to $48,000 each year. At the time, Cohn's total annual support from Harvard, the Rockefeller Foundation and the Office of Scientific Research and Development amounted to more than $140,000 annually. This new proposal thus projected a drastic retraction in total support of Cohn's scientific activities. If activated, it meant the exclusion of all but fundamental studies from the work of the Department of Physical Chemistry.

Seven years later, Cohn characterized this retraction in planned support as a fundamental error on his part, attributing it to preoccupation with the development and control of methods of plasma fractionation. Since he rarely admitted making a mistake, this was probably his way of acknowledging that he had been overloaded and exhausted from the tremendous burden he had carried since 1940. Whatever the case, it was his view that this error in judgment was a source of embarrassment to Harvard as well as to himself, since the space and funding for the work of his department subsequently proved to be less than his vision of what he needed.

In March, Hanson and George W. Gray of the Rockefeller Foundation paid a three-day visit to Boston to discuss with Cohn and the medical school authorities Cohn's plan for pure research after the war. At this time, Cohn outlined his ideas for keeping the principal people of his group together until the flow of graduate students through his department could produce another group of young well-trained men, whereupon he would be glad to see the present older group leave, one by one, as they are called to other institutions. They would be replaced by the best of the first crop of graduate students. In due course, Cohn hoped that several new crops of trained men would emerge during the fourteen or fifteen years before he retired.[5] The visit culminated in a private meeting of the visitors with Dean Burwell during which possible mechanisms for meeting Cohn's needs for the remainder of his career were discussed. Following the visit, the Dean took the matter up with President Conant and forwarded a letter to the Foundation requesting a lump sum grant to support Cohn for the remainder of his working career.

In late May, President Conant was notified that the Rockefeller Foundation had appropriated $350,000 to Harvard as an outright grant for research under the direction of Edwin J. Cohn "with the understanding that Harvard is assuming and will assume responsibility for the additional funds needed for Cohn's work."[6] The new grant was paid in full on September 1, 1945. With this generous action, the Rockefeller Foundation recognized Cohn's scientific stature and underwrote his future scientific endeavors to a substantial degree. At the same time, the receipt of this award tacitly foreclosed any possibility that Harvard could seek additional funds from the Foundation to support Cohn's studies.

Post Transfusion Hepatitis

For many years, reports appeared in the medical literature describing outbreaks of jaundice, a yellowish pigmentation of the skin symptomatic of liver disease, following the use of convalescent serum or plasma to treat infectious disease. In the 1930s, similar episodes were reported in England and Brazil, in each case associated with the inoculation of convalescent human plasma or serum. In late 1941, Beeson, Chesney and McFarlan, in a report from the American Red Cross-Harvard Field Hospital Unit in England, had reported an outbreak of hepatitis in a British training regiment being treated with convalescent mumps plasma in which over forty-four percent of the men were affected.[7] A year later, the National Research Council Subcommittee on Blood Substitutes had received a cable from the British advising of an outbreak of jaundice among subjects transfused with human serum, with most cases developing jaundice between sixty and 120 days after inoculation. Similar cases in Russia were known to members of the Subcommittee. At its October 19, 1942 meeting, the possibility that outbreaks of jaundice could result from the widespread use of dried human plasma or of products obtained by the fractionation of plasma had been discussed by the Subcommittee. It was noted that the then current practice in producing dried plasma involved the pooling of plasma from as many as twenty-five donors before drying, the implication being that a single infected unit of donated blood could contaminate all the plasma in the pool under those circumstances.

In 1943, Beeson, then at Emory University in Atlanta, reported seven cases of jaundice occurring one to four months after the transfusion of blood or plasma. According to Beeson, "the explanation which has generally been favored is that the jaundice is caused by a virus which happened to be present in the body fluids of the donors and which, after a long incubation period, produced a hepatitis in the recipient."[8] In response, the Subcommittee suggested that Red Cross reject all blood donors who had a history of jaundice within the previous six months. In view of the possibility that the products produced from large pools of plasma by the Cohn alcohol-water process might lead to post transfusion hepatitis in recipients, Cohn was asked to investigate whether the alcohol and low temperature environment used to prepare albumin from human plasma might have an effect on viruses. However, it was not until well after the end of the war that the causative relationship between post-transfusion hepatitis and infusions of pooled dried plasma began to be widely realized. [9]

Post transfusion hepatitis, variously called serum hepatitis, homologous serum jaundice and, more recently, hepatitis B (HB), went on to become the major infectious agent in the national blood resource for the next thirty-five years, being outranked in order of public health importance only by HIV, the retrovirus which causes AIDS, in the 1980s. Several factors contributed to this situation. Since the onset of the clinical symptoms of hepatitis B in infected individuals occurred so long after transfusion of the infectious agent, it was very difficult to associate the disease with the event so long before, especially in time of war when military personnel were dispersed around the world. Transfusion of a single unit of blood or plasma drawn from a single individual transmitted the virus only if the donor had carried the virus. However, as noted, the likelihood of viral transmission increased dramatically if the plasma of many donors was pooled, as in the preparation of dried plasma. In army hospitals in the U.S., the incidence of post-transfusion hepatitis was reportedly relatively low, perhaps because the dried plasma that was used domestically came from pools of fewer than eight donors. In contrast, the dried plasma used in the North African theater and in the European theater had been prepared from pools of fifty or more individuals.

Infectious Hepatitis

Concerns about the risk of transfusion transmitted jaundice took another turn in 1945 with the realization that there was a second type of hepatitis, called infectious hepatitis or hepatitis A (HA). Under a Florence, Italy dateline in 1945, *The New York Times* reported "Army Discovers Jaundice Cause . . . Jaundice, it may now be said, is a filth disease transferred from one person to another through the agency of flies, polluted water, and other means in much the same way as dysentery."[10] At a NRC Conference on Plasma Fractionation in March 1945, Stokes reported the results of a study of the effectiveness of gamma globulin in preventing infectious hepatitis. He prefaced his remarks by stating that "we are dealing with a material, the antibody content of which is unknown against the agent of hepatitis. We are dealing with a disease caused by an unidentified virus without a method for determining the susceptibility of exposed individuals." For these reasons, he found it necessary to conduct empirical studies with gamma globulin on large groups of subjects in order to obtain information of value.[11]

In a preliminary note published in the *Journal of the American Medical Association* in 1945, Grossman, Steward and Stokes reported that immune serum globulin, or gamma globulin, offered some protection in an outbreak of epidemic hepatitis in a summer camp.[12] On learning of that work, Colonel S. Bayne-Jones had sent Stokes overseas to test gamma globulin in an epidemic of hepatitis then being experienced by the 15th Air Force in Italy. Stokes began his study by conducting an initial test in a small Bombardment Group. Globulin was administered to three groups, each with a thousand men, while two other groups, the controls, were not injected. Among the treated groups, only one case developed hepatitis while there were twenty-one cases of hepatitis in the two control groups. Encouraged by these results, a larger study was conducted on a cohort of 12,000 men in the 5th Army and similar results were obtained. The incidence of jaundice among soldiers receiving gamma globulin was significantly lower than it was in those who did not receive gamma globulin. Stokes concluded that "the results obtained by the use of human immune serum globulin both in the 15th Air Force and in the 5th Army suggest that this material was a potent agent for the protection and attenuation of infectious hepatitis of the type occurring in an epidemic wave in the fall and early winter throughout MTOUSA."[13]

This new study provided an important answer to another question: could gamma globulin itself transmit hepatitis to recipients? The use of large pools of plasma to isolate gamma globulin had been a source of concern to Cohn, since it could be assumed that the plasma from only a few infected donors would contaminate the pools. Indeed, he had initiated a search for a stabilizing agent that would enable a heat treatment for gamma globulin analogous to that being used for human serum albumin. However, efforts in that direction were not successful and had been abandoned. The results reported by Stokes—only one case of hepatitis in 12,000 men who received gamma globulin—provided a clear answer in the negative. Confirmatory evidence was later reported by Janeway who found only one case of hepatitis among nearly 1,200 children who had received gamma globulin and had been followed for more than three months.

Late in 1944, Cohn was the guest speaker at the presentation ceremony in New York by the American Pharmaceutical Manufacturers Association of an award recognizing the important role of the National Research Council in making available to government agencies the knowledge and advice of civilian experts in the medical sciences. In his talk, Cohn predicted that "the control of infectious diseases by passive immunization with Gamma Globulins may well be the largest need of a civilian population for a blood derivative and one to which a civilian population can be expected to contribute in the interests of the modification and control of a children's disease such as measles until such time as the immunity of each growing generation is achieved."[14]

There was considerable interest in administering gamma globulin by the intravenous route, primarily because this could bring about a higher circulating level of antibodies in recipients than levels following intramuscular use. Unfortunately, attempts to administer gamma globulin intravenously had met with a high incidence of rather serious reactions. In one study, out of nineteen injections of nine different lots of the product, there had been nine severe reactions and two moderate reactions, a totally unacceptable level of reactions. At the time, the labels for immune serum globulin, the official name, carried the statement: "Not for intravenous use." (A method for treating gamma globulin to make it safe for intravenous use was developed several years later.)

Further Improvements in Subfractionation of Fraction II+III

In March 1945, Cohn reported at a NRC Conference in Washington substantial progress by Oncley and his associates in improving the subfractionation of Fraction II+III.[15] With a newly modified procedure called Method 8, the isoagglutinins were obtained in substantially increased yields and greater purity. Moreover, the use of Method 8 increased the yield of gamma globulin antibodies from 50% with Method 7 to 80% with Method 8. Given the rapidly mounting importance of the antibodies in treating measles, and its possible effectiveness against infectious hepatitis, this represented a substantial improvement. However, it would be necessary to test this new product during a measles epidemic so as to ascertain that the increased yield represented additional measles antibody.

Fibrin Foam and Fibrin Film

Fibrin foam was being produced in large quantities by the spring of 1945. Its use in general surgery was well established, and its use was being extended to dental and urological surgery. Total production of fibrin foam for use by the armed services exceeded 100,000 packages.[16] Production of Fibrin Film for general use had been delayed by technical difficulties in large scale sterilization of the finished product in the pharmaceutical firms. A solution to the problem had been found late in 1944 by the group at Armour Laboratories.[17] Fibrin Film was officially accepted as a dural substitute by the NRC Subcommittee on Neurosurgery on May 17, 1945.

Inactivation of Possible Viruses in Human Serum Albumin

The formulation of the human serum albumin that was first used at Pearl Harbor in 1941 consisted simply of a 25% solution of the protein in dilute sodium chloride. In those early days, the possibility of dispensing human albumin as a stable concentrated solution was particularly attractive to the Navy, constituting a definite advantage over the use of dried plasma. Cohn gave frequent attention to increasing the stability of albumin solutions. Indeed, he frequently referred to the worst case specification that albumin withstand transport in a "tank in Tobruk." Not surprisingly, a frequently employed test of albumin in the laboratory simulated that condition. Vials of sterile albumin solution were submerged for prolonged periods of time in constant temperature hot water baths. Initially, the vials were viewed visually for turbidity, but as the stability of albumin solutions improved in succeeding months, a sophisticated optical instrument known as a nephelometer was used. With foresight, Cohn had the good fortune to enlist the collaboration of J. Murray Luck in improving the stability of albumin solutions. Using this approach, the first substantial improvement had been introduced in 1943 simply by doubling the sodium chloride concentration in the 25% albumin solution. In 1944, Luck discovered that sodium acetate was a better stabilizer than sodium chloride.

This set off a concatenation of gains in albumin stabilization which Cohn later recounted in the following lengthy footnote in a chapter on the History of Plasma Fractionation.

> Dr. Luck (Leland Stanford University) found that acetate gave a greater stability than the same amount of sodium chloride; Dr. Edsall (Harvard University) suggested trying propionic acid; Dr. Luck found that this was more efficient, and that longer-chain aliphatic acids, from butyrate to caprylate, and some aromatic acids were very much more effective, and recommended phenylacetate or mandelate; Dr. Clarke (Columbia University) suggested the sodium salt of acetylphenylalanine as more closely related to the natural constituents of the plasma; Dr. Strong (Harvard University) suggested the salts of the acetyl derivatives of the amino acids in which albumin is deficient for nutrition (tryptophane and isoleucine).[18]

In this way, sodium acetyltryptophanate became the first stabilizing agent for Human Albumin. Having come that far, Scatchard predicted that it should be "possible to determine conditions such that the albumin will remain stable while bacteria and viruses will be destroyed." The first step in achieving that goal, actually the pasteurization of human albumin, was accomplished when Scatchard, Strong, Hughes, Ashworth and Sparrow showed that "albumin stabilized with 0.04 molal acetyltryptophanate may be heated for ten hours at 60°C. and remain more stable than the present albumin," a finding that was published in 1945.[19]

With such progress in stabilizing albumin, it was inevitable that the question would be asked, how much heat would it take to inactivate the virus responsible for post-transfusion hepatitis? The answer to that question was provided in 1947 by Gellis, Neefe, Stokes, Strong, Janeway and Scatchard.[20] Using an albumin solution known to contain the virus, two different heating regimens were tested in human volunteers: in the first, a sample of the solution was heated for ten hours at 60°C; in the second, a sample was heated for ten hours at 64°C. The virus was completely inactivated under both conditions. Thereafter, every vial of albumin produced in the United States was subjected to the 60°C treatment for ten hours, which became the standard for inactivating the virus in albumin. It resulted in the remarkable record of safety of human serum albumin to the present day. In the intervening years there has never been a need to use the more rigorous treatment at 64°C. As a result, human serum albumin is probably the safest biologic ever produced. Years later, it was found that the AIDS virus is inactivated under these conditions. The investigations of Luck, Scatchard and the others thus made possible something like a quantum leap ahead in stabilizing albumin. Since 60°C is equivalent to 140°F, 10 hours at 60°C is indeed vigorous treatment. This illustrates Cohn's attention to detail and his openness to the suggestions made by individual collaborators. It is also an example of how important decisions were arrived at during the wartime program.

A further step in the modification of human serum albumin solutions for clinical use was introduced after sodium acetyltryptophanate had cleared most of the hurdles for acceptance. Clinical studies at the Peter Bent Brigham Hospital in 1945 by Janeway on the use of albumin in patients with hypoproteinemia, and on the use of albumin in patients with kidney disease, led to a desire to reduce the amount of salt and mercury in albumin, since both salt and mercury were contraindicated for the treatment of kidney disease. Thus a requirement insisted on by the U.S. Public Health Service that merthiolate be added to the standard Army-Navy albumin solution was finally expunged. Cohn had never forgotten that Soma Weiss' recommendation against adding merthiolate to albumin had been overruled by Veldee of the Public Health Service. This time, the Service went along with the Subcommittee's recommendation. The new albumin solution, which temporarily bore the name "normal human serum albumin, salt-poor," replaced the original high-salt formulation at the end of the war.[21] This formulation is used for all albumin produced today.

Antihemophilic Globulin

Hemophilia is an uncommon hereditary hemorrhagic condition of variable severity, seen only in males. It is marked by impaired clotting that can result in life threatening hemorrhages. In Boston, many of the hemophiliacs relied upon George Minot's service at the Boston City Hospital for their care during hemorrhagic crises. It was therefore quite natural that samples of the Cohn plasma fractions had been made available to Minot's associates for testing. Their tests revealed that intravenous infusion of Fraction I reduced the clotting time of hemophilic blood in the laboratory. By 1945, injection of solutions of Fraction I in hemophiliacs reduced the clotting time to normal in most individuals, making it possible in several instances to extract teeth without untoward loss of blood.[22] Despite these initial favorable results, it was several years before a satisfactory product could be produced and made available generally.

Victory in Europe and Japan

Allied forces crossed the Rhine into Germany on March 7, 1945 by dramatically capturing the bridge at Remagen just as the Germans were about to blow it up. A few days later, Patton began crossing the Rhine at Oppenheim. As German resistance crumbled and victory appeared certain, the Allied world was plunged into mourning by the news that President Roosevelt had died. After the unconditional surrender of Germany was signed by German General Jodl on May 7, 1945,[23] the airlift of blood to ETOUSA was discontinued. By then, more than 181,000 units of blood had been airlifted from the USA. Fighting in the Pacific Theater continued unabated following VE day. The first atomic bomb was dropped on Hiroshima on August 6, 1945. Three days later, the second bomb was dropped on Nagasaki. Emperor Hirohito surrendered unconditionally. The surrender documents were signed on the deck of the battleship Missouri by General MacArthur and Japanese General Umezu on the second day of September 1945. [24]

The Subcommittee on Blood Substitutes met for its last official meeting on May 18,1945, less than two weeks after the unconditional surrender of Germany.[25] Most of the National Research Council wartime committees and boards were discharged at that time. In October 1945, citing the responsibility of the Red Cross to its volunteer blood donors, Maj. Gen. Norman T. Kirk, Surgeon General of the Army, ordered the return of all excess Army dried plasma to the Red Cross. The actual transfer took place over a period of many months, with nearly 1,000,000 "large" and over 500,000 "small" packages of dried plasma returned.[26]

In December 1945, the Red Cross convened a new Committee on Blood and Blood Derivatives with Charles Janeway in the chair. The other members of the committee were Edwin Cohn, Alfred Blalock and Robert Loeb. After discussing the difficulties being experienced by some Red Cross chapters, as well as by state and community organizations in establishing civilian blood programs, the committee recommended that the Red Cross actively encourage the development of civilian blood and blood derivatives programs on a national basis through its chapters. Where financial assistance was needed, the committee recommended that the Red Cross authorize the use of national funds to assist chapters to establish civilian blood programs, including the training of personnel, and that chapters should be authorized to assist in financing the processing of blood and blood derivatives in civilian programs in which Red Cross chapters participate. This committee was also charged with developing plans and policies for dealing with the surplus dried plasma being turned over to the Red Cross by the Armed Services. A plan for "American Red Cross Distribution of Surplus Dried Blood Plasma" was published in the *Journal of the American Medical Association*, with the note that its use had been presented to and concurred in by the Association of State and Territorial Health Officers, the American Medical Association and the American Hospital Association.[27]

More Honors

Edwin Cohn was the recipient of numerous honors and received extensive mention for his work in the national press. In April 1945, he was the William Henry Welch Lecturer at the Mt. Sinai Hospital of New York. Established in honor of the great Johns Hopkins pathologist, Cohn thus followed in the footsteps of Simon Flexner, Theobald Smith, Sir Henry Dale and Harvey Cushing as Welch Lecturers. Cohn gave two lectures—on the "Separation, Concentration and Characterization of Blood Derivatives," and the "Natural Functions and Clinical Uses of Blood Derivatives." The lectures were black tie affairs at 8:30 p.m., the customary hour for formal functions in the New York medical community.

Cohn also received the first Passano Award from the Passano Foundation of Baltimore "in recognition of his fundamental and fruitful investigations of blood derivatives and of their application to the preservation of life and the alleviation of suffering."[28] The presentation was made on May 16, 1945 at the Medical and Chirurgical Faculty of the State of Maryland in Baltimore. Among the attendees were Morris Fishbein, the editor of the *Journal of the American Medical Association*, Major General W.S. Rankin, who was a member of the award jury, Alan M. Cheney, Dean of the Johns Hopkins Medical School, and William Mansfield Clark, Professor of Biochemistry at Hopkins. Introduced by Brigadier General Stanhope Bayne-Jones, Cohn talked on "The Chemical Separation and the Clinical Appraisal of the Components of the Blood." Following the lecture, he was presented with a $5,000 check by Edward B. Passano, chairman of the board of the Williams and Wilkins Co., medical publishers in Baltimore. *Newsweek Magazine* framed its story of the award by the words of Mephistopheles in asking Faust to pledge his soul in blood: "Blut is ein ganz besonderer Saft."[29]

In the following June, Cohn was honored by Columbia University when President Nicholas Murray Butler awarded him an honorary Doctor of Science degree for his "productive research in the physiology and chemical composition of the blood, the results of which have not only greatly increased fundamental scientific knowledge, but have saved many thousands of lives among the wounded in the war." A few days later, he sat with a distinguished company of colleagues and graduates assembled in Harvard's Sever Quadrangle where he received another Doctor of Science degree, this time from James Bryant Conant. The citation credited Cohn as "Resourceful and energetic driver of a scientific team; his years of fundamental study of the proteins now enable him to draw new materials from blood to aid the sick and the wounded." Among other honorees that day at Harvard were the physicists Frederick V. Hunt and Frederick E. Terman, Alexander Fleming, the discoverer of penicillin, Admiral Ernest J. King and A. Baird Hastings. Hastings was cited as "one of the Harvard commuters to Washington where he guides wartime medical research."

During the summer of 1945, Cohn was invited by the New York Academy of Medicine to participate in the Academy's eleventh series of *Lectures to the Laity*, but he declined the invitation with the comment that he "was not a prophet." Shortly thereafter, however, Alan Gregg, of the Rockefeller Foundation, confessed to Cohn that he was "somewhat responsible, for I urged that they invite a protagonist rather than a spectator to talk on medical research. Furthermore, I can see advantages in creating at the Academy the kind of lectures which students and interns from various New York schools and hospitals can profitably attend. There would be an economy of effort in reaching a larger audience and it would tend in the direction of forming habits of attending solid lectures for young New York physicians as they progress from school to practice."[30] This letter placed the invitation from the New York Academy of Medicine in a completely different light. Cohn promptly accepted with the understanding that he would eschew prophecy and attempt an historical analysis. Projecting himself as "an amateur in the history of medical science," Cohn began his Laity Lecture with a discourse on the forces that shaped the course of research in the medical sciences since the Renaissance. He concluded by stating that "The new tools would appear to be on hand for the isolation as chemical substances of natural products such as the vitamins, hormones, enzymes, and the proteins, which are the structural elements of the tissues, and to yield fundamental knowledge regarding their functions and interactions and thus of the mechanism which control bodily processes."[31]

The editor of the British journal *The Lancet* forwarded personal copies of two editorials to Cohn. The first, published in March 1945, bore the title: "Globulin in the Control of Measles." Its lead sentence read: The physicochemical researches on human plasma carried out by Professor Cohn and his colleagues at Harvard have culminated in discoveries of immediate practical application in medicine, and one of the protein fractions he has isolated in almost pure form—gamma globulin—is likely to prove of great value in the control of measles.[32] In the second editorial, entitled "Therapeutic Value of Blood Proteins," the editor wrote:

> One may assume that every part of the blood performs an important natural function. Clearly, therefore, it is an advantage to make available as many as possible of its diverse cellular, protein and lipoid components in separate and concentrated form, for selective use in appropriate conditions. This concept suggests a brave new world of transfusion which has been brilliantly initiated and exploited by the Harvard Medical School under the leadership of Edwin J. Cohn, who, with the support of the National Research Council, has had the integrated help of workers skilled in every aspect of the subject.[33]

In October 1945, A. N. Richards, the Chairman of the OSRD Committee on Medical Research, advised Cohn that he and his colleagues at OSRD held him and his associates in their highest esteem for the quality and spirit of their services to CMR, and that, in recognition of their confidence, the Harvard contract (OEMcmr-139) was to be continued until such time as the opportunity "shall have arisen for consideration of its long-time significance by a national scientific research foundation which Congress is expected to create." In response, Cohn complimented OSRD for demonstrating that scientific knowledge could be applied by sufficiently wise and well-trained scientists in the national interest in time of stress, afterwards telling that

> I am by no means certain . . . whether the national scientific research foundation which Congress is expected to create will in fact permit government participation without incurring government domination of scientific work. In-sofar as a new foundation may make possible a more intense scientific development at the level of untrammeled, long-range scientific research, not only at the applied but especially at the creative and imaginative level, we will be honored to

> accept its support for some of the work which we believe, as a result of observations made in the course of our wartime experience, must be carried forward into the peace. This work will, of course, involve some colleagues in the natural and some in the medical sciences, but with even more emphasis on acquiring the fundamental knowledge, which must now be gained in the natural sciences if we are to make more far-reaching contributions to medicine and the public health. Until such time, therefore, as the pattern of organization and principles of administration of the new foundation have been further developed, we shall be happy to continue under Committee on Medical Research sponsorship, as suggested in your letter.[34]

Somewhat later, Richards telegraphed Cohn soliciting his views regarding the administrative objectives and organizational pattern of government support, and on the interrelations between the natural and the medical sciences. Cohn responded at length, warning that the direction that the national economy was taking was making it necessary that there be more support for scientific research than was available in the pre-war period. The objectives of the research carried on in government laboratories in the pre-war period were, he believed, for the most part, limited, for inevitably the quality varied considerably with the director. However, the organizational pattern and the atmosphere of such laboratories could at their best be compared with industrial laboratories rather than with institutes created and maintained for the exploration of the phenomenal world. He pointed out that government support of science during the war was aimed at specific objectives with success of many of the undertakings depending on the fact that both the members of the committees and the responsible investigators were scientists with a long tradition of the freedom necessary to ensure the exploitation of even a tactical advance, and who were able to communicate with each other in a common language. When the effort is made to increase efficiency or administrative responsibility by interposing scientific administrative officers who are peers neither of the committee members nor the investigators, he believed that the system degenerates. "Creative scientific research depends upon the imagination and technical resourcefulness of the investigator. He can be chosen by a committee competent to judge of his quality. He cannot be directed."

Recognition of the implication of observations was the essential element in the creation of a scientific structure, he believed. By supplying the funds to make administrators or applied scientists of creative investigators, the government could stultify and impede the national economy. A national foundation which derived its funds from Congress should in the long run be responsible to it. The Congress has no experience, however, and cannot be expected to acquire judgment in the wise expenditure of funds for scientific research without decades of experience. The inevitable tendency would be the selection of problems rather than of men, a tendency which leads, as has already been seen in many private institutions, to a pattern which represents the past but not the future. "The basis of the stability of our government is the balance between the Congress and the Executive." An administration which was not responsible to a representative government unquestionably can operate more expeditiously and with greater independence for a limited period of time. This has been demonstrated by the independent agencies that have functioned during the war. Freeing the administrators in independent agencies from oversight by a body competent to review their decisions should be terminated now that the emergency is over. The Congress should determine how scientific research should be supported in the postwar world.

Cohn observed that Congress created the National Academy of Science during President Lincoln's administration to advise the government in precisely this area. If the National Academy of Sciences has become unpopular in certain circles in Washington, this depended in part upon the functions that it has thus far had to perform, but in part also upon its having maintained a tradition of freedom from political pressure. He concluded by saying, "It is my hope that if a national science foundation is created by the Congress, nomination of its officers to the President and review of policy may be vested in the National Academy of Sciences as the body historically charged with the responsibility of advising the government on scientific matters."[35]

In a 1945 letter to Paul Buck, the Dean of the Harvard Faculty of Arts and Sciences, Cohn, called attention to the intermediate positions of the basic medical science departments at Harvard—Anatomy, Physiology, Pathology, Biochemistry, Bacteriology and Pharmacology—with respect to the departments in Cambridge and the clinical departments in the teaching hospitals. For these departments, he stressed the importance of maintaining balance between the training of graduate students, on the one hand, and of medical students, on the other. It was his observation that the forces relating the preclinical and clinical departments in the Harvard Medical School had grown stronger in twenty-five years. He attributed the growth of these departments, in part at least, to the trend toward appointment of full time faculty in the clinical departments, many of whom had greater experience in a basic science. In his opinion, "this increased intellectual intimacy has become apparent at the level of medical research where extremely close collaboration between clinical and preclinical departments has established a pattern for the attack upon disease, which has been amply tested and found effective during this war." At the same time, he advocated revision of the curricula for training of graduate students, pointing out how in many of the wartime advances, the physicist, physical chemist and organic chemist worked in close collaboration, without intermediaries, not only with the bacteriologist, pharmacologist, pathologist, anatomist, but also with clinicians and surgeons. Pointing out that the barriers between the disciplines had all but vanished, he stressed the importance of incorporating these experiences into professional educational programs.

In October 1945, anticipating the return to more normal academic pursuits following the war, Cohn recommended that Buck appoint a committee representing the Division of Medical Sciences and the closely related departments in Cambridge to work toward revisions of the curriculum for graduate instruction in the medical sciences in order "to take advantage of our wartime experience and undertake the training of graduate students of the highest competence." Pointing out that some of the wartime advances by physicists, physical chemists and organic chemists came from working in close collaboration without intermediaries, he noted that the barriers which might have separated scientists in individual fields, whether at the level of munitions or of medicine, had all but vanished. In his opinion it was necessary to train men who have been disciplined in the bodies of knowledge which proved so important for research in the medical science by incorporating this experience into the educational programs for future graduate students.[36]

Between 1940 and 1945, OSRD funds supporting the work of the Department of Physical Chemistry, excluding contracts to other universities for collaborations with the Harvard contracts, totaled $750,950. The other major sources of support were $100,300 from Harvard University and $82,500 from the Rockefeller Foundation. In December 1945, the U.S. Patent Office awarded Patent Number 2,930,074 to Cohn for his invention relating to the fractionation of proteins. Technically, this new patent was a "continuation in part" of the application that had been filed in December 1940 following the development of the first alcohol-water fractionation procedure.

The Red Cross blood donor service ceased operating on September 15, 1945 after collecting more than 13,000,000 units of blood, of which over 10,000,000 units were processed to dried plasma, and over 2,000,000 units were fractionated to yield human serum albumin, immune serum globulin, fibrin film and fibrin foam. Plans for a greatly expanded supply of blood for use in an invasion of Japan were abandoned. The U.S. Navy had airlifted a total 181,000 units of whole blood from Oakland California to Guam and Leyte in the Pacific Theater. The flying time was about five days, including stopovers for reicing of the containers. Once the airlift to supply blood to the Pacific was instituted, an adequate supply of blood was available at all times. With a dating period of twenty days from the time of collection, this resulted in approximately two weeks for use before becoming outdated in the field. The Army, in contrast to the Navy, used very little serum albumin because of its satisfaction with plasma. At times, however, the compact size of the serum albumin package was a distinct advantage. One medical officer, for instance, related how he and some of his corpsmen, after they lost all their plasma when their landing boat was sunk off the Normandy beaches, filled their pockets with packages of serum albumin and administered it to many seriously wounded men, most of whom lived to be taken aboard ships on which they could receive definitive medical care.[37]

13 A Peaceful Interlude

Cohn emerged from the war with greatly enhanced scientific stature and enormous power. In less than six years he had solved an abstract problem of no immediate practical value and had applied it to an endeavor of urgent national importance in a time of great crisis. In a quarter of a century at Harvard, he had progressed from an entry-level post in an obscure field of science to a position of international stature. He was held in great respect at high levels of governments. Although his achievements turned out to have great importance to the war effort, "they represented nothing more than a step forward in a quest that was at once inexorable and only dimly charted" at the personal level.[1] Among his peers, he was treated with respect, although for some, the awareness of his power could have a chilling effect. Others, familiar with his violent temper, even hesitated to engage him in the ordinary give and take of academic life. At the same time, he was inherently modest. While communicating a sense of authority, his manner of discourse tended to be serious. There were very few light moments. He shunned use of the first person singular.

According to Charles Janeway, Edwin Cohn believed in living life to the full. Knowing that he had hypertension, he preferred to continue his normal life, confident that when he worked hard and effectively and played his extremely good game of tennis, he was more serene and relaxed than he could ever be if he devoted his energies to worrying and "taking care" of himself. He felt a creative force within himself which could not be denied and often spoke of the keen sense of exhilaration that was his during the post-war years, when, without the terrible pressure of responsibility which he felt during the war, he could see that each new idea emerging from the work of the "kids" in his laboratory was taking its place in the overall picture of protein structure which was beginning to emerge in his mind.[2]

Socially, he was full of himself. He took great pleasure in organizing dinner parties and enjoyed them to the hilt. He loved to dance. There was an enthusiasm, an exuberance that characterized everything he did, which found expression in his "con brio" vitality. With his imposing bearing and formality, he was attractive to women, even to the point of being fawned upon. Marianne did not share his exuberance. She felt increasingly uncomfortable about the effects this new power and fame were having on him. She even felt that the war had done him a disservice. He had too much power. She saw it as potentially destructive. His interests and pleasures diverged from hers. He became difficult to live with. Among her circle of close friends in Cambridge she could not help but learn about his dalliances. Their marital relationship was strained.

In 1945, following the end of hostilities in Europe, the Cohn's elder son, Ed, left Washington on an assignment with the Mission for Economic Affairs in the U.S. Embassies in London and The Hague, the objective of which was the economic reconstruction of Europe. Less than a year later, however, Ed resigned that post and returned home, in so doing destroying the basis of "what little independence from his father" he had managed to achieve. In Cotuit, Ed was seen by Howard Armstrong who had sailed over from Edgartown with his wife to spend a weekend with the Cohns. It was Stanley Cobb, the Harvard professor of Neurology, and a Cotuit neighbor, who recommended that Ed should immediately enter the Austen Riggs Center in Stockbridge for evaluation and treatment. Both Edwin and Marianne were terribly upset by this totally unexpected development. Edwin became very angry, denying the gravity of his son's illness and resisted the medical advice. Nevertheless, Ed Cohn was admitted to Stockbridge. Marianne Cohn was deeply concerned for her son. She felt a tremendous sense of guilt over his misery. She saw it as an indication that she had not adequately protected the boys against their father's dominant personality while they were growing up. Ed Cohn was discharged from Stockbridge in 1947 and entered Columbia University as a graduate student in Economics for the fall term. [3]

The end of the war signaled a series of staff changes in the Department of Physical Chemistry, of which the most important was the departure of several colleagues who had worked with Cohn since the early 1940s. Howard Armstrong, John Ferry and Larry Strong left to accept appointments in other institutions. Howard Armstrong was appointed Professor of Medicine at the University of Illinois School of Medicine in Chicago. Ferry became Professor of Chemistry at the University of Wisconsin. Strong became Associate Professor of Chemistry at Kalamazoo College, later moving to Earlham College as Professor and Chairman of the Chemistry Department. Charles Janeway was appointed Thomas Morgan Rotch Professor of Pediatrics at Harvard Medical School and Physician-in-chief at Children's Hospital. George Scatchard, still never officially appointed at Harvard, left for Berlin where he served as Chief of the Scientific Research and Control Branch of the American Military Government. His principal responsibility involved advising General Lucius Clay on how to reestablish German science. In that position, Scatchard often had to do verbal battle with the Russian members of the quadripartite committee concerning the treatment of German scientists.[4] Only John Edsall, Oncley and Hughes remained with Cohn in the Department of Physical Chemistry. During the same period, the much larger group of about twenty-five doctoral level associates, the backbone of the wartime laboratory, left Harvard and moved on to careers in other universities, government and the pharmaceutical industry. Their contributions to the wartime achievements of the laboratory were permanently recognized by their inclusion as co-authors in the published scientific scientific papers from the laboratory.

The embargo against publication of the wartime work of the laboratory was completely lifted in 1945. It had applied primarily to the "methods" papers, the purely chemical articles describing the development of the plasma fractionation and subfractionation methods, the derivation of the physical chemical equations for osmotic pressure of Scatchard, and the characterization of specific proteins. Cohn chose to publish these "methods" papers in the *Journal of the American Chemical Society* as a series under the general title, "Preparation and Properties of Serum and Plasma Proteins." The first postwar paper described the fractionation of human plasma to obtain the major fractions: I, II+III, IV-1, IV-4 and V. Subsequent papers dealt with the subfractionation of the major fractions, as well as with physical chemical studies characterizing the properties of specific proteins. In all, thirty-four papers were published in this series. Publication of the papers in the "Chemical, Clinical and Immunological Studies on the Products of Human Plasma Fractionation" series in the *Journal of Clinical Investigation*, begun in 1944, was completed with the publication of thirty-nine papers by 1949. Taken together, these two series provide a permanent scientific record of the great work of Cohn and his colleagues arising from the wartime studies.

The first methods paper was entitled "Preparation and Properties of Serum and Plasma Proteins. IV. A System for the Separation into Fractions of the Protein and Lipoprotein Components of Biological Tissues and Fluids." The authors were E. J. Cohn, L. E. Strong, W. L. Hughes, Jr., D. J. Mulford, J. N. Ashworth, M. Melin and H. L. Taylor.[5] The initial sentences of this landmark paper were phrased in broad terms.

> Natural products rarely exist in a state of maximum purity or maximum concentration. They are generally found in plant or animal tissues or fluids in the presence of other natural products, and often in small amounts, or in inactive physical states in which they are stored as biological reserves. Natural function may demand the liberation of the active component in but a small, constant concentration. The greatest value of each active component, however, whether as reagent in therapy or in chemical technology, is often in a highly purified, stable and concentrated state.

The authors then told that the present series of reports describe the development of a system for the separation of the protein and lipid components of a biological tissue, first into a small number of fractions in which the major components are separated, and then into a large number of subfractions into which they are further concentrated and purified. The tissue that had thus far been most thoroughly investigated was human blood. The methods that had been employed were, however, general and could be applied to give comparable inclusive fractionations of other biological systems in the interest of obtaining many valuable components as nearly as possible in their natural state. The separation of the many protein and lipid components of a biological fluid or tissue was accomplished by controlling their relative solubilities in a multi-variable system. The larger the number of components and the more nearly alike their physical chemical properties, the more variables, each under accurate control, were needed in order to achieve conditions in which sufficiently large differences in solubility permitted satisfactory separations. In practice, they observed, conditions must be determined so that the protein to be separated has

> a high solubility when most other components of the system have low solubilities, or a low solubility when most of the components of the system have high solubilities. Solubilities of 0.01 to 0.1 g. per liter generally suffice to separate fair yields of a component as a precipitate. Solubilities of 10 g. or more per liter are generally adequate if other components of the system are to be precipitated. Conditions such that the solubility can be varied from a hundred to a thousand fold should thus be known if satisfactory fractionation and purification are to be carried out. The larger the variety of conditions under which a given protein in a mixture can be maintained in either the higher or the lower solubility range, the greater is the chance of effecting sharp separations from other proteins whose solubilities are influenced by the same variables.

In the course of the wartime work, the initial fractionation scheme, known as Method 1, was modified five times. Method 6, the most advanced method, yielded five major fractions. Of these, Fraction I contained approximately 5% of the plasma proteins; its major component was the protein fibrinogen. The second fraction, II+III, contained 25% of the proteins in plasma, of which the major components were beta and gamma globulins. A third fraction, IV-1, included 6% of the proteins. Fraction IV-4, about 6% of the plasma proteins, was comprised mainly of alpha and beta globulins. Fraction V, which included slightly more than half the proteins in plasma, consisted of almost pure albumin.

The authorships of these papers deserve comment for they reflect Cohn's style of leadership and direction of the wartime enterprise. Inclusion as an author of a scientific paper constitutes an important credential for scientists and physicians. Customarily, the person who led the team that did the work and wrote the paper was listed as the first, or senior author. Surprisingly, Cohn's name appeared as senior author on only four of the twenty-one papers in the "Preparation and Properties" series. The first was the 1940 paper describing the fractionation of plasma with ammonium sulfate.[6] The second was the paper in which the ethanol-water system was described for the first time.[7] The third was the 1946 paper, "A System for the Separation into Fractions of the Protein and Lipoprotein Components of Biological Tissues and Fluids," cited above.[8] The fourth was a paper on the "Crystallization of Serum Albumins from Ethanol-Water Mixtures" that was published in 1947.[9] Among the thirty-nine papers in the "Chemical, Clinical and Immunological Studies" that were published in the *Journal of Clinical Investigation*, Cohn was the senior author of only one paper entitled, "The Characterization of the Protein Fractions of Human Plasma."[10]

A group of seventeen new postdoctoral fellows arrived at the Department of Physical Chemistry during the summer of 1946 for stays of one or two years. They came from Argentina, Australia, Belgium, Canada, Chile, China, England, Finland, France, Sweden, Switzerland and the United States. Four were Harvard Medical School students. Their introduction into the laboratory followed the pattern that had worked so well during the war for training visiting scientists and pharmaceutical industry executives who became the managers of the plasma fractionation plants. They began by working in the Harvard pilot plant. This offered training in the practices of the laboratory in handling proteins, including such basic things as the measurement and adjustment of the pH of protein solutions, adding ethanol and other reagents to cause the formation of precipitates, centrifugation, and the drying of products from the frozen state. By its very nature, the work in the pilot plant included periods during which the fellows could gather around a blackboard and discuss scientific matters with Cohn's colleagues. In this way, the new fellows learned about the currently active research projects of Cohn and his associates and began to consider opportunities for more intense endeavors during their stays in the laboratory.

Three factors present in the laboratory environment at Harvard

Medical School were particularly important for the growth and development of the new fellows. These were, first, the air of openness and informality among Cohn's associates who willingly answered questions and helped the newcomers learn new experimental techniques; second, the extensive sharing of common space, equipment and instruments within the laboratory; and third, the weekly luncheon meetings which provided the best opportunities to experience Cohn's style of leadership. Cohn seldom met with individuals; thus the luncheon meetings provided the main forum for scientific communication in his presence. Within two or three months of such exposure to the Department, most of the fellows had linked up with Cohn or one of his associates and were hard at work.

In the absence of the National Research Council in an active role, the new American Red Cross Committee on Blood and Blood Derivatives was soon recognized as the focal point in Washington for issues concerned with blood and related matters. The Committee strongly endorsed Red Cross plans to develop civilian blood programs, even urging that the Red Cross authorize the use of national funds to assist chapters in establishing such programs. It also warned of possible abuses of the Red Cross offerings as they became available free of charge, advising the Red Cross to "make it clear that its responsibility ceases when blood and blood derivatives are distributed to hospitals and physicians." It advised the Red Cross to comply with the minimum requirements of the U.S. Public Health Service at all times.[11]

The Red Cross administration in Washington faced a number of serious challenges. First and foremost was the sudden termination of Red Cross collections of blood for the armed forces. There was no program to take its place. It was not simply a matter of continuing and expanding the splendid Armed Services wartime blood collection program. The wartime program had focused Red Cross blood collection activities in certain urban areas of the country with favorable proximity to the pharmaceutical plants with Army and Navy contracts, leaving a void in other areas of the country where existing Red Cross chapters had never collected blood. At the same time, stimulated by the return of physicians newly discharged from the armed services who had seen how the transfusion of airlifted whole blood benefited wounded soldiers, independent blood programs were being established by community hospitals. Sensitive in its public relations, the Red Cross adopted a policy of not competing with these developing community blood banks. Instead, the Red Cross followed a more deliberative process. It set out to secure the endorsement of its blood program by national entities, including the American Public Health Association, the Catholic Hospital Association, the American Medical Association, the Veterans Administration, the Association of State and Territorial Health Officers and the Surgeon General of the U.S. Public Health Service. This proved to be a time-consuming process.

Meanwhile, the Red Cross national headquarters staff was preoccupied with the distribution of the huge inventory of surplus dried plasma being returned by the Army. Unfortunately, pursuit of that task was plagued by mounting concerns that the dried plasma might be transmitting the agent that caused homologous serum jaundice. Until the cessation of hostilities, no measures prohibiting the use of dried plasma had been taken by the Army.[12] However, when the Red Cross began distributing packages of surplus dried plasma to state health departments, some surprising reactions were encountered. Citing concerns about the risk of transmission of jaundice, New Jersey health authorities flatly rejected the proffered dried plasma. In Minnesota, questions were raised, although in the end, the state medical society voted to receive shipments of dried plasma. Nevertheless, the Red Cross soon halted further distribution of dried plasma and referred the matter to its new committee for advice.

The possibility that the surplus dried plasma might transmit the virus of hepatitis was discussed at a meeting at Red Cross Headquarters in Washington in March 1946. In addition to the members of the Committee, those present included Generals George Callender and Bayne-Jones, and Colonel Michael DeBakey representing the Army Surgeon General, Edward Churchill of the Massachusetts General Hospital, and representatives of the Navy, U.S. Public Health Service, American Medical Association and National Research Council. General Bayne-Jones presented a letter from the Army Surgeon General to Red Cross President Basil O'Connor stating the Army's conclusion that the benefits of dried plasma "far outweigh the risks." Nevertheless, in the ensuing discussion, it emerged that it was difficult, if not impossible for the Army to determine from the histories of returned transfused wounded service men whether they had received only blood, only plasma or both, data that would be necessary to estimate the risk. Col. DeBakey pointed out that records of hospitalized soldiers were not always complete and the use of dried plasma was not always recorded. Further, many cases may have been missed because the incubation period of hepatitis was not accurately known.

Churchill noted that the Mediterranean theater where he served was an area of epidemic infectious hepatitis. Only dried plasma had been used. Some blood transfusions were also administered using blood collected and processed in the field. Based on records that were accurate after September 1944, an estimated 51,094 cases of trauma had been hospitalized of whom 153 had developed hepatitis. This corresponded to an incidence of jaundice of less than 1%. In a British study, 7.1% of almost 1,200 transfused soldiers had jaundice. Among untransfused soldiers in the same study, only 0.1% of almost 5,000 men developed jaundice. Following the discussion, the committee recommended that the American Red Cross continue to distribute Army dried plasma as planned; that the Surgeon General of the Army be asked to prepare a statement regarding the danger of transmission of homologous serum jaundice by dried plasma for possible publication over the signatures of those present at the meeting; and that every effort should be made to obtain scientific evidence on the incidence of jaundice following use of Army surplus dried plasma.

Before adjourning, the committee discussed the options available to the Red Cross should the suspected contamination of the dried plasma with the virus of homologous serum jaundice be confirmed. The most attractive option involved the salvage of clinically useful proteins from the dried plasma. Assuming that the proteins in the dried plasma had not been altered in the process of drying the plasma—a reasonably safe assumption—the chances were good that the immune serum globulin could be salvaged. As for excess human serum albumin, Cohn knew that it would be contaminated with mercury from the mercurial preservative that had been added to the plasma before drying. However, he suspected that it would be possible to remove the mercury during the fractionation process. Lacking experience on the effectiveness of this option, the committee advised the Red Cross to arrange for a batch of dried plasma to be fractionated at a pharmaceutical firm, and that the human serum albumin and immune serum globulin so obtained be subjected to chemical, biological and clinical testing.[13]

By October 1946, the Red Cross had distributed over 400,000 packages of dried plasma for civilian use and had received requests for an additional 200,000 packages. Statewide Red Cross blood donor programs had been established in Michigan, North Dakota and Massachusetts. A statewide plasma fractionation program was in operation in Massachusetts; and Michigan was recruiting a director for its new fractionation program. Blood was also being collected in Los Angeles and Detroit, as well as in certain communities in Louisiana, New Hampshire and Maine. The development of blood programs had been authorized by Red Cross chapters in New York City, Kansas, New Jersey, New Hampshire, California, Arizona and Iowa.[14]

The results of the first large scale fractionation runs with 175,000 units of reconstituted surplus armed services dried plasma were reported to the Red Cross Committee in April 1947.[15] The antihemophilic activity of the Fraction I from this plasma was surprisingly strong. The immune serum globulin in Fraction II tested satisfactorily for antibody titres. For the human serum albumin, there was an important bonus; in Cohn's words, "the brilliant experiment of Dr. Walter L. Hughes Jr., of our laboratory solved the problem of the removal of mercurials which had been added to the dried plasma, and thus made possible the salvage from the dried plasma of mercury-free, salt-poor albumin."[16] Cohn and his associates thus opened the way to salvage useful products from the huge wartime Army stockpile of dried plasma. This achievement assumed even greater importance once valid clinical evidence was in hand indicating that dried plasma was now safe from the transmission of virus diseases.[17] Since a less expensive substitute had subsequently been developed, Fibrin Foam and Thrombin was no longer in demand, and because Fibrin Film proved to be difficult to produce from dried plasma, it too was abandoned.

The American Red Cross national blood donor program was formally approved by the Red Cross Board of Governors on June 12, 1947.[18] In July 1947, Admiral McIntire, now retired from the Navy, was appointed Administrator of the Red Cross National Blood Program, with Edwin Cohn as an unpaid Chemical Consultant. Late in 1947, Cohn suggested to Louis K. Diamond, the Chief of the Blood

Grouping Laboratory at Children's Hospital in Boston, that he to go down to Washington, study McIntire's plans for the Red Cross National Blood Program, and offer him "advice on how best to operate the numerous Red Cross regional centers that were to be opened in the future." When Diamond got off the train in Washington, he was met by a Red Cross driver who whisked him off to Red Cross National Headquarters. There he received a VIP reception, beginning with the Admiral, who in due course passed him on to G. Foard McGinnes, the Red Cross Vice President for Medical Affairs, and then to Basil O'Connor, the Red Cross President. What Diamond did not know was that Cohn had alerted McIntire, suggesting that Diamond be taken on immediately as Technical Director of the Red Cross Blood Program. The following morning Diamond was busy at Red Cross headquarters working on plans for the operation of the first Red Cross blood program scheduled to be opened in Rochester, New York, on January 12, 1948.

Diamond turned out to be the right man for the task; on December 15, 1950, he opened the thirty-sixth Red Cross center in Columbia, South Carolina. Later, there were fifty-nine Red Cross Blood Centers. Cohn accepted invitations to be the guest speaker at opening ceremonies of many of the centers, stressing repeatedly the need for blood and expounding on the importance of separating the blood into parts that could benefit more than one recipient. Cohn and Diamond saw each other frequently at Red Cross meetings, at the National Institutes of Health, and at the dedications of blood centers. Diamond characterized this hectic period as "a liberal education, scientifically, medically and even gastronomically, for with respect to the last of these, Dr. Cohn was a true gourmet. And dinner conversation was rarely strictly business but rather about the personalities of the great, the near-great and the not-so-great, and about medical philosophy and philosophers."[19]

During this period, Cohn was contacted by several colleagues in Europe. In one instance, he wrote a long reply to a letter from Professor E.J. Bigwood, at the Free University of Brussels, a Councilor to the Belgian Government, setting forth his views concerning the provision of blood in a free society.[20] "I am in complete agreement with your original position that the collection and distribution of human blood and blood derivatives can best be carried out either by the Red Cross or by government public health agencies," Cohn assured Bigwood. Cohn also stated his aversion to the commercial collection and distribution of blood and blood products primarily because it "implies the paying of blood donors and the cost of the products becomes prohibitive." With respect to dried plasma, he advised Bigwood that

> Many of us begin to doubt its place in blood programs of the future . . . both from the point of view of safety and economy . . . Evidence from England and the United States has rendered it certain that infectious hepatitis can be transmitted by human plasma. Although the statistical incidence remains to be determined, it will always be higher than in whole blood since the latter is not pooled. There is no evidence thus far of virus transmission by a product of plasma fractionation. This must not be construed as evidence that conditions (during fractionation) are such as to assure virus destruction during processing. That is not the case. Indeed the general methods of protein fractionation . . . can be employed not only for the concentration of antibodies and antitoxins but also for toxins, toxoids and viruses. It is for this reason that we have recommended that blood programs be associated with (state) antitoxin laboratories so that the advances that have been made in plasma fractionation may be applied also to the most diverse natural products, of plant or animal origin. [In a subsequent article, it was told how this was accomplished.] It has been our constant aim in developing methods for plasma fractionation to eliminate the risk of virus transmission. In the case of... human serum albumin...the stability is so great in the presence of certain non-polar anions, that the salt poor human serum albumin now being produced can be heated in the final container for ten hours at 60°C.[21]

The latter statement referred to findings from a new study that had not yet been submitted for publication, which showed, by injection in man, that the jaundice virus is destroyed by heating under these conditions.

In June 1946, Cohn was invited by C. J. Van Slyke to serve as a consultant to the National Institutes of Health concerning a new program of NIH Research Grants. Before accepting, Cohn wrote a letter to Van Slyke setting forth, with unusual gentility, some of his strongly held views. He opened by telling Van Slyke, "I find it rather difficult to give the kind of opinion that I should like to without knowing more about the policy being adopted," and he asked,

1. Is policy in making grants to be based entirely on the quality of the man rather than on his description of his project?
2. May I assume that no supervisory activity at the administrative level will be exercised by the U.S. Public Health Service?
3. May I assume that the object of these grants is to increase fundamental knowledge and that therefore they are to be made to laboratories directed by well-trained original investigators, capable of making advances in the direction of their choice as effectively as other laboratories in other parts of the world.[22]

In particular, Cohn had reservations about the first point, for he was a "conscientious objector to naming [i.e. describing] projects in advance." Although the subsequent support of basic research by the National Institutes of Health closely adhered to all but the first principle that Cohn advocated, he quietly acquiesced to NIH regulations and became a charter member of the NIH Hematology Study Section, the panel that reviewed NIH grant applications relating to blood.

Shortly thereafter, Dean Burwell submitted to Rollo Dyer, the Director of the National Institutes of Health, a draft of a grant application on behalf of Edwin J. Cohn that had been "prepared in very general terms... without a detailed budget."[23] This brought a response from Ernest Allen at the NIH Research Grants Division, who stated that he did not believe that "the Division will be able successfully to process the application." While conceding "that no investigator would receive more immediate consideration than Dr. Cohn," Allen explained that "in view of the close scrutiny of project proposals that may be given by the Bureau of the Budget...we would be in an indefensible position if we did not have definite budget proposals included in all applications for grants in aid." Allen suggested that anticipated expense for Cohn's grant be broken down into five categories, "with assurance from this office that return mail approval can be given" for transfers that Cohn might later desire within budget categories.[24] In this enlightened way, Ernest Allen persuaded Edwin Cohn to accept strictures that bordered on violating Cohn's strongly held principles.

More Honors

Edwin Cohn was the John Phillips Memorial Medallist at the annual meeting of the American College of Physicians in Philadelphia on May 14, 1946. His lecture before the College was entitled, "The Separation of Blood into Fractions of Therapeutic Value."[25] In his introductory statement, Cohn stated that "Medicine is concerned with understanding, in the interests of control, all of the component parts of the body and their functioning." He believed that anatomy and physiology had contributed to this understanding, as had bacteriology and immunology, while alchemy had contributed both to pharmacology and to chemistry, and an unfolding chemistry had contributed a knowledge first of the elements, then of their relation to each other in the organic molecule and then of intermolecular reactions. The level of understanding, at any time, had reflected the new knowledge in the natural, as well as in the medical sciences. A central portion of his lecture was devoted to the development of the physical chemistry of proteins, attributing to Sir William Hardy the correct interpretation of early observations on the behavior of proteins in electrical fields as due to their dissociation as electrolytes. If proteins were electrolytes, it follows that proteins were in equilibrium with the other ions of the body, and there must be a zone in which they bore no excess of positive or negative charges, i.e. were in an isoelectric condition. Cohn gave credit to the two "great" papers by Hardy and by Mellanby in the *British Journal of Physiology* in 1905 dealing with the interaction of proteins with salts. He believed that there was no satisfactory approach that could be applied to our understanding of solutions of ions, dipolar ions and uncharged molecules until 1923, when Debye developed his electrochemical theory of solutions that accounted for the interaction of ions with each other as well as with organic molecules. "In terms of electrostatic forces," Cohn continued,

> the solvent action of salts upon globulins depends not only upon the number of electrical charges on the protein, but also upon their spatial distribution on its surface . . . The value of blood in the prevention or treatment of disease depends but rarely upon the whole tissue. Rather it depends upon one or more of the cellular, protein, or smaller organic constituents of blood for which there is great need.

> This need is for quite different constituents of the blood in hemophilia, edema, shock or hypoproteinemia. The number of specific interactions between the small organic molecules and the large protein molecules that occur in blood and in other tissues is far greater than has heretofore been demonstrated.

As a result, Cohn believed that the separation of proteins as natural protein complexes renders it possible to study the interacting physicochemical forces, as well as to explore these natural physiological mechanisms and to determine their possible value in therapy.

In this lecture, Cohn used two figures to illustrate the diversity of the plasma proteins. The first of these was a chart showing the relative molecular dimensions of hemoglobin and several plasma proteins as calculated from ultracentrifuge data (Fig. 5). Viewed in this way, hemoglobin and albumin, despite almost identical molecular weights, 69,000 vs. 68,000, had quite different shapes. Similarly, gross differences are evident in the shapes of albumin, β_1-globulin (metal combining globulin), χ-globulin, α_1-lipoprotein, fibrinogen and β_1- lipoprotein. To offer perspective in interpreting this chart, Cohn pointed out that if the model labeled "fibrinogen" were a yard long, the human red blood cell would be about the size of the Harvard Stadium. He liked to point out the physiological importance of the shape of fibrinogen, since one of fibrinogen's physiologic functions during the coagulation of blood resulted from its ability, following treatment with the enzyme thrombin, to polymerize end-to-end, thus forming long fibrin strands.

The second image added another dimension to understanding the diversity of proteins in plasma. Known in the laboratory as a "Plasma Pie," this figure depicted graphically the relative proportions of each of the major proteins in normal plasma (Fig. 6). These two images—the size and shape of the major proteins, and their proportions in plasma—were used in many of Cohn's public lectures in the postwar period. In concluding, Cohn identified a new principle that was emerging from the post war studies in his laboratory. This was the realization that some of the plasma proteins serve as vehicles for the transport of low molecular weight substances in the blood stream by a process involving the reversible binding of small molecules or ions to specific sites on certain proteins. Examples that he cited included the binding of estrogens and vitamin A by lipoproteins, the binding of chloride ions and other stabilizers to albumin, and the newly discovered binding of iron by the metal combining globulin.

In its wisdom, the program committee of the American College of Physicians had arranged for Joseph Stokes, George Minot and Charles Janeway to follow Cohn's lecture by presenting new clinical findings that stemmed directly from his work. Stokes focused on the public health significance of gamma globulin, stressing how the pooling of plasma from large numbers of blood donors resulted in surprisingly uniform and reproducible titers of antibodies against infectious agents. He stated that gamma globulin was the most satisfactory biological developed up until then for the passive-active immunization of children exposed to measles.[26]

George Minot identified two problems concerning the potential clinical utility of antihemophilic globulin in treating hemophilia.[27] These were the substantial variation in the potency of the crude products then being tested; and the observation that some cases of hemophilia failed to respond to Fraction I, the fraction which carries the antihemophilic activity. He speculated that the defect in hemophilia might not be a single factor. Later, Minot's reasoning proved valid when it was discovered that there are two types of hemophilia, A and B.

Janeway dealt with physician concerns stemming from the fact that most of the studies on albumin had been driven by military considerations, e.g. the dispensing of albumin as a 25% solution. Although postwar albumin was still being dispensed in that formulation, Janeway pointed out that it did not have to be used in that form; this referred to an allusion still held by some, that 25% albumin could dehydrate patients. He then described a number of conditions in which albumin had great value as a therapeutic agent in diseases of the liver and the kidney.[28] Human serum albumin subsequently was made available as a 5% solution, the form in which it is used today.

Cohn was honored by the Belgian American Foundation late in 1946 and was the recipient of the Medal of the Free University of Brussels. He received honorary degrees from the University of Bern and the University of Geneva early in 1947 and was Visiting Lecturer of the American-Swiss Foundation for Scientific Exchange, giving lectures in

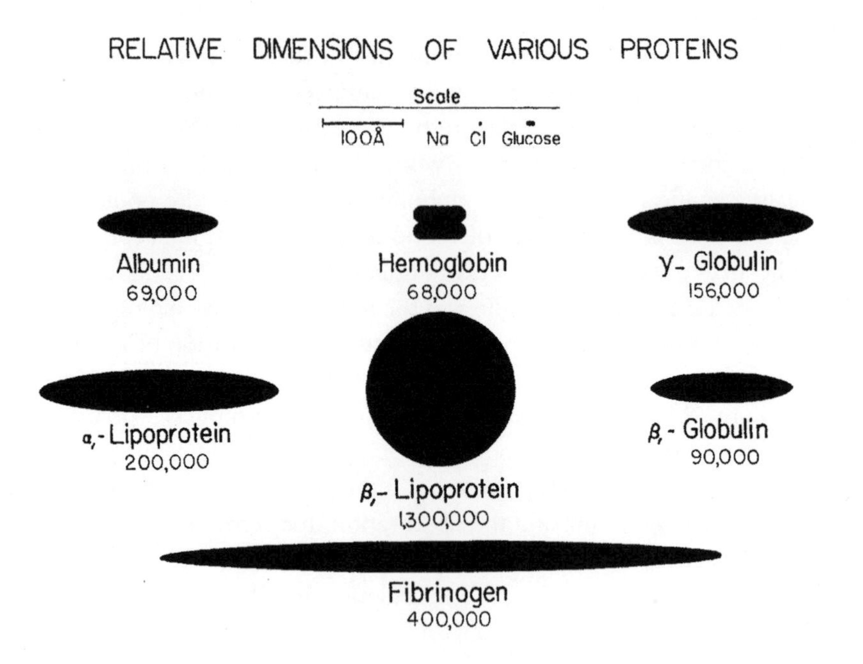

Fig. 5. Relative dimensions of various proteins. Revised by J. L. Oncley, 1945.

Bern, Zürich, and Geneva in February and March. The substance of the Swiss lectures was later published as a scientific paper in the Swiss journal, *Experientia*.[29] An unusual feature of this paper was a table in which he listed the properties of all the proteins then known to be present in human plasma. The summary of these lectures offers Cohn's overview of the wartime program.

> Starting with the assumption, which must for the present remain an assumption, that every part of human blood performs an important natural function—that the amount of any plasma component that may wisely be used in replacement therapy is the amount that will tend to re-establish rather than to derange the normal concentrations of the various physiologically significant plasma components—

PLASMA PROTEINS
THEIR NATURAL FUNCTIONS AND CLINICAL USES
AND SEPARATION INTO FRACTIONS

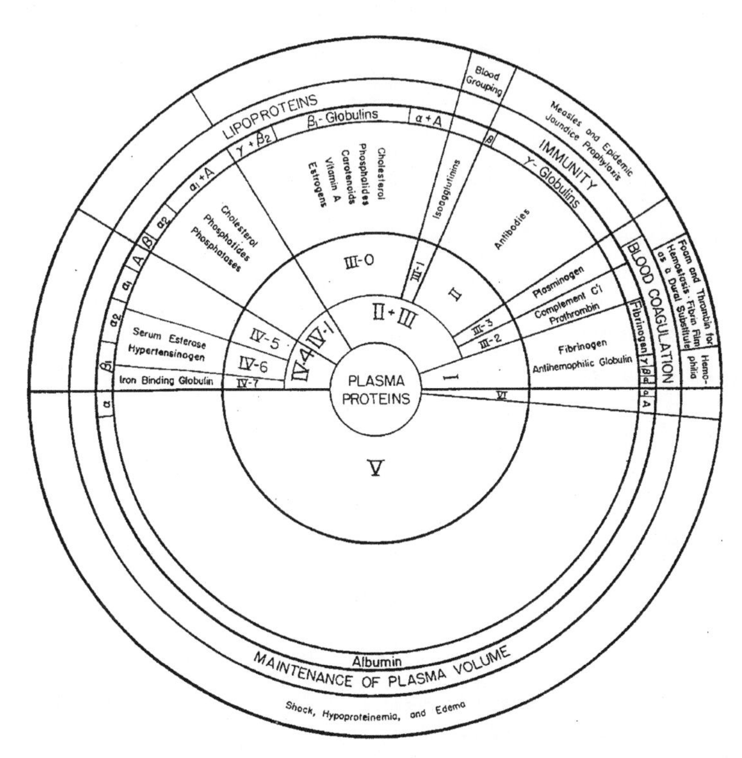

Fig. 6. Plasma proteins, their natural functions and clinical uses, and separation into fractions.

> we have attempted to make available as many as possible of its diverse protein components, separated and concentrated as specific therapeutic agents, of value in different conditions, in the interests of the most effective and economical use by a society of the blood which it contributes.

Cohn frequently received requests for samples of the plasma fractions from other scientists. Typically, an investigator might have found a new biological activity in plasma and wanted to see if it could be identified in one of the fractions. Samples of dried fractions were mailed off along with an expression of interest in hearing the results of any findings. In this way, important new leads were fed back to the lab-

Table II

Protein Components of Human Plasma separated and concentrated in Diverse Fractions

Protein Component		Estimated Amount in 100 g. Plasma Protein gms	Concentrated in Fraction[4]	Approximate Isoelectric Point	Specific Chemical Interaction
Fibrinogen		4	I-2	5·3	Thrombin
Non-Clottable Protein, insoluble at low temperature		0·15	I-1		
Antihemophilic Globulin[1]			I		
Antibody γ-Globulins	Diphtheria Antibodies[1]	(0·001)			
	Measles Antibodies[1]				
	Mumps Antibodies[1]				
	Streptococci Antitoxin[1]	11	II	7·3	Antigens
	Influenza Antibodies[1]				
	Pertussis Antibodies[1]				
	Typhoid "H" Agglutinins[1]				
Antibody Englobulins	Typhoid "O" Agglutinins[1]		III-1		Antigens
Isoagglutinins	Anti-A, Anti-B[1]	(0·03)		6·3	
	Anti-Rh Antibodies[1]		III-1		Incompatible Red Blood Cells
Complement Components	C'1	0·4	III-2[2]		Antigen-Antibody Complex
	C'2		IV[2]		
Enzyme Precursors	Prothrombin	0·3	III-2		Thromboplastin
	Plasminogen		III-3		Streptokinase
Serum Enzymes	Thrombin[1]		III-2	4·8	Fibrinogen
	Plasmin[1]		III-3		Proteins
	Amylase[1]				Starch
	Lipase[1]				Lipid
	Peptidase[1]		IV		*l*-Leucylglycylglycine
	Phosphatase[1] (alkaline)		IV[2]		Phosphoric Acid Monoesters
	Esterase[1]	0·02	IV-6	4·5	Acetyl Choline, Ethylbutyrate
Metal-Combining β_1-Pseudoglobulin: crystallized		2·5	IV-7	5·6	Iron and Copper
High Molecular Weight β_1-Globulins (Lipid-poor)	S = 7	2	III-0		
	S = 20	1	III-0		
Iodoprotein[1,2]			IV-6		
Thyrotropic Hormone[1]			IV-4		
Glycoproteins	α_2-Glyco-pseudoglobulin	0.7	IV-6	4·9	
	α_2-Mucoid Globulin	0·5	IV-6	4·9	
Lipoproteins	β_1; 75% Lipid-Containing "X"-Protein	5	III-0	5·6	Estriol, Carotenoids, and other Steroids
	α_1; 35% Lipid-Containing Protein	3	IV-0	5·2	Steroids
Blue-Green Pigment α-Globulin			IV-2		
Bilirubin-Containing α_1-Globulin[3]		0·05	V-1	4·7	Diazo-reaction
Albumin: Crystallized with Mercury			V	4·9	Mercury, Decanol
Albumin: Crystallized with Decanol		50	V	4·9	Fatty Acids, Bile Salts, many Dyes and Drugs

[1] These components represent but small proportions of the fraction and subfraction, and their properties cannot, therefore, be deduced from those of the concentrates in which they have been separated.

[2] These components have not been tested for since revision of the fractionation process.

[3] Albumin binds more bilirubin than the bilirubin pigment globulin in Fraction V-1 and more iodine than has been found in Fraction IV-6.

[4] When purified chemical components have been separated from fractions they have not been given new fraction numbers. In that case, the fraction number refers to the starting material for the separation of the component.

9 Exper.

Fig. 7. Protein components of human plasma separated and concentrated in diverse fractions.

oratory and occasionally a new plasma protein was identified. One such request came from Arthur L. Schade, a bacteriologist at the Overly Biochemical Research Foundation in New York. Schade had discovered that the growth of the bacterium, *Shigella dysenteriae,* which requires iron for its metabolism, was inhibited by egg white and by human plasma.[30] He showed that the growth inhibiting effect of human plasma was in fraction IV-4, the fraction that precipitated just before albumin in the main plasma fractionation procedure. By then, studies were well underway on the subfractionation of Fraction IV-4 at Harvard. Samples were forwarded to Schade who promptly found the growth inhibiting effect in a subfraction known as IV-7. Schade also discovered that a faint salmon red color developed when iron was added to solutions of Fraction IV-7.[31]

During the war, Navy contracts for the production of albumin specified that fractionation be initiated within seventy-two hours from the time of collection of blood from the blood donors. The setting of this specification was intended to provide a reasonable time for all the steps involved—prompt cooling of the individual bottles of blood at the blood donor center, refrigerated shipment of the blood to a pharmaceutical laboratory, centrifugation, drawing off the plasma from the red cells, and pooling it into the batch for processing, while at the same time minimizing deterioration. Some proteins were denatured by the conditions of fractionation. The hormone precursor called hypertensinogen was found by Lewis Dexter to be present in good quantity in the supernatant solution following centrifugation of Fraction IV-1, but had been completely inactivated at the high ethanol concentrations needed to precipitate the next fraction—Fraction IV-4. In view of the high priority attached to the production of albumin, Cohn could not consider any change in the specifications or in the fractionation procedure during wartime. But when the war ended, he became interested in looking into this problem. He wanted to obtain every plasma protein in its state in nature.

Division of Medical Sciences of the Faculty of Arts and Sciences

For many years the primary purpose for the establishment of medical schools in the United States was the training of physicians and the award of M.D. degrees for the practice of medicine. Research programs aimed at the eventual elimination of the causes of disease led, over the years, to the development of research and training programs in the basic medical sciences including anatomy, bacteriology, biochemistry, pathology and physiology, for which the Ph.D. degree was awarded. Historically, this degree had traditionally been administered by the Faculty of Arts and Sciences at Harvard, where it led to the creation of a Division of Medical Sciences in the Faculty of Arts and Sciences which was composed primarily of faculty members in the medical school who participated in training men and women for the Ph.D. degree.

Edwin Cohn served as Chairman of this Division between 1936 and 1949. With the termination of hostilities at the end of the war, interest in courses devoted to the study of the medical sciences offered at the Harvard Medical School attracted considerable interest. During the 1948–49 academic year over forty graduate students were registered for the Ph.D. degree in Anatomy, Bacteriology, Biochemistry, Comparative Pathology, and Physiology. For the most part, these students planned to spend their careers in the expectation of contributing to new knowledge in the medical and biological sciences in laboratories of universities, industry and governments.

In a report dated October 28, 1947, Cohn stated "there was a dearth of scientists whose training in both organic chemistry and pharmacology rendered them especially fitted for the synthesis and appraisal of new drugs and other chemical compounds; of those trained both in physical chemistry and bacteriolocy and thus especially fitted for the preparation of natural products destined either for prophylaxis or therapy; of those trained both in physics and physiology and thus especially fitted to analyze and to protect against the influence of high altitude." Cohn continued, "With these thoughts in mind, the Harvard Committee on the Curriculum has recommended the introduction of new courses especially destined for those taking higher degrees under the Faculty of Arts and Sciences."[32]

Marianne Brettauer Cohn 1893–1948

Marianne Cohn took her own life on January 20, 1948. Her funeral was held in the Harvard Memorial Church. She was buried in Mt. Auburn Cemetery in Cambridge.

14 THE COLD WAR

THE COLD WAR began with a series of strokes and counterstrokes on the part of the Russians and the Western Allies in 1947 and 1948. Czechoslovakia fell to the communists in February 1948. The Marshall Plan was enacted by the Congress in April 1948. In June, the Russians blockaded the free zones of Berlin. American and British air forces immediately launched the Berlin Airlift and operated it until Stalin ended the blockade in May 1949. However, the conflict would continue for the next forty years.

A New National Research Council Committee on Blood and Blood Derivatives

Collections of blood by the American Red Cross for the armed forces had been terminated in 1945. Lacking an organized civilian blood program of any magnitude, the United States was in a vulnerable position if large volumes of blood suddenly were to be needed. A new Research and Development Board of the National Military Establishment replaced the wartime Office of Scientific Research and Development. On January 8, 1948, the National Research Council resumed its wartime role. A new Committee on Blood and Blood Derivatives replaced the old Subcommittee on Blood Substitutes. Its roster included Edwin Cohn, Elmer DeGowin, Everett I. Evans, Frank W. Hartman and I.S. Ravdin, with Charles Janeway as its chairman. The ties to the armed services were reactivated, with representation from the Office of the Army and Navy Surgeons General and with Louis K. Diamond representing the American Red Cross.

The NRC Committee on Blood and Blood Derivatives met in Washington in April 1948 and adopted a revised set of principles: There is no substitute for whole blood. The wartime American Red Cross blood program should be reactivated and expanded as rapidly as possible. As by-products of such a program, new stockpiles of virus-free plasma and albumin should be accumulated for supplementation of available supplies of whole blood. Since the expansion of the program on the scope necessary to meet any possible emergency might take considerable time, the Committee recommended the accumulation of stockpiles of osseous gelatin by the Armed Forces, for use only if the supplies of blood, plasma or albumin were insufficient to meet needs. Osseous gelatin was the closest to being an acceptable blood substitute. Existing facilities for the production of plasma, plasma fractions, and osseous gelatin were then being dismantled; the Committee strongly urged that these facilities be kept in readiness for production in case of emergency. Furthermore, support should be given to studies of methods for further prolonging the life of red blood cells in storage.[1]

The Committee immediately reviewed the Red Cross plans for developing a national blood program. Admiral McIntire's blueprint called for establishing over sixty regional Red Cross blood centers in the next several years. The new program would not only have to meet the needs of the armed services, but also of the civilian population, in peace as well as in case of nuclear war. Many new problems had to be dealt with. Isadore Ravdin, the surgeon who had seen the casualties following the 1941 attack on Pearl Harbor, characterized the task ahead as requiring that the Red Cross be ready to face a national tragedy that no private blood bank can meet. While the Red Cross had posted a splendid record in disaster relief and aid to military families during and after the war, the operation of a national blood program would require the Red Cross to provide a basic medical service on a day-to-day basis throughout the year. In each new blood center, operating personnel—physicians, nurses and technicians—would have to be recruited and trained. New policies governing the distribution of blood and derivatives to hospitals would need to be developed. During the war, difficulties in organizing blood collections in large cities such as Los Angeles and New York had not been resolved and would undoubtedly recur. Meeting the needs for blood for civil defense added to the challenge. By 1948, about a fifth of the hospitals in the United States had established private blood banks that depended on a variety of means to attract blood donors from their communities. In some places this included payment of blood donors. These hospital blood banks and the new organization that represented them—the American Association of Blood Banks—posed strong potential competition for the Red Cross.[2]

Following a visit from Red Cross Vice President G. Foard McGinnes, Cohn wrote to him in forthright terms, forwarding copies of his letter to Navy Secretary Forrestall, Admiral Swanson and Admiral McIntire. "I am convinced that the Red Cross must not only collect and distribute whole blood, but also create a stockpile of stable and safe blood derivatives for use in emergencies." He urged that the Red Cross begin by establishing new pilot plants and negotiating contracts with commercial firms for the production of new blood derivatives that had been developed during the war at Harvard and elsewhere.[3] This was strong talk to an executive of an agency that, at the time, was not collecting blood and had none to distribute. The Red Cross's only tangible asset was the stockpile of outdated armed services dried plasma and a limited supply of surplus gamma globulin that had been returned by the Armed Services.

Less than a month later, Cohn dispatched another letter to McGinnes in which he stated that "the impossibility of judging, in less than a decade, the complex roles that protein components of the blood play in their interactions with each other and with the smaller substances with which they are in labile equilibrium and which they transport to and from the tissues, seemed to me seven years ago to preclude the serious consideration of the use of blood substitutes." He said that he could not deviate from that conviction at this time. It was because of that conviction that he had long since directed his efforts to the further exploration of the unknown, or little known, components of the blood, as well as to the multiple physiological functions of the major components of the blood.[4]

Homologous Serum Jaundice

In the meantime, credible new information had revealed that the incidence of homologous serum jaundice among patients receiving army surplus dried plasma ranged between 3% and 7% in Massachusetts, New York State and in England. This news effectively terminated further civilian interest in dried plasma.[5] At the same time, this gloomy picture was balanced somewhat by new proof that heating serum albumin solutions for ten hours at 60^0 C. inactivated the hepatitis virus.[6] Back in 1945, when he authorized that study, and, without waiting for the results, Cohn had ordered that all human serum albumin be subjected to the new heat treatment in its final containers. That step insured that a safe supply of human serum albumin was already in place. A second positive development came from the brilliant work of Hughes in devising an elegantly simple method for removing the mercury preservative contained in dried plasma. This could be accomplished while the dried plasma was being fractionated to salvage its albumin and immune serum globulin.[7] The Red Cross Committee on Blood and Blood Derivatives promptly urged Red Cross officials to "push forward with the greatest possible speed" to fractionate all the dried plasma that had been returned by the Army at the end of the war. Although this would take more than a year to accomplish, it would also assure that valuable technical resources in the plasma fractionation industry would be in usable condition, in the national interest.

A Letter from Lieutenant General Leslie Groves

In July 1948, Vannevar Bush, now Chairman of the Research and Development Board of the National Military Establishment, forwarded to the Board's Committee on Medical Sciences a letter he had received from Lt. General Leslie Groves, the former Director of the Manhattan Project, concerning the need for stockpiling large quantities of whole blood to be used for the treatment of atomic warfare casualties.[8] Groves wrote, in part:

> If an atomic bomb is ever used against us, the matter of therapy for the victims will be of the utmost importance. The best medical opin-

> ion is that the most essential part of this therapy will be the use of whole blood. It is true that during World War II dried plasma was used with remarkable success in treating the injured. However, plasma itself will not be sufficient for an individual who has been injured by ionizing radiation. Whole blood will be given until such time as the patient's bone marrow has a chance to recover. We are not aware that any process has been developed for the storage of the cellular elements of the blood for long periods.

Groves pointed out that during World War II, some 10,300,000 pints of blood were collected from which 5,150,000 units of plasma were made and delivered to the Armed Services. Of that amount, 3,500,000 units of plasma, the equivalent of 7,000,000 units of whole blood, were actually used. He said that it had been estimated by medical authorities that if adequate blood transfusion therapy had been used on all the injured at Hiroshima and Nagasaki, it would have required as much blood as was used in World War II. He concluded that "it seems obvious that some system must be devised within the next few years whereby a large amount of all the elements of the blood will be prepared and stored to be readily available in use in case of an atomic war."

In response, the Executive Director of the Committee on Medical Sciences of the National Military Establishment's Research and Development Board wrote to Lewis Weed at the National Research Council on July 16, 1948, as follows:

> In a recent letter addressed to the Research and Development Board, Lt. General Leslie Groves brought up the need for stockpiling large quantities of whole blood to be used for the treatment of atomic warfare casualties, should such an eventuality occur. General Groves also pointed out the need for methods of storage of whole blood for prolonged periods with preservation of the cellular elements, if successful stockpiling were to be accomplished.
>
> This matter was brought up for discussion at the second meeting of the Committee on Medical Sciences of the National Research and Development Board that was held on July 9, 1948. The facts noted above were confirmed.
>
> As a result of action taken at this meeting, I have been instructed by the Committee to address an official request to the National Re-

> search Council to convene its Committee on Blood and Blood Derivatives in the near future to review this problem and to make specific recommendations for such programs or studies considered desirable.[9]

This matter was referred to the NRC Committee on Blood and Blood Derivatives at a meeting on October 2, 1948, with the request that it organize a national conference and review actions that should be considered. Following discussion within the Committee, its chairman, Janeway, "asked Dr. Cohn to serve as chairman of an Ad Hoc conference to discuss the preservation of whole blood and of blood elements as well, urging that all members of the Committee as well as outside experts attend." To meet adequately the full purpose of this study, the assignment was interpreted in terms which appeared solvable, namely, that the need was for "The Preservation of the Formed Elements and of the Proteins of the Blood." [10]

Given the high priority assigned to the conference by the Research and Development Board, January 6, 7 and 8, 1949 were set as the dates for the conference, which would be held in Boston. Cohn moved swiftly to develop the agenda for the conference. With only three months at hand to complete the arrangements for the conference, Cohn plunged into a series of meetings in Washington and Boston, supplemented by telephone consultations with a range of experts. He relied heavily on Patrick Mollison of the Blood Transfusion Research Unit in London, Orville Denstedt in Montreal, Kenneth Brinkhous of the University of North Carolina and Charles Doan of Ohio State University.

The Formed Elements Group

At the same time, Cohn initiated a series of steps that typified his ability to grasp an important scientific problem and arouse the enthusiasm of associates to work toward its solution. He gathered an informal group of associates from the Boston area to meet in his office on Saturday morning, October 23, 1948. This group included John G. Gibson, 2nd, whose pioneer work had contributed substantially to introducing ACD for the preservation of red cells, Edward Buckley, a young physician working with Gibson, Carl Walter, the Peter Bent Brigham Hospital surgeon who had collected the first human plasma for Cohn's plasma fractionation work in 1941, and Charles P. Emerson, a hematologist at Boston University Medical School and the co-author, with Richard Ebert, of the Ebert-Emerson study of blood volume in shock. Others participating in this effort were Robert Pennell, a chemist at Sharpe and Dohme Laboratories in Philadelphia, Eugene Rochow, a Harvard Professor of Chemistry, Dwight Mulford of the Massachusetts Antitoxin and Vaccine Laboratory, and Janeway, Diamond, Hughes, Scatchard and Surgenor. After briefing us on the plans being developed for the meeting, Cohn challenged the group to devise a method for separating the formed elements—red cells, white cells and platelets—from freshly collected whole blood. He had in mind a tour de force that could be reported to those attending the January meeting.

On the following Saturday, Gibson and Buckley returned with some favorable results. A key problem lay in the difficulty of separating white cells from red cells, whose densities were not very different, thus making it almost impossible to separate them by centrifugation. Gibson and Buckley's initial approach involved causing the red cells to reversibly stack up like coins, called rouleaux, which, by virtue of their size, rapidly sedimented under the influence of earth's gravity. This left the white cells in suspension, which could then be siphoned off. Finally, the rouleaux could be disaggregated, yielding suspensions of discreet red blood cells.

This group came to be known as the Formed Elements Group. By December 21, 1948, the group had arrived at a relatively simple laboratory procedure for separating the red cells, white cells and platelets. The group met at least monthly for the next two years, with some changes in membership over time. It became a gathering place for the development of new ideas and suggestions from a stream of experts who participated on an ad hoc basis. Under Cohn's leadership these participants worked together voluntarily, without formal organization or authority, and without special funding, until August 1951.[11]

In December 1948, just before the planned conference, J. E. McCormack, the Deputy Executive Director of the National Research and Development Board Committee on Medical Sciences, wrote Cohn summarizing how the Research and Development Board viewed his assignment in organizing the conference. He pointed out that it should be apparent that the NRC committee in its discussion evidenced enthusiasm to have the whole field of blood preservation explored. Work in this area had been brought nearly to a standstill at the end of the war by the disbanding of OSRD. One member had particularly urged that the stockpiling of huge amounts of plasma or an acceptable substitute was equally as important as the availability of whole blood or erythrocyte suspensions. The problem of thermal burns at Hiroshima and Nagasacki was a great one, making it apparent that disturbance of the osmotic relationships of the blood and extravascular fluid was in many respects as serious as the destruction of blood cells. Since the relationships to other problems considered at the meeting, such as sterilization of blood and plasma and standardization of transfusion equipment, was apparent, he told Cohn that:

> It is expected that the symposium which you are arranging will (a) arrive at a summary opinion of the present state of our knowledge in the field of blood preservation, (b) survey the known current research endeavor which is related to this over-all problem in order that it might be possible to (c) make recommendations as to a probably effective program of research designed to prolong the period over which blood cells can be preserved. We confidently entrust the program entirely to your own good judgment. From your knowledge of the history, the trends and the probable developments in this field, the symposium may be so oriented that there will emerge a reasonable opinion as to how blood in quantity can best be collected, handled, and preserved and what derivatives may be stockpiled in quantity.[12]

Additional Honors are Bestowed

On February 2, 1948, Cohn was awarded the "Medal For Merit" by President Truman for his "contribution of inestimable value in the field of medicine in the production of a pure human albumin and in packaging it in such a manner as to prevent deterioration under various conditions of climate. As a result of his research, progress has been made in obtaining gamma globulin for use in the prophylaxis and treatment of measles and probably of infectious hepatitis."[13] Cohn also was the recipient of the Theodore William Richards Medal by the Northeastern Section of the American Chemical Society in May 1948. The presentation, a black tie affair, took place in Room 10-250, the large lecture hall under the Great Dome of the Massachusetts Institute of Technology before a capacity audience that included many of Cohn's colleagues, associates, students and friends. The occasion gained a certain sparkle from the presence of many ladies, some of whom were in evening dress. The medallist was formally introduced by George Scatchard, who began by stating:

> I need not tell you how great Dr. Cohn is, but only how he is great. He is the most meticulous man I know, willing to spend a huge amount of time to see that each experiment, each sentence is exactly right. He has known better than most of us what he was going to do . . . but he has had to know even more surely what he was not going to do . . . He makes no pretense to being a Jack-of-all-trades, but consults an expert for each problem and does not waste time on any but the first rate. His prime interest is in ideas. The kind of teamwork inspired by Dr. Cohn makes it often impossible to tell with whom any scientific idea originates.[14]

In accepting a medal that had been awarded on only eight occasions in the previous sixteen years, Edwin Cohn held to a serious, formal comportment. There were no light moments and no anecdotes. He began by admitting to the satisfaction which comes from recognition of one's efforts by those with whom one has been most closely associated; by those who have known one's failures as well as one's successes, and whose judgment thus depends less on the more tangible results of one's investigations, which so often are favored by chance, than on the significance of the problems which one has attempted to understand.[15]

He acknowledged his great pleasure in being chosen to receive the Richards Medal because of his sense of deep gratitude to, and very happy memories of Professor Richards. He noted that it was then almost a third of a century since he had come under the influence of Richards as a graduate student working in the Wolcott Gibbs Memorial Laboratory under Lawrence Joseph Henderson. In that environment it was clear that the same standards of accuracy and reproducibility that had been characteristic of Dr. Richards' atomic weight, compressibility and other physical chemical studies had also to be brought to bear in any physical chemical attacks on biological problems.

He reminded his audience that it was in order to learn about the chemistry of sea water, as the environment of marine life, that he had come to Harvard in 1915. It had become apparent in his earliest studies that the efforts that he and others were making to understand the behavior of unicellular organisms in terms of their environment could not proceed effectively until more was known about the chemistry of the environment. But as he studied the chemistry of sea water, especially the equilibria between the weak acids and bases that were concerned with the neutralization of the carbon dioxide produced by organisms under a variety of conditions, it seemed that this first approach was very indirect, primarily because the same environmental variables appeared to influence all organisms. This held not only for organisms that lived in the sea, but also for the more complex multicellular animals whose blood stream, in the terms of Claude Bernard, could be considered as their "internal environment." But if these environmental variables influenced all cells, in mammals or in invertebrates, they clearly influenced them to do different things. Oxygen lack, an acid concentration in the environment that was greater than normal, a salt concentration greater or smaller than normal, influenced all cells—however much the cells might differ from each other. The influence on the motility of the sperm cell was quite clearly different than the influence on the maturation of the egg cell. The influence of these same variables upon the enzymes of various systems had also begun to be studied. One could describe the changes in rate, but not the nature of specificity, in terms of the environmental variables which fascinated so many of his contemporaries and teachers. Without their contributions the next scientific steps could not have been taken.

In order to understand biological specificity, he reasoned, the problems had to be carried to the interior of the cell; they had to be attacked directly rather than through investigations at the periphery. This had brought him to the study of the proteins which were, he reminded his audience, "of the first importance," which, in turn, led him to conclude that "many phenomena of importance in biology and medicine depended upon protein interactions. Moreover, the study of such interactions could not readily be investigated until satisfactory preparations of undenatured proteins were available and until their molecular properties and their electrochemistry were understood."

X-Ray Crystallography

During the summer of 1948, Barbara W. Low was among an arriving class of six postdoctoral fellows. Although the onset of the war had disrupted Cohn's 1940 plan for X-ray crystallographic studies on proteins under Fankuchen and Warren at MIT, his determination to mount an X-ray project as an integral part of the studies on proteins in his laboratory had never flagged. Low had completed her training at Oxford in chemistry and physics and had become an expert crystallographer in the laboratory of the distinguished British crystallographer Dorothy Crowfoot, who, during the war, had played a major role in determining the structure of penicillin. Low had then spent a year at Cal Tech in Pasadena working with Linus Pauling. During the course of searching for a crystallographer, Cohn had consulted with a number of experts about the approaches to studying protein structure. In particular, he had become intrigued with the idea of labeling the vast protein molecule with a single heavy metal ion as a tag that would be "visible" on X-ray diffraction.

He had explored this idea with John von Neumann, the Princeton mathematician, who, after making some calculations on the back of a menu at lunch one day, had agreed that it was a feasible approach. As a result, Cohn expected that investigations along these lines would yield fundamental knowledge about the three-dimensional structure of proteins and contribute important information linking the structure of proteins to their physiological functions. In preparation for Low's arrival, a special cold room had been installed to accommodate the X-ray and optical equipment and enable the study of proteins in the cold under optimal conditions. This facility was one of the first of its kind, and the first in the United States.

Other new fellows arriving in the laboratory in 1948 included Walter B. Dandliker, Robert Davis, a Harvard medical student, Margaret J. Hunter, a Scottish postdoctoral fellow, J. Morton Gillespie, an Australian, Frederic W. Kahnt, a Swiss research chemist at CIBA, and two new graduate students, Howard M. Dintzis and Frederic Richards. With forty three attendees, Edwin Cohn opened the first Department luncheon meeting in September 1948 by reminiscing about the history of the Department of Physical Chemistry.[16] He told them that the Department was organized to engage in fundamental research on proteins, with few teaching responsibilities; furthermore, when President Conant visited the laboratory a year before, he had remarked that not one method then being used in protein research had been in use during the first years of the laboratory: the Clark hydrogen electrode now being used had just been developed; the Tiselius electrophoresis apparatus arrived in 1930; and the ultracentrifuge came in 1936. At the outset, horse plasma proteins had been studied, but had to be abandoned after a year of difficulties in separating pure proteins. Their investigations had then turned to casein and determined that ethanol-water mixtures helped with precipitation. Vegetable globulins, such as edestin, were easy to work with, since they can be found in nature as crystals. The first crystalline enzyme, urease, was isolated in 1926. After Northrop began in 1928, many more enzymes were crystallized. The hormone insulin was not isolated until 1923, and was crystallized several years later.

Thus, when this laboratory started in 1920, very few pure proteins were available. The policy of the laboratory, which was patterned on Osborne's work, was that one must prepare a protein in order to make it worthwhile to study; therefore, he told his new trainees, a part of this laboratory will always be devoted to preparing proteins. "This year there will be a continuation of investigations of human plasma proteins, with emphasis on the labile components," he informed the group. During the war, it was necessary to overlook many of these components "although we knew in many cases that we were denaturing them; for instance hypertensinogen, antihemophilic factor, prothrombin." He concluded, saying, "We know that one factor of importance was the time that elapsed between collection and processing of blood. One of the largest groups of unstable proteins was the lipoprotein group. As we developed our new fractionation process to replace the wartime process, it became apparent that we had never even seen the alpha lipoprotein before in its native state."

Method 6, the final version of the wartime plasma fractionation process, he continued his review, incorporated a revision of the plasma fractionation procedure at the stage following the removal of Fraction II + III. This began by precipitating a small fraction of lipid rich α-globulins into a new fraction called fraction IV-1. A second step, precipitation of Fraction IV-4, separated a group of albumin-like α- and β-globulins. This revised procedure continued to provide the major fractions of plasma—fibrinogen, gamma globulin and albumin—in good amounts, while providing access to certain other components of plasma in a native state.[17]

A separate study, begun after the war, led to the subfractionation of Fraction IV-4 into three new subfractions: Fraction IV-5+6, rich in α-globulins and serum esterase, Fraction IV-7, consisting primarily of β_1-globulins, and Fraction IV-8, containing small amounts of albumin.[18] Schade reported that the new Fraction IV-7 contained the iron binding protein component. Subsequent purification and crystallization of this β_1-globulin by Koechlin, a Swiss fellow in the laboratory,[19] confirmed that the crystalline product was the protein responsible for the iron binding capacity of serum that had been described by Schade and Caroline, who called this new protein Siderophyllin.[20]

Further studies at Harvard revealed that the newly identified plasma protein had a molecular weight of 90,000, slightly greater than the molecular weight of serum albumin.[21] It comprised three percent of the proteins in plasma. As isolated during the subfractionation of Fraction IV-4, the metal-combining globulin contained no iron. Studies of its interactions with iron revealed that each molecule was capable of binding two atoms of iron. The protein was also found to be capable of binding copper and zinc and perhaps iron as well. That copper was bound to the same site as iron was revealed from studies of the competition of iron and copper binding to the protein. These revealed that iron was capable of displacing bound copper. Based on its yield from plasma, it was estimated that two and one tenth grams of the protein were present in a liter of normal plasma, suggesting that it comprised slightly less than 3% of the proteins in normal plasma With essentially pure Fraction IV-7, a distinct color could be observed when iron was added. On these grounds, Cohn gave the protein the name "metal combining globulin."

The first clinical study of this protein was conducted at the Peter

Bent Brigham Hospital by Charles E. Rath and Clement A. Finch, he related. While confirming the chemical estimates of the iron binding capacity of plasma, they found that only a third of the iron binding capacity of plasma was normally occupied with iron. The remaining available unsaturated iron binding capacity readily bound iron when it was added, providing a way to measure the total iron binding capacity of serum simply by adding iron to serum until there was no further change in the characteristic color of the iron-transferrin complex. Rath and Finch extended their study to a series of patients with other illnesses. Patients with iron deficiency anemia were found to have a very low plasma iron level and an elevated unsaturated iron binding capacity. On the other hand, patients with hemochromatosis, a chronic disease characterized by deposition of excess iron, had no residual unsaturated iron binding capacity.[22]

A second study, conducted by George Cartwright and Max Wintrobe at the University of Utah, confirmed the normal degree of saturation of the β_1-globulin with iron, (about 35%) and total iron binding capacity corresponding to 3.6 mg of iron per liter of plasma. When infused, iron-free metal combining globulin was tolerated without reaction. It resulted in a transient predicted rise in iron-binding capacity, followed by a transient rise in serum iron. They concluded that the low serum iron often associated with infections was not the result of a reduced iron binding capacity. These studies filled in a missing link in the elucidation of the transport of iron in the blood stream.[23]

The ability of iron binding globulin to bind iron and copper proved to be yet another example of a general property of many proteins; small molecules, metallic ions and other substances were often bound specifically to them. The stabilizing effect of caprylate on human serum albumin, which made it possible to heat albumin solutions for ten hours at 60^0 C., is an excellent example. Another was Hughes's finding that mercury ions reacted with a single specific site on the albumin molecule. Cohn increasingly saw this binding property of certain proteins as specific examples of protein-small molecule interactions. This led to the general postulate that such binding involved a specific site on the protein to which the ion or small molecule was attracted. This postulate was confirmed in numerous other instances.

Pernicious Anemia Revisited

As the Hematology Study Section assumed its new responsibilities within the NIH framework, the members inevitably identified the prominent unresolved issues in the field. The nature of the factor in liver that cured pernicious anemia was one such unresolved issue. In discussing this later, Cohn said,

> When I left liver extract, I had the definite feeling that I had been throwing away a fair amount of activity in Fraction D. [Since 1940] we have been using alcohol[-water] mixtures at low temperatures so as not to denature the proteins. Using such fractionation...we could proceed on the working hypothesis that the active principle was in more or less labile combination with one or another of the coagulable proteins but had been liberated as a result of the denaturation of the protein[s] to which it had been bound. In such cases, the small molecules are liberated by denaturation of proteins, which would explain their presence in the filtrate.

He postulated that he could now take advantage of that knowledge to fractionate the proteins of the liver, isolating the protein-active principle complex by splitting off the active principle so as to detect its presence. So long as the active principle remained in a protein complex, then it was a problem that he would be interested in pursuing and that he felt well equipped to pursue. If the active principle was in the noncoagulable fraction, he would probably drop out of the field.[24]

Against that background, Cohn had been made chairman of a subcommittee on Pernicious Anemia, and the Hematology Study Section organized a national network to identify patients with pernicious anemia who could become the subjects for clinical testing of new Cohn fractions. Cohn immediately enlisted the assistance of Jules Porsche and James Lesh at Armour Laboratories in Chicago and work began. The starting material was bovine liver, which was thoroughly perfused with saline solution to minimize contamination with blood. The livers were then frozen, comminuted and fractionally extracted, first with 60% ethanol at pH5, then with 30% ethanol. The latter extract was rich in protein. In reporting on this experiment, Cohn stated "The active principle effective in pernicious anemia was not extracted with these proteins. The active principle was, however, extracted at lower ethanol concentrations. The active material reprecipitated and redissolved with the proteins of this fraction, though not without loss. The active fraction was still rich in enzymes, and it appeared of the first importance to separate the proteins from the enzymes that rendered these solutions unstable."

At this point in the investigations, the crystallization of vitamin B_{12} was announced. It thus appeared probable that this vitamin was bound as a prosthetic group to a protein in Fraction C (the third extract). This discovery that the active principle in pernicious anemia was tightly associated with a protein in the liver when isolated under controlled conditions that prevented denaturation was a source of satisfaction to Cohn.[25]

In his 1948 report as Chairman of the Division of Medical Sciences, Edwin Cohn recounted that because the observations of physiologists had often contributed to medicine through biochemistry, Harvard had recognized the need for collecting all of those concerned with this basic discipline by the creation of a Department of Biochemistry within the Faculty of Arts and Sciences by merging the existing departments of Biological Chemistry and Physical Chemistry at the Harvard Medical School, and of a new department to be created in Cambridge. This plan to bring together all biochemists in the University, wherever their laboratories, was undertaken in the interest of the best training of the biochemists of the next generation. Comparable relations should come into existence between other disciplines such as bacteriology and physiology, whether training be carried out in the Biological Laboratories or in those of the Medical School, in the interests of the best training of the next generation in these fields also.

Edwin Cohn and Rebekah Robinson Higginson

Edwin Cohn and Rebekah Robinson Higginson were married in Mrs. Higginson's home at 11 West Cedar Street on Beacon Hill in the early spring of 1948. The wedding was attended by members of their families. Rebekah had been the widow of the late Charles Higginson since 1936. She was quite well known in Boston and Cambridge. She was a tall and willowy lady, with a gracious manner and attractive personality. Prior to her marriage to Cohn, Rebekah had been a frequent "odd lady" at dinner parties in Boston and Cambridge, including some at 183 Brattle Street. She was also known for her thirty minute radio program that was broadcast on weekday mornings by one of the Boston stations. Speaking with a pleasant Boston accent, she dealt with a range of topics generally related to foods, varieties of food, history of specific foods and trends in the use of foods. She continued her broadcasts after her marriage to Edwin Cohn.

15 The Need for a National Blood Program

Over 125 participants registered for the Conference on "The Preservation of the Formed Elements and of The Proteins of the Blood" at the Harvard Medical School on Thursday morning, January 6, 1949. The conference was called at the Request of the Committee on Medical Sciences of the Research and Development Board of the National Military Establishment by the Committee on Blood and Blood Derivatives of the National Research Council in collaboration with the Panel on Hematology of the National Institute of Health, and by the Committee on Blood and Blood Derivatives of the American National Red Cross."

An eighteen-page program identified the list of government agencies and public organizations and their official representatives who were participants, beginning with Francis C. Blake, Chairman of the Committee on Medical Sciences of the Research and Development Board of the National Military Establishment. Other distinguished guests were Lt. Col. Howard A. Van Auken, representing Major General R.W. Bliss of the Army, Captain Lloyd R. Newhouser, representing Rear Admiral C.A. Swanson of the Navy, Shields Warren, representing the Atomic Energy Commission, Lewis H. Weed, Chairman of the Division of Medical Sciences of the National Research Council, Charles A. Janeway, Chairman of the NRC Committee on Blood and Blood Derivatives, Lt. Col. J. M. Matheson of the British Joint Services Mission, Rollo E. Dyer of the National Institute of Health, Charles A. Doan, Chairman of the NIH Hematology Study Section, Basil

O'Connor, President of the American Red Cross and Ross T. McIntire, Administrator of the Red Cross National Blood Program.

Edwin Cohn had carefully framed the detailed program of the conference so as to address the urgent purposes of the Research and Development Board, while at the same time interpreting the assignments to the contributing participants in terms that were scientifically and logistically achievable. The opening Thursday morning session was devoted to presentations concerning "Methods of Estimating Viability and Survival of the Formed Elements of the Blood." Following lunch in Vanderbilt Hall, the afternoon session dealt with "Properties and Functions of the Proteins and Protein Enzymes of the Plasma and of the Formed Elements of the Blood."

After dinner, an evening session was devoted to presentations on "Blood Preservation in World War II and Present Red Cross Programs." Among the topics dealt with were: "Wartime Practice in England during World War II," by Patrick Mollison; "Wartime Practice in Germany during World War II," by Kurt Reisseman; "Research on Blood Preservation in the United States during World War II," by George M. Guest; "Experiences of the Armed Forces with Whole Blood in the Pacific during World War II," by Captain L. R. Newhouser; "The Canadian Red Cross Post-war Program," by W. S. Stanbury; and the "American Red Cross Post-war Program," by Ross McIntire.

On the second morning, the program dealt with "Modification of the Chemical Environment of Blood Cells." This was followed by an afternoon session devoted to simultaneous conferences on three different topics:

1. "Influence of Low Temperature and of Freezing upon the Metabolism of Cells with Reference to the Formed Elements of the Blood."
2. "Influence of Ultraviolet Radiation, Cathode Rays and Chemical Reagents upon the Viruses and Proteins of the Plasma and of the Formed Elements of the Blood."
3. "Factors Concerned in Blood Cell Formation and Maturation." This topic extended to the morning of the third day. It dealt with the metabolism of bone marrow, protein metabolism, nucleic acid metab-

olism, iron metabolism, metabolic governors, and factors inhibiting red blood cell formation and maturation.

The final session, on the morning of the third day was devoted to the "Use of Protective Interfaces in Blood Collection and Storage."

Visits were also made to the Harvard Pilot Plant and to the new Blood Processing Laboratory of the Massachusetts Department of Health which was then serving as a Pilot Plant for the Red Cross National Blood Program. The conference stimulated a wide array of responses within the national community. Within the biomedical community, it also served to identify and orient investigators to areas in which new knowledge was urgently needed. Within the highest levels of government, it stimulated the identification of policy questions to be addressed if the country was to be brought to a better state of preparedness of its blood supply against emergencies.

After the conference, the NRC Committee on Blood and Blood Derivatives met with Dr. Blake to plan its next moves. Blake explained that the recommendations of his Committee to the armed services would relate to the broad strategic plans of the Army General Staff. They should reflect the priorities of the different research undertakings emerging from the conference. He expressed the hope that the NRC Committee on Blood and Blood Derivatives would make concrete recommendations regarding improvement of the methods for collecting and preserving blood and its components.

Before departing Boston, the NRC Committee adopted a report for forwarding through channels to Blake after careful deliberation of the problems confronting the armed forces and civilian populations. This recommended to the Committee on Medical Sciences of the Research and Development Board of the National Military Establishment the following steps in the interest of developing, having available for current needs, and accumulating for an emergency, the most satisfactory stores of blood and blood derivatives:

A. The agency charged with the responsibility for supplying blood to our armed forces and to our civilian population in the event of an emergency, should establish pilot plants, so that whatever advances are made as a result of free research in universities or foundations or as a result of government contracts or grants, may be reduced to practice as rapidly as possible and incorporated in improved procedures of blood collection, preservation and distribution.

B. The conference on the Preservation of the Formed Elements and of the Proteins of the Blood indicated that far-reaching improvements in blood collection, processing and preservation could be achieved by vigorous, adequately supported, fundamental research, as well as by development in the following fields:

1. Studies of methods for the collection of blood under such conditions as to assure maximum stability and preservation of the proteins of the plasma and of the formed elements of the blood.

2. Studies of methods for the separation of the formed elements and of the plasma proteins of the blood at the time of its collection, or as soon thereafter as possible, in order to achieve further improvements in their stabilization and preservation and to render possible their accumulation as reserve stores.

3. Studies of the surfaces of all equipment designed for use in blood collection, preservation and transfusion, in order to achieve the maximum stabilization and preservation of the proteins of the plasma and of the formed elements of the blood.

4. Studies of the enzymes of cells and plasma in order to provide the essential information necessary to prevent destruction and preservation of the proteins of the plasma and of the formed elements of the blood.

5. Studies of the environmental conditions which affect the stability and preservation of whole blood and its separated components, including studies of the effects of electrolytes, carbohydrates, lipids and proteins, at controlled pH and temperature; especially temperatures below O° C. since there have been many previous investigations at higher temperatures.

6. Studies of the sterilization of blood, its separated cellular and protein components, by all possible methods; especially by the use of the nitrogen mustards for the destruction of viruses.

7. Studies of the value of human serum gamma globulin for the control of diseases in warfare; especially of its value in the prevention of homologous serum hepatitis.

8. Studies on the physical and chemical characteristics, the physiological and immunological properties, and the clinical uses of the separated formed elements of the blood, and of the proteins of the plasma; especially such combinations of them as may achieve the maximum stability and preservation of the proteins of the plasma and of the formed elements of the blood.[1]

The Committee on Blood and Blood Derivatives also set in motion plans for two important NRC conferences. The first of these, a "Conference on Blood Substitutes," was held in February 1949. In introducing the topic of the conference, the representative of the Army Surgeon General warned that a future national emergency would raise a new set of problems requiring the stockpiling of preserved whole blood, blood derivatives and blood substitutes. Moreover, the equipment for transfusing these agents—bottles, filters and intravenous sets—would have to be standardized and stockpiled around the country. He warned that, given the possibility of atomic bomb attacks on large centers of population, large numbers of burn and blast injuries could be expected. Moreover, the army would have an added responsibility in meeting civil defense requirements. At the conference, presentations had been made concerning gelatin, periston, dextran, intravenous hemoglobin and human globin as possible blood substitutes. However, since none of these were considered to be ready for serious consideration, the NRC Committee used the new conference to stress the importance of encouraging work in this field in the academic community.[2]

A second February conference reviewed progress toward standardizing blood collecting and transfusing equipment, including the specifications for surfaces coming into contact with blood. It also heard a report from Louis Diamond concerning Red Cross blood collections. Diamond announced that nine Red Cross blood centers had become operational during the first half of 1948, while ten more centers had been activated in the second half. However, total Red Cross collections of blood in 1948 had amounted to less than 80,000 units.[3] At this point, the Committee forwarded a stern report to Blake. Its opening sentence, "There is no substitute for blood," set the tone. Stating that top priority should be given to studies concerning the separation, preservation and stabilization of the formed elements and the plasma proteins, it advised that adequate stockpiles of stable blood derivatives could be built up only if the American National Red Cross were designated as the official agent to collect and stockpile blood and its derivatives in the U.S., not just for the civilian population, but for the armed forces as well. In a hint of mounting frustration, Blake was reminded that a recommendation to that same effect had been forwarded in October 1948, and was known to have been transmitted to the Secretary of Defense in November following. However, no action had been taken. "Delay in acting upon this recommendation and thus stimulating the Red Cross to expand its program...may create a deficit in defense preparedness which cannot be compensated by any known blood substitutes." The report concluded by stating that "there is no evidence known to the Committee that any other material offers greater promise as a blood substitute."[4]

The incidence of homologous serum jaundice among patients receiving the surplus Army dried plasma that had been returned to the Red Cross was an issue of increasing concern in the discussions about the national blood program. This had stimulated investigations of ways to inactivate the presumed agent, and the NRC Committee on Blood and Blood Substitutes invited a group of experts to a conference in Washington in July 1949 to discuss the problem. Three possible approaches to killing the presumed viral agent were being evaluated at the time: irradiation with ultraviolet light, use of chemical additives such as nitrogen mustard, and by heating. Scatchard posed three questions to be answered for any blood sterilization procedure: 1) will it sterilize? 2) is it damaging to the formed elements or to proteins? And 3) can it be used to treat dried plasma, the only product that could then be stockpiled in large quantities? In response, the Committee on Blood and Blood Derivatives issued the following statements:

1. Research on the application of bactericidal and virucidal agents to the sterilization of blood and its derivatives should be prosecuted as vigorously as possible.

2. Ultraviolet irradiation of fresh plasma appears to be a reasonably safe and effective procedure for sterilization, although further studies of its effect on the chemical integrity of the proteins and field studies of its effectiveness in preventing homologous serum jaundice should be pursued.
3. Existing stocks of dried plasma should be reconstituted and fractionated to serum albumin and other stable blood derivatives because of the proven stability and safety from the standpoint of homologous serum hepatitis of serum albumin so produced.
4. A revolving stockpile of serum albumin and of plasma sterilized by an approved method should be set up as an integral part of the National Blood Program of American Red Cross.

Use of the term "revolving stockpile" in these discussions reflected the seriousness of planned preparedness for nuclear disaster. The Army was counting on building up a revolving stockpile of Red Cross blood, i.e. blood that had been collected but not yet transfused, as well as other blood substitutes.[5]

University Professor

Shortly after the January 1949 conference, President Conant notified Edwin Cohn that he had been recommended for promotion to University Professor. Coming just four years after being awarded an honorary degree by Harvard, this new distinction would place him in an elite group of Harvard's scholars. University Professorships had been established at Harvard in 1935 "for men of distinction not definitely attached to any particular department...It was proposed to reserve these new chairs for men working on the frontiers and in such a way as to cross the conventional boundaries of the specialties."[6] Up until then, four Harvard Professors had been named University Professors: Roscoe Pound in law, Werner Wilhelm Jaeger in classics, Sumner Huber Slichter in economics and Ivor Armstrong Richards in linguistics. Cohn would thus be the fifth member of the University faculty and the first scientist to be so recognized. The new appointment was to become effective on July 1, 1949.[7] However, "remembering that the work which had won him this supreme recognition of Harvard Scholarship had been shared with other members of the Department of Physical Chemistry," Cohn wondered how his elevation to a University Professorship would affect his colleagues. When this question was raised, President Conant agreed that it was a pertinent issue and called for full consideration. During the next several months, there were periodic conferences between the President and Edwin Cohn relative to the continued academic careers of Cohn's colleagues. The outcome of these deliberations was an agreement to transfer Cohn's associates from the Medical School to the University, reconstituting Cohn's unit as the University Laboratory of Physical Chemistry Related to Medicine and Public Health." The formal announcement of Cohn's University Professorship was not made by Harvard until February 1950.

At this time, G.W. Gray, the Rockefeller Foundation staff member who periodically wrote reports for the Rockefeller Trustees about Foundation supported programs, paid another visit to Cohn. In answer to a question about factors that had influenced his career, Cohn identified five men. "The first was Jacques Loeb, not only a great scientist but a sympathetic friend when I was floundering in my youth and continued to help me all his life. Second was C.M. Child, Professor of Biology at the University of Chicago during my student days there. His book, *Senescence and Rejuvenation*, which I read avidly, and his seminar, which I attended, were important influences in molding my biological thinking." It was in Child's seminar, he told, that he became acquainted with Arrhenius's idea of quantitative laws in biology and knew at once that here was a chemist whose methods he had to master. The third was Lawrence J. Henderson, whose philosophical insight into complex systems was unsurpassed, and whose encouragement was invaluable. The approach to chemical experimentation of the fourth, Thomas Barr Osborne, was simpler. Half a century ago, Osborne was convinced that proteins were respectable chemical substances, and his forthrightness convinced those who studied with him that his respect could only be maintained by continuing in the path that he had made easy. Finally, there was Thomas Hunt Morgan, who Cohn said he did not know at the time. Dr. Morgan, whose approval of Cohn's approach to cytological problems was noted in Chapter 1, provided great encouragement to the young investigator at the outset of his career.[8]

A New Department of Biochemistry in Cambridge

As soon as Cohn's appointment as University Professor was in process, Harvard Provost Paul Buck found himself, as Cohn later put it, "in a unique position to proceed with the immediate activation of the Department of Biochemistry of the Faculty of Arts and Science."[9] On May 18, 1949, the Provost had advised Cohn that "the President will announce at the next Faculty meeting your appointment as Chairman of the Department of Biochemistry of the Faculty of Arts and Sciences. You may, therefore, proceed to plan with your colleagues in the Department the development of this vital and new program which has so great a promise."[10] In this manner, the President and the Provost indicated their intention to resolve a long effort to establish a new Department of Biochemistry in the Faculty of Arts and Sciences in Cambridge.[11] The Provost also committed funds for two new faculty posts in the new Department, one at the Assistant Professor level and a second at the tenure level.[12]

As planned, the new Department was to be composed of the existing departments of Biological Chemistry and Physical Chemistry at the Medical School, and a new department to be created in Cambridge, with laboratories in the biology buildings. A search for a head of the new department had been disappointing. By late 1947, separate approaches to three distinguished American biochemists, including a Nobel laureate, had failed.[13] Under the circumstances, Conant and Buck had then turned to Cohn, their strongest internal candidate, who had the advantage of being in a Harvard Division that possessed extensive experience in training graduate students in Biochemistry. Since the founding of the Division of Medical Sciences at the Medical School in 1909, eighty-five Ph.D. degrees had been awarded under the Division, of which thirty-seven were in biochemistry. In the 1949 academic year alone, twenty-four out of forty-two enrolled students in the Division were candidates for the Ph.D. in biochemistry.[14]

Cohn wasted no time. Since the academic year was rapidly winding to a close, he immediately sent off a series of letters to colleagues in the Biology and Chemistry Departments in Cambridge, as well as in the Medical School Departments of Biological Chemistry and Physical Chemistry, informing them of his impending appointment. He also named three subcommittees for the new department with representation from each of the participating departments involved to seek candidates for the new posts. In addition, he designated a curriculum committee to initiate plans for implementation of courses under the new department.

On learning of Cohn's proposed new appointment, Edward Doisy who earned his Harvard Biochemistry Ph.D. in the Division of Medical Sciences, and Carl Cori, both Nobel Laureates, promptly expressed their satisfaction over Cohn's new appointment. Cori wrote, "My congratulations to you on your appointment as University Professor and Chairman of the Department of Biochemistry. It seems to be an excellent and as it now appears quite obvious solution of a knotty problem, excellent certainly from the standpoint of Cambridge, but how will it affect Harvard Medical School?"[15]

With the end of the academic year fast approaching, less favorable responses emerged from those more directly involved in Cambridge. Following a meeting of the Department of Biology, its chairman, J.H. Welch, advised Cohn that "the unanimous opinion of the members present was that, although his work and that of his group was of unquestioned distinction, it was of special and restricted nature that is considerably removed from the kind of general and comparative biochemistry that was so much needed by the Departments of Biology and Biochemistry. Moreover, he noted that the act of transferring Cohn's center would not increase the overall resources in biochemistry at Harvard, as would the addition of new men."[16] Sensing other opposition to the Cohn appointment, the Provost called a meeting. However, Louis Fieser, chairman of the Provost's earlier committee to present recommendations for the establishment of the new Department of Biochemistry in Cambridge, was unable to attend the meeting.

In a letter he sent to Cohn on June 6, 1949, Fieser explained that the Provost had asked him to present his views in the form of a letter that could be read at the meeting he had called. Fieser related that he had composed such a letter a couple of days ago and had showed it to the people in Chemistry and Biology who would attend the meeting. In order to remove any air of conspiracy, and in the hope of eliminating unpleasant debate from the discussions, he was now sending Dr. Cohn an advance copy. "There is no question of any antagonism or lack of confidence," he assured Cohn. "The thing boils down to a difference of opinion...But I think you will find that Wald, Thimann, Bartlett and Woodward agree with me and that there is no point in debating this point any further." In the letter sent to the Provost to be read at the meeting, Fieser stated,

> The policy of the subdepartment of biochemistry should correspond to that of other departments under the Faculty of Arts and Sciences, including the departments of chemistry and biology. Thus the chairmanship should be exactly what the name implies, and not a headship or dominating directorship patterned after the European tradition. The latter is assuredly the practice in most biochemistry departments associated with medical schools, but all of our thinking was along the lines of something that would be an improvement over this tradition. We envisioned a department including not only the members of the full-time staff but also part-time members from Chemistry and Biology, and we assumed that all such members would be charged with joint responsibility for operation of the department and would vote on all major issues. It would appear difficult to secure effective cooperation from part-time members under any policy structure different from that obtaining in their own departments.
>
> I think we felt further that, although at the outset the new subdepartment [sic] might be staffed by a group imported from a medical school and operating as a coordinated research unit, definite provision should be made for future development of a decentralized department including completely independent individuals or research groups of diversified interests and objectives.
>
> These, I think, are the basic points of policy that our committee regarded as of key importance.
>
> The plan to call Dr. Cohn to the Chairmanship of the subdepartment at first seemed to me to offer great promise. However, his initial acts and statements have not indicated a desire or intention to follow the basic policy outlined, and hence I feel that plans for the department should be reviewed.[17]

Cohn explained in a letter to the Provost a few days later that he had acted out of a belief that "abstract discussion was less likely to resolve these differences than effective collaboration in solving the concrete problems that must be faced in expanding, without ignoring, the excellent record of the Faculty of Arts and Sciences in the training of biochemists, including two winners of the Nobel Prize."[18]

On June 7, Warren Weaver, Director of the Division of Natural Sciences at the Rockefeller Foundation, visited Cohn at the Medical School and later had dinner with the Cohns in Cambridge. His notes read as follows:

> E. J. Cohn's home on Brattle Street proves to be pure Marquand. He has remarried, his wife (younger, attractive, two children) being a Higginson. WW [Warren Weaver] has on an ordinary business suit, and EJC turns up with a summer tuxedo, cummerbund and all.
>
> The evening is devoted to several hours of exceedingly devious conversation. EJC has recently been made a "University Professor"—the first such appointment of a scientist. He labors over many long stories of historical background, his relations with Conant, the position of his group in the medical school; and talks in very confused and vague ways about "meeting this great new opportunity," "what is really best for biochemistry," "this problem is much bigger than he is"—etc., etc.
>
> About 11:30 WW finally tumbles to what this has all been about. Some time ago Harvard (Conant?) decided to found and develop a general university (arts and sciences college) department of biochemistry. They tried to get [3 candidates mentioned]...Failing in these, Conant has offered the post to Cohn. Should he take it? And EJC almost surely had in mind, possibly as his primary question, would the RF object?
>
> C's institute has, in sober fact, never become integrated or absorbed into Harvard Medical. When C retires, it is very problematical what will happen to the institute, the work, and the other men. If he shifted them to a general university department of biochemistry, the problem of stability and of succession would almost surely be easier. Furthermore, WW happens to believe that biochemistry deserves and requires a general setting. Thus he is inclined to think this is a sensible move, but is by no means prepared to advise C to make it.[19]

Cohn's appointment as Chairman of the new Faculty of Arts and Sciences Biochemistry Department was never announced. Months later, in a long letter to President Conant, Cohn wrote: At the time . . . we believed it feasible that our group could make the desired contribution of creating a teaching department in Cambridge . . . However, conferences that followed rendered it most probable, (happily before this experiment was tried,) that the definitions of function, shared by the senior members of our group, could in all probability not satisfy the aspirations of some of the members of the Departments of Biology and Chemistry on the Provost's committee.[20] With Commencement and the end of the academic year at hand, the issue faded into the background, to be left, as so often happens in academic life, to a future date.

A Visit by Frederick Sanger

The guest at the last Department of Physical Chemistry luncheon of the 1948-1949 academic year was the English chemist Frederick Sanger. In introducing Sanger, Cohn stressed the enormous importance of Sanger's current work which was aimed at determining the amino acid sequence of the hormone insulin. An organic chemist, Sanger was conducting a painstaking study of the arrangement, or sequence, of the amino acids in the peptide chains of insulin. He gave a fascinating report on the present status of his work. Reasoning that the amino group ($-NH_3+$) of the terminal amino acid(s) of peptide chain(s) should be reactive chemically, Sanger devised a method for "tagging" this group by reacting it with a fluorinated derivative of dinitrobenzene (FDNB). When, following such a reaction, insulin was hydrolyzed to its constituent amino acids, the terminal amino acid(s) could be identified as dinitrobenzyl derivatives. After working on this problem for several years, Sanger had established that the insulin molecule was comprised of four peptide chains. In two of these, the N-terminal amino acids were glycines; in the other two, phenylalanines. Remarkably, the four chains accounted for all the N-terminal amino acids in insulin.[21]

Sanger's work confirmed earlier conclusions that proteins like insulin behaved as true chemical compounds, always having the same four peptide chains and N-terminal amino acids. On the other hand, the insulin of different species of animals (pigs, cows, sheep) differed slightly in amino acid composition. Sanger did not complete his epoch-making work until 1953, when he worked out the complete sequence of amino acids in insulin by an iterative process involving reacting the N-terminal amino acid with FDNB, splitting off the resulting dinitrobenzyl derivative, identifying the amino acid involved, and repeating the process stepwise down through each peptide chain. Sanger won the 1958 Nobel Prize in Chemistry.

As the 1949 academic year ended, Edwin and Rebekah flew to Anglet, near Biarritz on the French Atlantic coast, where they had taken a villa. They had a long holiday that alternated between swimming, enjoying French cuisine, tennis, visits to Romanesque churches in France and Basque churches in Spain. They even went to a bullfight in Spain.

Homologous Serum Jaundice

Cohn did not attend the NRC conference that was held in July 1949, which addressed the disturbing incidence of homologous serum jaundice among patients receiving surplus Army dried plasma. To stimulate investigations of possible means to inactivate the presumed viral agent, the NRC Committee on Blood and Blood Derivatives had invited a group of experts for a review of the problem. In addition to the use of heat, three possible approaches to killing putative viral agents were being evaluated at the time: irradiation with ultraviolet light, irradiation with cathode rays, and treatment with new viricidal agents such as nitrogen mustard. Both ultraviolet and cathode ray irradiation had the potential of being safe and nontoxic, but considerable additional work would be needed concerning their effects on the chemical integrity of proteins. Nitrogen mustard was known to be effective in killing viruses, but carried with it risks of undesirable side effects. Nevertheless, since these were the only agents then under serious consideration, the Committee advocated additional research. Unfortunately, the effectiveness of any candidate viricidal agent in preventing homologous serum hepatitis transmission would have to be evaluated in human clinical tests with human volunteers. Under the circumstances, Joseph Stokes remarked that "the scarcity of volunteers for testing meant that only candidate treatments which showed the greatest promise could be tested."

In executive session following the conference, Diamond reported that the Red Cross had ordered the return of all dried plasma, although many physicians wished to continue using it. The Committee urged that existing stocks of dried plasma be fractionated to yield albumin and immune serum globulin to take advantage of their proven stability and safety with respect to transmission of homologous serum hepatitis. It also urged that serum albumin be stockpiled as an integral part of the National Blood Program of the American Red Cross.[22] As of June 1, 1949, dried plasma equivalent to 170,000 liters was on hand and awaiting fractionation at three pharmaceutical firms. It was estimated that, once fractionated, this would yield approximately 200,000 100 ml bottles of 25% human serum albumin.

Method 10 of Plasma Fractionation

During the war, scientific observations had occasionally suggested simplifications of the plasma fractionation procedures. However, since almost any change, even a minor one, would have meant a costly interruption in the delivery of serum albumin and gamma globulin, Cohn seldom considered them seriously. After the war, however, he urged Dora Mittelman to pursue one such suggestion, based on an observation of Harold Taylor during the war, that might increase the yield of albumin. Mittelman, a clinical chemist from the Institutum Campomar Para Investigaciones in Buenos Aires, had come to work with Cohn in the late summer of 1946. Working with small volumes of plasma, Mittelman discovered a set of conditions (pH 5.8, 18% ethanol at −5°C.) for separating plasma into two major fractions. Under these conditions, a precipitate was formed that contained one third of the proteins, including all the gamma globulin, $\beta_1$lipoprotein, isoagglutinins, plasminogen, cold insoluble globulin, fibrinogen, and prothrombin. The supernatant solution contained two thirds of the proteins, including all the albumin, transferrin, serum esterase, α_1 lipoprotein, α_2 glycoproteins, α_2 mucoproteins, choline esterase and phosphatase. Cohn and Mittelman described this separation at a meeting of the American Chemical Society in New York in September 1947.[23] It was unexpected that the gamma globulins would be insoluble under those conditions, or, for that matter, that the $\alpha_1$lipoproteins would be soluble. The explanation lay in finding that the proteins in the precipitate under these conditions had formed insoluble protein-protein complexes involving salt-like linkages between proteins of opposite net electrical charges at pH 5.8. The proteins that remained in solution under these conditions had electrical charges of the same sign.

In 1948, Mittelman had returned to Argentina. Continuing her line of research, J. Morton Gillespie, an Australian postdoctoral fellow, contributed another important observation. Gillespie discovered that the addition of small quantities of zinc acetate (0.02 molar) to Mittelman's supernatant solution quantitatively precipitated the albumin and other soluble proteins contained therein. Taken with Mittelman's work, Gillespie's observation meant that almost all the proteins of plasma could be rapidly reduced to two insoluble precipitates without going more acid than pH 5.8 and without requiring more than 18% ethanol. Cohn was attracted by this observation because it offered important advantages over the wartime plasma fractionation procedure in which successive fractions were precipitated and removed, one by one, from solution. It also took advantage of the greater stability of proteins in the insoluble state rather than in solution.

Cohn believed that the mild conditions involved in these new methods would result in the survival of many additional protein components of plasma that might have been adversely affected by the conditions inherent in the wartime methods, which could now be recovered in their natural state. In 1948, this set in motion further studies that ultimately involved several visiting postdoctoral fellows and led within a year to what Cohn called a "new process" of plasma fractionation that he named Method 10. Method 10 was published in the *Journal of the American Chemical Society* under the title, "A System for the Separation of the Components of Human Blood: Quantitative Procedures for the Separation of the Protein Components of Human Plasma."[24] Method 10 offered the potential of doing for the plasma proteins what the

Formed Elements group was doing in adapting newly emerging technologies for harvesting the formed elements of blood. However, as published, Method 10 did not offer subfractionation procedures to obtain gamma globulin and other purified proteins from the initial globulin precipitate.

Cohn saw one further advantage: Method 10 opened up the possibility of continuously processing blood to obtain fresh, potent products for clinical use. In a state of excitement, he consulted Warren K. Lewis and Edward Gilliland in the Chemical Engineering Department at MIT in the hope of identifying a local firm that could help him in designing and fabricating processing equipment with which further to exploit the new process commercially. Lewis and Gilliland suggested Artisan Metal Products, a Waltham firm owned by James Donovan, as ideally suited to deal with the problems being posed. Arrangements were promptly made whereby two engineers from the firm began working with Cohn's associates.

Prior to publication of Method 10, Cohn had forwarded copies of the manuscript to Vlado Getting, the Massachusetts Commissioner of Public Health, as well as to officers of the pharmaceutical firms engaged in plasma fractionation. Cohn wrote,

> Whereas we are quite satisfied that the new procedures in Method 10 are far superior to those introduced under pressure during World War II, a period of clinical trial will be necessary regarding each new product—whether previously available or prepared for the first time by the new method—before it can safely be recommended for use in man. Whereas, we cannot authorize your substituting the new procedures or introducing new products at this time, we thought, as a matter of courtesy, that you should be aware of them before publication of our new paper. Should you wish us to, we will forward to you detailed procedures for the preparation of each product and would be happy to have you cooperate, on a completely voluntary basis, in the preparation of these materials during a period of clinical trial.[25]

Taken together, these combined efforts of the Formed Elements Group and the protein scientists in the Department of Physical Chemistry, working under Cohn's leadership, comprised a daring initiative to jump-start a new national initiative on blood.

An Incident Regarding Faculty Appointments

As the 1949–1950 academic year got underway, Cohn forwarded a routine set of recommendations for appointments of six postdoctoral Research Fellows to George Packer Berry, the newly appointed Dean of the Harvard Medical School. In a brief covering letter to Berry, he noted that "it has always been customary to give nominal titles to those who come to us with fellowships ... It is my impression that these do not necessarily go through the Faculty, but may go directly to the Corporation. However you will know this better than I."[26] Unfortunately, Berry misunderstood the intent of Cohn's letter, writing, in response,

> In your letter you give your impression, if I interpret your statement correctly, that these appointments of yours need no longer come before the Medical Faculty. I have explored this point and agree that you are quite correct. Having been signally honored by the President and Fellows of Harvard College, who have elected you to a University Professorship, you are now responsible only to the President. Obviously it would be entirely inappropriate for me as Dean of the Medical Faculty, of which you are no longer officially a member, to express either approval or disapproval of your recommendations. I am, therefore, returning the data...without comment, save to say that I have greatly enjoyed looking over their records and take the present opportunity to extend my congratulations to you on this impressive group of men who are continuing their scientific training under your guidance.[27]

Cohn was taken aback at being so brusquely informed that he was no longer a member of the Medical Faculty. Indeed, in all the excitement of becoming a University Professor, he may not have considered the implications for his old department and the associates on whom he relied so heavily. Ignoring the balance of Berry's gracious comments, he immediately drafted a response that began "The point that I had in mind...was not related to my new status in the University." Then he thought better of the idea; Berry's letter was filed with a handwritten note: "Not answered, but discussed with President Conant." In its place, he wrote a long letter to President Conant, reviewing the history of the Department of Physical Chemistry, its teaching activities, and the men and women who were trained there. He also reiterated his great satisfaction to be able to report directly to Conant on the rather intricate problems which had to be considered in order to bring his research group to its present state and to develop essentially new interrelations to government, to industry, and to the various departments of this and other universities that he noted. The appointment of a scientist to a University Professorship would have value only if the group of investigators, with whom a natural organic relationship had developed, were considered as well as the individual,[28] Conant agreed, and undertook to consider matters further.

Commission on Plasma Fractionation and Related Processes

The Commission on Plasma Fractionation and Related Processes held its annual meeting on November 30, 1949. In the interval since its founding in 1944, over 500 lots of human serum albumin and almost 400 lots of immune serum globulin had been submitted for testing and release. At the meeting, Edwin Cohn stated that "the Commission's experiment in the control of plasma products in the public interest is clearly beginning to be understood." In his view, there was no reason to alter the position that caused its establishment in 1944: to assure the public of products of the highest possible quality, potency and safety. He also stated that "there would seem to be agreement in Washington on the principles established by the Commission, and it is, indeed, conceivable and desirable that some of the functions of the Commission may one day be taken over by National Agencies."[29]

This referred to what had been Cohn's unwritten, but openly held objective for the Commission. He intended to use the Commission's licensing and control of the products of plasma fractionation as leverage to force the government to adopt more rigid standards. If successful, the functions of the Commission in controlling the quality, potency and safety of products made under the Cohn patents would no longer be necessary. Control by the government would have risen to the level of control by the Commission. In keeping with this objective and the openness of Commission affairs, Norman Topping, Associate Director of the National Institutes of Health, was elected to the Commission, and attended the next meeting. Roderick Murray and William Workman, senior members of the staff of the NIH Division of Biologics, were appointed to the Advisory Committee on Serum Albumins and the Advisory Committee on Serum Gamma Globulins, respectively.[30]

As an indicator of the quality of products being produced by the licensed pharmaceutical firms at the time, only twenty-four out of the 505 lots of human serum albumin had been rejected by the Commission. Of those, twelve were rejected because of unacceptably low stability, eight because of clinical reactions in man, and four because of unacceptably high mercury content. During the same period, 384 lots of gamma globulin (immune serum globulin) had been tested, of which only one lot was unacceptable.

By 1949, the Commission had detected no change in the potency of the oldest lots of gamma globulin. As a result, the dating period was increased to a total of eleven years from the time of manufacture, as follows: up to five years of refrigerated storage as dried powder, up to three years refrigerated storage of the final product by the producer, and three years storage after distribution for use. From the outset, the Commission Specifications for human serum albumin had required the conduct of a pyrogen test in humans. The Minimum Requirements of NIH mandated an animal pyrogen test (in rabbits) that was much easier to perform, albeit with less reliability than the human test. At the time, without differing with the Commission on the validity of the human test, the NIH Division of Biologics had reviewed the matter and decided to take no action pending completion of a study and development of a definitive (animal) test.

The Commission and the NIH Division of Biologics also differed with respect to their ground rules in dealing with licensees. Under Federal law, communications between producing firms and the NIH Division of Biologics Control with respect to products submitted for approval were strictly confidential. The Commission on Plasma Fractionation and Related Processes followed the same practice, with a minor difference. Without divulging the details of its actions with respect to specific producers, aggregated comparative information concerning the testing outcomes of each producer, by product, were distributed at annual meetings to which all licensed producers and NIH representatives were invited, individual firms being identified only in code. As a result, representatives of each firm present at the meeting could only identify data from their own firm. This turned an annual review of the products being produced by the firm into an educational experience in which each firm could view the qualities of its products in relation to those of other firms licensed by the Commission. Even more important, this made possible discussions between producers and Commission Advisory Committee members about related topics as, for example, the value of specific tests.

The dialogue between the Commission and its licensees contributed in other important ways to the quality of the products being produced.[31] While gamma globulin was being developed in 1943 and 1944, samples from almost every lot had been tested clinically for safety. This not only assured the integrity of the product but also contributed valuable data for arriving at a standard dosage for its use in measles. Thus, for prevention of measles, the dosage was set at 0.1 ml per pound while the dosage for modification of measles was set at 0.025 ml per pound. The aggregate experience with gamma globulin also made it possible to reduce the frequency and cost of testing. In 1949, only twenty-three out of 135 lots of gamma globulin were subjected to a clinical safety test.

1949 Symposium on Blood Preservation

On December 2, 1949, the NRC Committee on Blood and Blood Substitutes held a Symposium on Blood Preservation to hear follow-up reports of progress made since the January Conference in Boston. According to McIntire, thirty Red Cross regional blood centers were then in operation, with another ten scheduled to be running by July 1950. In the four month period prior to the conference, 150,000 units of blood had been collected. At that time, nearly twenty percent of the 6,600 U.S. hospitals were receiving blood from Red Cross, an indication of the long road that lay ahead in building the Red Cross blood program. The total peacetime need for blood at the time was estimated to be 3,000,000 units a year. Diamond reported that contracts then in force called for the fractionation of 480,000 units of dried plasma. Serum albumin, in short supply, was being distributed only for special cases of hypoalbuminemia due to liver or kidney disease. During the 1948-49 season, 600,000 two ml units of gamma globulin had been distributed to state health departments for the control of measles. Diamond stressed the inadequacy of the supply of fractions, indicating that there was little likelihood that a backlog of gamma globulin could be built up. Even more disturbing, the supply of plasma for fractionation would soon run out, raising the possibility that it would be necessary to obtain plasma from current blood collections.[32]

At this meeting, the Committee once again discussed the need for a National Blood Program. By then, the NRC Committee on Surgery had joined the Committee on Blood and Blood Derivatives in calling for a blood program that would be national in scope and capable of very rapid expansion in case of emergency. To that end, Janeway had invited Richard Meiling of the Research and Development Board to attend the meeting. Meiling pointed out that direct responsibility for defense was vested in the President of the United States. Only he, as Chairman of the National Security Council, could initiate a national blood policy. He noted that a fact-finding group representing the three Armed Forces was then at work. When that group had the needed information, he expected to be in a much better position to request recommendations from the Committee.

In the discussion that followed, Edwin Cohn summarized the efforts of the old Subcommittee on Blood Substitutes in repeatedly recommending the use of whole blood to the Armed Forces. Those recommendations had not been acted upon until, following a period of heroic attempts at blood collection in the field by the Army, reports arrived from the African campaign pleading for blood. It was only when the Office of Scientific Research and Development came into the picture that work was begun on the preservation of whole blood. He reminded those present that he had suggested to NRC as early as 1943 that the gamma globulin being produced in surplus quantities under military contract be made available to the civilian population. To overcome difficulties in divided responsibility and ownership, an agreement had been adopted under which the Red Cross re-acquired the surplus product.

Cohn proposed that this wartime agreement be amended to reflect the changes that had taken place since 1943. In his view, no national blood program could work without the full cooperation of all groups and agencies concerned. During the war, no decision regarding blood products had been made by the NRC Subcommittee without the approval of the NIH Division of Biologics Control. He suggested that the NRC would be a natural liaison group between the Armed Forces, NIH, Veterans Administration and other agencies. Any rivalry between the procurement agencies would be disastrous for a national program. A single agency should be designated. He even went so far as to note that the new Committee on Blood and Blood Substitutes had not ruled out the possibility that some agency other than the Red Cross might be given the responsibility to collect blood. If another agency were designated, he would advise the Red Cross to withdraw. He stressed that "it is absolutely necessary to have a single national program to make products available anywhere for an emergency." As a result, the Committee unanimously adopted a set of recommendations that amended the old Armed Forces-Red Cross agreement and called for immediate steps to prepare recommendations aimed at development of a National Blood Program.[33]

A few days later, the American Medical Association Committee on Blood Banks, together with the Red Cross Executive Committee and the Committee on Blood and Blood Derivatives of the Advisory Board on Health Services of the American Red Cross, held a joint meeting. Among the guests in attendance were Meiling and General Marshall, who had just been elected President of the American Red Cross. Ernest L. Stebbins, the Chairman of the Red Cross Advisory Board, chaired the meeting. Janeway reviewed the operation of the blood program for the armed forces during the war, in particular, stressing the important role of the technical experts on the Subcommittee on Blood Substitutes. This was the pattern his committee felt should guide any truly National Blood Program. He stressed the importance of research, pointing out that research to extend the preservation period for whole blood had been halted at the end of the war. He stressed the view of his committee that the blood program should be officially recognized by the government against the possibility of a national emergency, since "only by having a unified integrated national blood program in operation could we be prepared for such an emergency."

As the meeting neared adjournment, Stebbins asked General Marshall if he would like to comment. Marshall asked two questions: does the Red Cross National Blood Program meet a public demand that otherwise could not be met? And, to what extent is the National Blood Program approved by the American Medical Association? Based on the responses offered by those present, General Marshall stated that the coming together of professional men in that meeting reassured and fortified him. It was a lengthy meeting, but after working out the exact wording, it was unanimously agreed that

> The committee believes that, as never before in peacetime, it is our solemn duty to encourage the development of, and participate in, a plan which will insure an adequate supply of blood and blood derivatives, in the event our country is suddenly, and possibly without warning, plunged into another world war. The committee is convinced that because of its pattern of nationwide organization and its experience and record in the procurement and processing of blood during the last world war, the Red Cross is the logical agency to assume this responsibility.[34]

Edwin and Rebekah Cohn enjoyed a pleasant life at 183 Brattle Street. Breakfast came up on the dumb waiter, one tray at a time, and was taken in Edwin's study on the second floor. Rebekah had many interests that took her on a round of appointments and other activities, many of which were related to her daily broadcast schedule. On most days her broadcasts were done "live" from the station, although some programs could be recorded in advance. At the end of their day, over a drink and dinner at home, they enjoyed comparing their experiences. Rebekah liked to stop at Postar's, a wholesale antique emporium on Market Street in Brighton that periodically brought in shipments of furniture from Europe. At dinner one night, she described an unusual piece of furniture that had caught her eye. Edwin listened to her description and asked her a few questions. There was a pause. Then, with a tinkle in his eye, he asked if she had bought it. She had! He told her that it could only be a Queen Anne sideboard.

Weekends were spent in Cohasset where they often worked in a set of tennis with friends. Thanksgivings were informal family occasions with the Higginson offspring and their grandmother in the big house. Rebekah might roast a goose on a spit before the fire. One or two of

Edwin's younger associates and their families might have been invited. After a long and pleasant dinner, there would be a walk down by the water. For quiet weekends during the winter months, rather than open the main house, they lived in a smaller, separately heated wing of the house.

16 University Professor

Although Edwin Cohn's promotion to University Professor had become effective on July 1, 1949, it was not until February 1950 that the University announced the appointment and the creation of a new university laboratory to be called the University Laboratory of Physical Chemistry Related to Medicine and Public Health. The University announcement stated that the new laboratory would "make fundamental studies upon the constituents of body fluids and tissues and extend the research upon the physical chemistry of proteins and other biological substances which are characteristic of all living matter and systems. Physical and chemical theory and methods will be brought to bear on problems related to medicine and public health. Research will center on the study of proteins, the major structural and functional elements of plant and animal cells and tissues."[1] The announcement also stated that the Harvard Corporation had approved the construction of a new University Laboratory building in Cambridge for Dr. Cohn and his group as soon as new funds are available.[2] The new laboratory was the first to be associated with a University Professor. It was free to continue cooperation with groups within the University and elsewhere whose interests and activities converged upon the problems being investigated.

The roster of the new laboratory as of July 1, 1950 included Edwin J. Cohn as Director, John L. Oncley and John T. Edsall as Members, and Walter L. Hughes Jr., Douglas M. Surgenor and Barbara W. Low as Associate Members, and three Research Associates and eleven Research

Fellows in Physical Chemistry. Academically, the group functioned as the new Department of Biophysical Chemistry within the Division of Medical Sciences of the Faculty of Arts and Sciences. The Department of Physical Chemistry at the Harvard Medical School was dissolved. For the time being, however, the new University Laboratory continued to use the facilities of the former Department of Physical Chemistry at the Medical School.

It was hoped that the name chosen for the new facility would ease certain sensitivities among physical chemists in the Harvard Chemistry Department in Cambridge. Unfortunately, the *Harvard Gazette* omitted the last six words in the title in announcing a series of seminars to be held in the new laboratory a few days later. Cohn promptly sent a letter of apology to George Kistiakowsky, the senior physical chemist in the Chemistry Department.[3] Thereafter, Cohn was careful to use the full name at all times.

To mark the occasion, a remarkable hundred page booklet was published in March 1950 under the title:

UNIVERSITY LABORATORY OF PHYSICAL CHEMISTRY
Related to
MEDICINE AND PUBLIC HEALTH
HARVARD UNIVERSITY
1950
Department of Physical Chemistry
Harvard Medical School
1920–1950

Its Foreword, written by President Conant, told that by vote of the President and Fellows of Harvard College on June 19, 1935, a plan was adopted for the "establishment of new professorships for men of distinction not definitely attached to any particular department" and these were to be known as University Professorships. It was proposed to reserve these new chairs for men working on the frontiers, and in such a way as to cross the conventional boundaries of the specialties, and Professor Edwin J. Cohn was the first scientist to hold a University Professorship. This pamphlet, which described the history of Cohn's activities over the last thirty years, made it evident why this appointment to a University Professorship was peculiarly appropriate. His Department of Physical Chemistry, organized originally as part of the Faculty of Medicine of Harvard University, soon became a focal point for the application of physical methods for the study of proteins. The men he has had working with him over this period of time and the students he has trained represented a wide variety of interests. It was therefore quite fitting that this undertaking should now bear the title of University Laboratory of Physical Chemistry Related to Medicine and Public Health of Harvard University. Those who were concerned with the history of Dr. Cohn's work and were curious as to the function and opportunities provided by a Harvard University Professorship would find this pamphlet of more than usual interest.[4]

Intended by Cohn for use in fund raising, the pamphlet provided a unique record of the scientific work of Cohn and his collaborators between 1920 and 1950. In the introduction, Cohn related that Universities had always served more than one function. The faculties of universities had the responsibility of teaching existing knowledge to the next generation, and additionally assumed the responsibility of acquiring new knowledge, and of preserving the accumulated knowledge of the past in university laboratories and university museums. In the past, the responsibility of acquiring new knowledge had generally been assumed by the professor rather than by the University. Indeed, in the United States, until the founding of Johns Hopkins University in 1867, the professor was supposed to do research, should he care to, in his leisure time. The epochal contribution of Daniel C. Gilman, the first President of Johns Hopkins University, was to diminish the amount and the kind of teaching. Fewer undergraduates to teach, and more advanced students to work with surely would be conducive to carrying on more effective research.

The logical conclusion to lightening the teaching load had, however, been achieved in the independent research institution from which formal teaching had been banished. Even the title of "professor" had vanished and been replaced by naming the investigator a "member" of the research institute. Implicit in the creation of research institutes outside the framework of universities had been the assumption that research was more effectively pursued when investigators were segregated—when the investigator was freed of the teaching function and charged

Fig. 8. Device designed by E. J. Cohn to symbolize the mission of the University Laboratory of Physical Chemistry Related to Medicine and Public Health.

with the sole responsibility of advancing the state of knowledge. However, the secluded investigator, working under these conditions, had lost day-by-day contact with the generation in training to become teachers or investigators in university, government, research foundation or industrial laboratories. Appointment of a scientist as a university professor raised the question as to whether a new pattern was not possible, namely the building, both spiritually and physically, of a university laboratory in which a university professor and the group of collaborators associated with him could make their most effective contribution to research, in the field of their interest and to the university of which they are a part. The definition of a university professor did not preclude, he affirmed in closing, but rather suggested the desirability of making available the contributions of a university professor not only to a Faculty of Arts and Sciences, but to all professional schools for which his work, or the work of his group, has significance.[5]

The pamphlet included the names of the twenty members of the new laboratory with a description of their scientific interests. It provided a listing of all "Investigators who have worked in the Department of Physical Chemistry, Harvard Medical School, 1920–1950," with years in the laboratory and present position, as well as "Colleagues from other departments or institutions who were directly associated with the work of the Department of Physical Chemistry in its Investigations on Blood and who worked under contracts recommended by the Committee on Medical Research to the Office of Scientific Research and Development, 1941–1946." There was also a list of "United States Army and Navy Officers assigned to the Plasma Fractionation Laboratory, Harvard Medical School, 1941–1946," and "Representatives of Pharmaceutical Houses who came to the Plasma Fractionation Laboratory for training in procedures, at the request of the Surgeon General of the United States Navy, as well as Technical Assistants whose contributions were recognized by co-authorship in scientific papers." Also included was a "Historical Sketch of the Department" written by John Edsall, which listed 399 "Scientific Papers from the Department of Physical Chemistry, Harvard Medical School, published between 1920 and 1950." This pamphlet was identified as the "First Publication of the University Laboratory of Physical Chemistry Related to Medicine and Public Health." Interspersed in the pamphlet were a series of couplets written by Charles Morton, who came to America in 1686 to become the pastor of the First Church in Charlestown, bringing with him the set of manuscript outlines for his *Compendium Physicae* that appear to have been adopted at Harvard in 1687, of which the following is an example:

Where New appearance is before the Eyes,
New Suppositions thereupon arise.

Prominent on the title page of the pamphlet, was a device, (Fig. 8) Romanesque in shape, that Cohn had designed to symbolize the mission of the University Laboratory of Physical Chemistry Related to Medicine and Public Health. In a note at the end of the pamphlet Cohn stated that

> The device was evolved gradually from its origin: the essentially uncorrected pencil marks on the margin of an agenda of a not completely absorbing conference in the fall of 1948. In it, the convention

> of the chemist, the benzene ring and the crossed retorts, formed a background for the staff of Aesculapius. Phrases derived from the definitions of the status and functions of the University Professor surround the central design. The spectral lines were suggested by I. Bernard Cohen; the actual spectral lines are those of iron in hemoglobin. The shape of the device derives from the Romanesque rather than the Roman arch and was suggested by R.R.C. on recalling the twelfth century bas-reliefs behind the choir of St. Sernin in Toulouse. The conviction that work "on the frontiers of knowledge" must be based on the traditional function of the Faculties of Universities in preserving and disseminating existing knowledge seemed to give symbolic significance to the form, which was therefore immediately and gratefully accepted.[6]

This pamphlet could be more accurately characterized as a memoir of the old Department of Physical Chemistry at the Harvard Medical School rather than as a fund-raising instrument for the new University Laboratory. Its only allusion to fund-raising was in its final sentence, which read, "It is believed that the construction of a research laboratory dedicated to research in biophysical chemistry, in which the teaching of the next generation can also take place, has great advantages in assuring the most rapid fundamental dissemination of newly developed scientific theories and methods basic to medicine and the public health." Given the splendid record of past scientific achievements documented in the pamphlet, it might also have raised some questions on the part of prospective donors as to why, after thirty years of scientific discoveries, including substantial scientific contributions in the national interest during World War II, it had suddenly become necessary for Harvard to move Edwin Cohn and his colleagues from the Medical School in Boston to the Harvard campus in Cambridge.

With the establishment of the University Laboratory, a fund-raising goal of $1,750,000 was arrived at. This included $750,000 for construction of a new laboratory building in Cambridge on a site owned by Harvard, and an endowment of approximately $1,000,000 to support the activities of Cohn's associates in the laboratory. President Conant and Cohn set out to identify some individuals willing to help raise the funds. Among these were Eustace Seligman, a partner at the Sullivan & Cromwell law firm in New York, and an old friend of Cohn's.

Seligman was very interested in the idea but was pessimistic about finding donors who would be interested in Cohn's field of endeavor. He could identify only two potential donors, who "for various reasons cannot be solicited."[7] Joseph Barker of the Research Corporation in New York, and Victor Conquest of the Armour Company in Chicago were asked to help, and did so, but concerns were expressed about the propriety of seeking funds from pharmaceutical firms holding licenses under the Cohn patents.

Another prominent prospective donor, when cautiously approached by Cohn, indicated that he would want to talk to Alan Gregg at the Rockefeller Foundation about the project. Cohn arranged to brief Gregg in advance of the prospective donor's visit. Whether the prospective donor ever talked with Gregg is not known, but in his diary, Gregg noted that "the idea is in some ways interesting but it will also be interesting to see how it fares in the hands of Cohn and Conant in the face of some considerable apathy or disagreement among others in the field of chemistry and other parts of the University, notably the Medical School."[8] President Conant went so far as to sound out Chester I. Barnard, the President of the Rockefeller Foundation, about obtaining a grant from the Foundation toward construction of the new laboratory, but without success. The fund-raising effort continued in earnest until the summer of 1950, when it was interrupted by the United States involvement in the Korean War.

During this period, General George C. Marshall was elected President of the American Red Cross. He immediately wrote to Cohn: "it would be of the greatest value to the Red Cross, and to me personally, if you would agree to serve as advisor and consultant to the president of American Red Cross on matters relating to the Blood Program. This would enable me to call upon your great talents and long experience in observing and guiding the future development of a vitally important program."[9] In accepting Marshall's invitation, Cohn confirmed the importance of the Red Cross Blood Program's relationship to medical science and the desirability of collaborating with other national foundations, societies or agencies, private and governmental, concerned with the diagnosis, prophylaxis, or therapy of various diseases.

Cohn and the combined Formed Elements and plasma fractionation groups were engaged in designing and testing a single inclusive system for the collection of human blood and the separation and preservation of the formed elements and plasma proteins in stable states. Cohn's plan was to combine and integrate the new knowledge gathered and new discoveries made since the 1949 Harvard Conference sponsored by the Research and Development Board of the National Military Establishment. In the planned new system, Cohn was determined to incorporate the advantages inherent in the new principles developed since that conference. These included the elimination of wettable surfaces, the removal of calcium by passage of donor blood over an exchange resin, the separation of red cells by sedimentation from the white cells and platelets, the removal of white cells and platelets by continuous centrifugation, the adsorption of prothrombin and serum prothrombin conversion accelerator from plasma by barium sulfate, the separation of fibrinogen and antihemophilic factor, and the isolation of a globulin fraction and an albumin fraction. His overall objective was a series of investigations, carried out in more than one center, that would yield new knowledge important for national health and security, while at the same time benefiting medical science.

The Korean War

In June 1950, the North Koreans swept down upon an unprepared South Korea, expecting no opposition from the United Nations. It was almost two months before the United Nations could mobilize a UN force under General Douglas MacArthur in an attempt to repel the invaders. Chinese troops then came to the aid of the North Koreans. In fierce fighting, the UN forces were driven deep into South Korea. General McArthur was relieved of his command by President Truman. Under General Matthew Ridgeway, the UN forces regained their original positions and Armistice negotiations were initiated in July 1951.[10]

On July 10, 1950, a meeting in Boston brought together General Marshall, representatives of the American Medical Association, the American Hospital Association and the American Association of Blood Banks with the National Research Council Committee on Blood and Blood Derivatives. Out of this meeting came the "Boston Agreement," under which the four organizations agreed to cooperate with each other in peacetime and with the National Security Resources Board in time of war.[11] The agreement called for the free exchange of blood between the organizations, the establishment of blood collecting agencies in communities not already served by the Red Cross, and a commitment to move toward the standardization of blood collection and transfusion equipment throughout the country. After the meeting, Cohn wrote to Marshall,

> The precedent of a most harmonious meeting of minds, both with representatives of agencies of government and with private associations, such as the American Association of Blood Banks, augurs well for the future. It also demands most careful consideration of the responsibility implicit in our [i.e. the Red Cross] offer to collaborate both with the government and with the medical associations which have been most fearful of governmental interference. The present situation demands compression of the time generally required for the completion of research before development begins . . . In the blood program, we are at the point reached during the early years of World War II . . . Some form of committee or agency responsible for research and development of blood and its derivatives must now, I believe, be established.[12]

In response to a question from General Marshall, Cohn pointed out the importance of greatly accelerating research and development in the interest of stockpiling. He cited opportunities to stockpile human prothrombin and platelets, success in separating red cells and white cells from platelets in a long traverse centrifuge, the potential of new plastic vessels for collecting and washing blood cells without transfer, and research on the optimal environment for preserving viable red cells. Shortly thereafter, General Marshall notified Cohn that Secretary of Defense Louis Johnson had designated the Red Cross as the official procurement agency for whole blood and blood derivatives to meet the needs of the armed forces.[13] In an important related step, W. Stuart Symington, Chairman of the National Security Resources Board, delegated the responsibility for funding research and development for the National Blood Program to the National Institutes of Health, through its Hematology Study Section. This had an immediate impact on blood research in the United States by activating NIH granting authority, including funding of the work at the Blood Characterization and Preservation Laboratory. Marshall responded on July 3 by making funds "immediately available through the Red Cross for continuance of the research that presently is yielding encouraging results." Cohn and Marshall agreed that the Red Cross should not support research, but in the present instance, should expect reimbursement from governmental sources "for funds advanced in preparing for an emergency." Later, in a letter to General Marshall written after Marshall's appointment as Secretary of Defense by President Truman, Cohn reported that the work that had been demonstrated in July was going forward rapidly. He referred in particular to a new "stable plasma protein solution," called S.P.P.S., which he hoped would replace serum albumin, dried plasma and liquid plasma.[14]

Large amounts of blood were needed in Korea. Initially, the blood was collected in South Korea and Japan. As the fighting intensified, an airlift of whole blood collected by the Red Cross was launched from the U.S. west coast in August 1950. In mid 1951, a shortage of blood developed and an Armed Forces Blood Donor Program was initiated by the Secretary of Defense to put blood donor recruiting on a stronger basis. The Navy contributed 50,000 vials of World War II human serum albumin to the Army. Although technically outdated by then, reports indicated that the albumin proved to be both safe and effective and was used widely. Packed in its tin can along with an infusion set, vials of albumin were particularly effective. Not only did they carry no risk of transmitting hepatitis, but they had an added advantage during the bitter Korean winter in that they did not freeze.[15] The Tokyo Blood Depot received a total of 303,000 units of blood, of which 76% was airlifted from the US.

A Mobile Blood Processing Laboratory

Cohn and the writer often drove to and from the laboratory together. In the morning, I walked over to 183 Brattle Street from my home. There I was admitted by a maid, and was asked to go upstairs to Cohn's study where, clad in a lounging jacket, he would be talking on the telephone or writing. Often, we began discussing the topic for the day. When finished, Cohn would return the papers to his briefcase and excuse himself while he donned his vest and jacket. He drove a new Chrysler Town and Country convertible with side and trunk panels covered with real wood. To accommodate his and Rebekah's luggage the spare tire had been removed from inside the trunk and remounted outside, atop the trunk cover. Cohn always parked his car at a fire hydrant in front of Building A of the Medical School on Shattuck Street.

In the late summer of 1950, two special projects altered the routine of our trips to the Medical School. Cohn decided to build the prototype of a Mobile Blood Processing Laboratory that was intended to accompany a Red Cross Mobile Unit and crew to the site of a blood donor drive and separate the formed elements and plasma components from each unit of blood immediately following collection. With this Mobile Laboratory, the procedures for separating the formed elements, and the new Method 12 for fractionating plasma, Cohn was confident that the plasma proteins could be brought to stable states even before leaving the blood collection site. The new laboratory was designed to process up to 200 units of blood in a single day.

Overseeing the construction of the Mobile Blood Processing Laboratory exposed Cohn to some new experiences which he greatly enjoyed. A thirty-two foot closed truck trailer was acquired, together with a tractor unit. There, over the next few weeks, with almost daily visits, a unique mobile blood processing laboratory took shape. The body of the truck was insulated. A diesel generator was mounted in a forward compartment to supply needed electric power. Stainless steel tanks, centrifuges, refrigerators, air conditioning, laboratory tables and instruments were installed. The fabrication and installation of the equipment was done at the workshop of the truck firm by a diverse crew of workmen recruited largely by Charles Harris, the wily Scotsman who had built cold rooms and refrigerated centrifuges for Cohn at the Medical School. Among the workmen were men with special talents. One of them, Fred Gilchrist, was an expert in shaping and forming sheet metal. Cohn found Gilchrist so valuable a collaborator that he was kept on for the rest of his career in the University Laboratory machine shop. Cohn drove everyone involved at a frantic pace, pushing to complete the project in time to demonstrate the Mobile Laboratory at the October 1950 meeting of the National Academy of Science on the campus of the General Electric Company in Schenectady, New York.[16]

The Blood Characterization and Preservation Laboratory

The second project involved the construction and equipping of a new Blood Characterization and Preservation Laboratory at the old Bussey Institute in Jamaica Plain. By this time, laboratory space at the medical school was seriously overcrowded. With the likelihood of NIH funding for work on the National Blood Program, Cohn searched for additional space and discovered an unused wing of a building that was owned by Harvard on the grounds of the Arnold Arboretum adjacent to the Massachusetts State Biologics Laboratory. The wing consisted of a large open gymnasium-like space, about the size of a tennis court, with a high ceiling and a basement. It had last been used as a dormitory for students at the Bussey Institute, a school associated with the Arnold Arboretum where apprentice gardeners were trained to work on gentlemen's estates. The main Bussey building, to which this wing was attached, was occupied at the time by a State Health Department laboratory performing testing for sexually transmitted diseases and related serological problems.

After arrangements had been made for use of the Bussey space, our daily visits in Somerville were extended to include stops at the Bussey site. The new laboratory was built by workmen from the Medical School buildings and grounds department on what could be called a "design/build" basis, with Edwin Cohn in a role that he particularly enjoyed: as the designer. The plan was simple. A new mezzanine floor was constructed in the open space. Together with the basement area, this meant that laboratory work could be conducted at three levels: basement, main floor and mezzanine. A central, vertical stack of cold-rooms provided access to cold rooms from each floor. At one end of the building, there was a paneled seminar room with a high ceiling, at the rear of which were glass enclosed offices underneath a balcony that housed a small library. The walls of the seminar room were inscribed with Charles Morton couplets. Around the frame of the blackboard was the famous statement from Goethe's Faust: "Blut ist ein ganz besondrer Saft."

A Public Relations Gesture

While these hectic activities were taking place in the fall of 1950, Cohn agreed to permit John Lear, a senior writer at *Collier's* magazine, to make an extended stay in the University Laboratory for the purpose of preparing an article. During a period of about a month, Lear was given access to meetings of the Formed Elements group of Cohn's associates. Lear's article, "You May Be Drafted To Give Blood," appeared in the March 10, 1951 issue of *Collier's*.[17] Its lead sentence read: "It may be necessary for us to conscript human blood along with manpower if we are forced to fight a total war against the Russians." While admitting that this was "a bold and extreme approach to a frightening problem which no responsible public official has ventured to suggest," Lear quoted a warning by General Marshall that we "might require more blood in a single week than all our requirements for a full year during the war." Lear described Cohn as

> the man in any scientific gathering who looks least like a scientist. The perfect fit of fashionable cloth across his shoulders, and the well-brushed sound of British phrasing in his speech all hint of a brisk and debonair investment banker-clubman. Within the research realm this Harvard professor operates on a scale that would be fully appreciated only in the world of industrial finance.

With respect to his associates, Cohn was described as making it "impossible to tell with whom any scientific idea originates." Lear saw Cohn as "not insensitive to any antagonisms his detachment might arouse," attributing it to an attitude that "the University Laboratory is his institution, and others can't expect to tell him how to run it." Furthermore, Cohn's associates were "never quite certain how much of the professor's apparent unconcern for them and their problems is real...Evidently feeling that doubt on this score is healthy, Cohn shuns administrative detail." As a result "he acquires freedom for himself to think out new concepts, and puts the staff on its mettle to devise unconventional implements. The results are sometimes as romantic as fairy tales." Lear characterized George Scatchard as "wise and witty" and Janeway as "crisp, boyish and hard headed." Edsall was "generally conceded to know more about proteins than any other living man" and Oncley was described as "a full professor whose earthy Wisconsin skepticism salts the tails of Cohn's high flying ideas before they hit the stratosphere." Lear characterized Hughes as a "dreamy theoretician," Surgenor as a "quick-triggered Connecticut Yankee who acts as the chief's executive officer," and Barbara Low as a "photogenic English X-ray genius."

In December 1950, Cohn met with the Formed Elements group for the last time and declared that the exploratory period was at an end. In less than two years, the group and his associates in the University Laboratory had devised a totally new system for collecting and processing blood. He now wanted to move forward into a new period of clinical testing of new products that were coming out of that work. In what turned out to be his final report to General Marshall, Cohn wrote,

> New procedures, far simpler than I had believed possible even two months ago, yield a Stable Plasma Protein Solution which will, I believe, replace serum albumin. Like albumin, it can be rendered safe from virus transmission by pasteurization. Neither alcohol, very low temperatures, drying equipment, nor radiation equipment for virus sterilization is necessary for its preparation. I will request field trials with this material to determine whether it should not replace all previously recommended blood derivatives as a basis for immediate emergency use or for stockpiling. The only thing that might turn up in the clinical trials is the possibility that a zinc complex forms which is a new antigen, or something like that. I will not willingly recommend a procedure that carries any risk of viral transmission.[18]

Alfred B. Cohn and Barbara Norfleet

Fred Cohn and Barbara Norfleet were married in Cape May, New Jersey during the summer of 1950. Barbara, whose home was nearby, was the daughter of a Naval Officer. Edwin Cohn wanted to attend the wedding, but Fred Cohn urged him not to come. Prior to their marriage, Barbara, a Swarthmore graduate, had worked in the Research

Center for Group Dynamics at MIT under Kurt Lewin, and was at the time a graduate student in Social Relations at Harvard. Barbara and Fred met in 1947 when both found themselves taking the same course in statistics at Harvard. She noticed him because he always sat in the front row. She sat at the back. Fred was then a student in the Law School.[19]

17 The Blood Characterization and Preservation Laboratory

The Blood Characterization and Preservation Laboratory in Jamaica Plain was dedicated on January 8, 1951 during some of the worst days for UN forces fighting in Korea.[1] At the time, the United Nations forces had been driven below the 38th parallel by a Chinese offensive in what Morison called the nadir of the Korean War. The guests included Francis G. Blake, the Chairman of the Committee on Medical Sciences of the Research and Development Board, Shields Warren, the Director of the Division of Biology of the Atomic Energy Commission and Professor of Pathology at the Harvard Medical School, William H. Sebrell, Jr., the Director of the National Institutes of Health, the agency that had only recently become a significant player in advancing medical research, Vlado A. Getting of the Massachusetts Department of Public Health, and Edward Reynolds, Administrative Vice President of Harvard University.

Vice President Reynolds welcomed the guests by describing the historic relationship between the Commonwealth of Massachusetts and Harvard University. He called attention to the succession of members of Harvard medical faculties who had served as directors of the Biologic Laboratories of the Commonwealth and stated that "It is our pride that during the last fifty years, these men, maintaining their active participation in the teaching and research in the University, have not only stimulated the development of prophylactic and therapeutic agents in the Biologic Laboratories of the Commonwealth, but have equally stimulated the research activities of their University colleagues in new lines of investigation which have led to new products of benefit to the community. And now the Federal Government has stepped into this very fruitful collaboration . . . by being of great help in the establishing of this new laboratory."

Francis G. Blake reviewed the events that led up to Cohn's January 1949 Harvard Conference on the Preservation of the Formed Elements and of the Plasma Proteins of the Blood. Drawing from the letter by General Groves that was quoted at that meeting, he stated that "some system must be devised within the next few years whereby a large amount of all the elements of the blood will be prepared and stored to be readily available in case of an atomic war." In Blake's view, the 1949 Conference had resulted in the fallow field being reseeded, "and from its recultivation has sprung not only much new knowledge about the collection and preservation of components of the blood but also this fine laboratory we dedicate today."

Shields Warren spoke in stark terms about the situation in Nagasaki, Japan, after being hit by the atom bomb: "It was the sense of overwhelming helplessness that our medical team faced at Nagasaki in September 1945, that first forced on me the feeling that whole blood, and particularly the formed elements of that blood, played an extraordinarily important role in this field of attempting to save atomic casualties. In the few cases in which we were able to give transfusions, the improvement that promptly appeared in the patient was very impressive." The analysis of the damage that is done by ionizing radiation shows that it hits very largely the formed elements of the blood, Warren told the guests. "As we follow through the symptoms, we find that the appearance of symptoms is very closely related to the changes in these formed elements." First, was the great susceptibility to infection when there are insufficient leukocytes; second the hemorrhagic manifestations when, along with other changes, there was a reduction in the number of blood platelets; and finally, the weakness due to anemia, appearing usually ten days or so afterwards and increasing with passage of time for eight to ten weeks while the effect of damage to the red blood cell progenitors was being registered.

"We have, then," he pointed out, "the problem of how to meet infection, the problem of how to control hemorrhage, the problem of how to provide sufficient oxygen-carrying capacity to the organism as a whole." These three aims could best be achieved, Warren believed, if, instead of starting with the initial building blocks from which these cellular elements come, we could bridge over the gap by using these cellular elements themselves, not wastefully, blindly, just as they come, but specifically to meet the needs that exist at the time those particular needs arise. In the cannibalizing of jeeps in the Pacific Islands, it was frequent to need only the cap of a distributor to get another jeep under way, and the rest of the jeep went to pot for a brief period until others came along and cannibalized other parts. "This whole blood program I visualize something like this. We can steal the whole jeep in the form of whole blood, or, if we need a rotor, we can leave the rest of the jeep for other people who have other needs, each using the distributor cap or the battery or whatever it may be for his specific purposes."

"I was fortunate enough twenty-eight years ago, as a medical student, to be able to sit in on some seminars which Dr. Cohn gave," where, Warren revealed, he had his first real glimpse as to how the field of physical chemistry, which was essentially an unknown at that time, might be applied to biologic problems. Although he did not have any part in carrying on work in this field, it had been extraordinarily gratifying to watch its development as applied to medicine, from one that was at first nebulous to one that through its particular application in this way concerned the lives of virtually everyone in this country. And this was only one of the many applications that had been opened up.

William H. Sebrell, Jr. spoke about the role of the U.S. Public Health Service with respect to the National Blood Program. "We face the possibility of war with attendant catastrophe and confusion. In this perilous situation, we must prepare for the eventuality of thousands, perhaps millions, of casualties, both military and civilian. In dedicating this new laboratory, we open the door to a new realm of possibilities for saving life and relieving human suffering." Noting that the Public Health Service had distributed over $50,000,000 since the establishment of its extramural programs in 1937, Sebrell admitted that it was only recently that the mission of NIH had been expanded to include the responsibility for supporting the research and development aspects of the National Blood Program, and he reported that NIH had already allocated $600,000 in grant funds as a contribution to the Blood Program's total requirements of $2,000,000 in its first year.

Edwin Cohn reminded those present that the original National Research Council body with responsibility for blood during World War II was the Subcommittee on Blood Substitutes that operated formally under the NRC Committee on Transfusions. He recounted how the surgeon scientist Alexis Carrel had returned from France in 1940, and, justifiably agitated, as others have been since, hastened to Washington to urge the need for immediately preparing 400,000 units of blood to be sent to France for transfusion. Before this end could be achieved, however, France had fallen. The Committee on Transfusions was convinced that the U.S. should resort to blood substitutes. Cohn credited Elmer DeGowin, a member of the Subcommittee, for insisting that whole blood be made available and that the Armed Forces begin to acquire the necessary equipment for the collection of whole blood. It was only after the American landing in North Africa late in 1942 that whole blood became available for the treatment of American casualties. According to Cohn, the most impressive achievements in making whole blood available in theaters of operations were those of John G. Gibson, 2nd and Lloyd R. Newhouser. As a result of their initiatives, refrigerated whole blood was flown across the Pacific and used in the final landings on those distant islands. The tragedy was that the stimulus for research and development for military, as well as civilian needs, was permitted to lapse after V-J Day.

As for blood substitutes, Cohn stated that it had been his misfortune to have evaluated one after another so-called blood substitute, an effort that failed to identify a single substance that could be approved. None of the blood substitutes that had been considered performed the respiratory, phagocytic, coagulative or transport functions of blood, leading to the conclusion that nothing justifies the current hope that efforts to collect and stockpile human blood can someday be abandoned. Consistent with that position, he urged that attempts to continue accumulating plasma substitutes should be abandoned so as to "avoid interfering with the willingness of the American public to donate blood." Plans to accumulate blood substitutes, like those to stockpile atom bombs, should be classified top secret, Cohn advocated. This tenet formed the basis of his current strategy of separating the formed elements of blood as the first step in their preservation. Noting that the rising threat of virus transmission was the greatest menace to any blood program, Cohn reminded those present of the narrow differences between the intensities of radiation that destroy viruses, on the one hand, and that destroy parts of the blood, on the other. Nevertheless, he conceded that active efforts to inactivate viruses in blood products must be continued.

Cohn used the dedication of the Blood Characterization and Preservation Laboratory to announce the development of Method 12, a new plasma fractionation method, and two new plasma products, stable plasma protein solution, (SPPS), and plasma globulin precipitate (PGP). This new procedure depended on the use of zinc to cause the precipitation of a plasma globulin precipitate that included gamma globulin, fibrinogen and clotting factors, and which could be removed by centrifugation. After removal of the zinc by passage of the supernatant solution over an ion exchange resin, the result was a 5.5% solution consisting of all the serum albumin, alpha-lipoproteins, glycoproteins, and the β-1 metal-combining globulin in plasma. SPPS included about 70% of the proteins in normal plasma. Neither alcohol, very low temperatures, drying equipment nor radiation equipment for virus sterilization was necessary for the preparation of these products. Moreover, SPPS could be heated for ten hours at 60° C to inactivate any virus present. Cohn saw SPPS as a substantial technological improvement over wartime human serum albumin. The principal design criteria for the blood collection system used in preparing SPPS included avoidance of exposing the blood to wettable surfaces, prompt decalcification of the blood by passage over an ion exchange resin, thus inhibiting blood coagulation, and prompt separation of the formed elements by centrifugation. The system yielded red cells ready for transfusion, with the option of obtaining white cells and platelets for research. In this way, plasma became available for use shortly after collection of the blood from the donor.

In a letter written to licensed producers of plasma products, Cohn expressed confidence that Method 12 of plasma fractionation would render obsolete all previous methods for fractionating plasma. The first lot of SPPS produced outside of Boston was completed at the Michigan Department of Health Laboratories in Lansing on Wednesday March 21, 1951, using ion exchange columns and plastic recipient sets developed by Carl Walter in Boston. As Cohn concluded, he

Illustration 13. Above: Rebekah and Edwin Cohn, accompanied by Rebekah's daughter Susan Higginson (center), boarding an airplane in New York for London in 1949. Below: General George C. Marshall (left) and Dr. Cohn talking to newsmen on the front steps of Building A of the Harvard Medical School in 1950.

Illustration 14. Above: Edwin Cohn showing an early version of his blood centrifuge to a visiting engineer. Below: A mobile blood processing laboratory enabled the fractionation of plasma to be initiated at Red Cross collection sites in 1950.

Illustration 15. Dedication luncheon of the Blood Characterization and Preservation Laboratory at the Bussey Institute in Forest Hills in 1951. Identified participants are: l, L. K. Diamond; 2, V. A. Getting; 3, K. M. Brinkhous; 4, W. H. Sebrell, Jr., 5, E. J. Cohn; 6, Captain L. C. Newhouser; 7. F. G. Blake; 8, D. M. Surgenor; 9, Edward Reynolds; 10, Shields Warren; 11, C. A. Dorn; 12, W. Workman; 13, Benjamin Alexander.

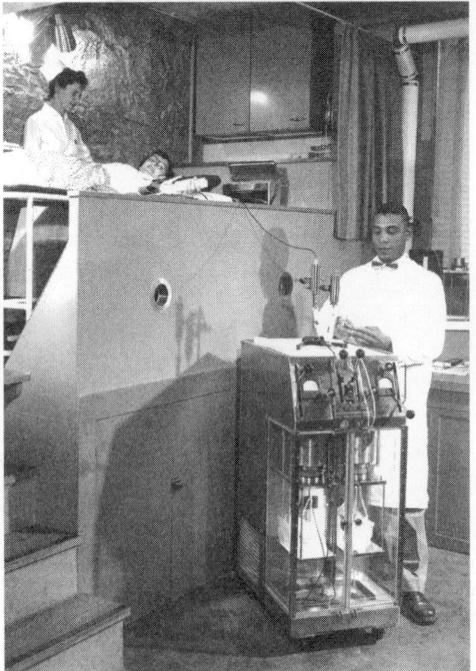

Illustration 16: Above: Edwin J. Cohn contemplating a bottle of suspended red cells in the post World War II era. Below: The Cohn centrifuge being operated by Robert Tinch during blood collection at the Blood Characterization and Preservation Laboratory. This was the centrifuge that was demonstrated at the Establissement Central de Transfusion et Reanimation de l'Armee in Paris in 1952.

Illustration 17: Massachusetts governor Christian Herter presenting the Massachusetts charter of the Protein Foundation, Incorporated to Chester I. Barnard, its chairman. Front row, Left to Right: Roger I. Lee, Governor Herter, Mr. Barnard; in back, Edwin J. Cohn (left) and Dr. Wilbur K. Jordan (right). The presentation took place on February 25, 1953 in the Governor's executive office in the State House.

Illustration 18: Above: Allen D. Latham, Jr., photographed in 1991. Below: Stages in the development of the design of the bowl for the Cohn centrifuge. Left to right, a stainless steel bowl with carbon-steel seal, a reusable plastic bowl, and a single use, disposable bowl developed by Mr. Latham.

stressed the importance of having equipment immediately available which could be thrown quickly into high gear if tremendous quantities of blood were to be suddenly needed. It would not be prudent for such matters of life and death to be left to chance.

The Cohn Centrifuge

Following the dedication of the Blood Characterization and Preservation Laboratory, the Cohns left for a brief stay at the Arizona Inn in Tucson. The few days in Tucson provided an opportunity to reflect at ease on the directions of the work in Boston. The developing program at the Blood Characterization and Preservation Laboratory was very much in Cohn's mind. He was excited by the prospect of replacing the wartime alcohol-water plasma fractionation system with the new system that was based on the use of zinc rather than alcohol. The simplicity of the new fractionation methods was startling. The key steps were carried out at physiologic pH and without dilution of the plasma. The process was susceptible to automation and yielded a solution that was ready for infusion without being dried. As he considered the situation, he realized that the weakest link in the overall process of blood collection and processing was at the very beginning, at the stage when the blood was collected and the formed elements were separated from the plasma proteins. Important time was being lost between the drawing of blood from the donor and the reduction of its component parts to environments that were conducive to their optimal preservation. What was needed was a small centrifuge into which fresh blood could flow from the vein of a blood donor and be rapidly separated into useful components. The more he toyed with this idea, the more excited he became.

Cohn returned to Boston with some sketches drawn on the backs of old envelopes. Within a few days, the first crude centrifuge bowls had been made from Erlenmeyer flasks, a common piece of glassware in chemistry laboratories, by a glass blower at the Macalaster Bicknell Company in Cambridge. The inverted bowl was mounted vertically above a stationary assembly glass trough. The bowl was caused to rotate around its vertical axis at a speed that produced the desired centrifugal force on its inner surface. Blood was introduced from below the spinning bowl through a vertical central feed tube, impinging on the flat upper inner surface of the rotating bowl where it was exposed to the centrifugal force within the bowl. This caused the separation of the blood into two vertical layers: the heavier red blood cells were forced toward the periphery of the bowl, while the lighter plasma formed an inner yellow layer. When the bowl was filled, clear yellow plasma began to overflow the bottom lip of the bowl, dropping into the collection trough and thence into a plastic collecting bag. When the desired volume of blood had been introduced, the flow of blood into the bowl was halted by a clamp and the needle in the donor's arm vein was removed. The plastic bag of plasma was then removed and sealed; and a new plastic bag was attached in its place. The rotation of the bowl was halted, permitting the red cells and some remaining plasma to drop into the second bag.

While the characteristics of the new collection system—speed of revolution, capacity of the bowl, etc.—were being explored, outdated bank blood was used. During the early phases of the study, the gap between the rotating bowl and the stationary collecting trough was left "open." Eliminating this gap was seen by some of Cohn's associates as an almost insurmountable problem, but Cohn was confident that a way of closing the gap would be found at the appropriate time. In the meanwhile, he devoted his efforts to studying the dynamics of the new centrifuge, working out the effects of changes in the shape of the glass bowl and the volume of liquid it retained while spinning.

Initially, Cohn drew upon the experience of his colleagues at the Medical School, members of the Formed Elements Group, and the expertise of Oliver Eckel and Fred Gilchrist at the Mundet Cork Corporation in Somerville where earlier the mobile laboratory in the truck had been built. Somewhat later, American Optical Company personnel were invited to participate, primarily because Cohn thought they could contribute in helping to record, by optical means, variations that might occur during the separations. As the project grew, more collaborators were added. The earliest units were fabricated in the Medical School shop. The first prototype machine was built at the American Optical Company in Southbridge, Massachusetts. Later, modifications were explored, including the treatment of the interior surface of the glass bowl with silicone to render it non-wettable, addition of a water cooled heat exchanger to cool the blood after leaving the vein of the donor, and the introduction of an ion exchange column to evaluate various ion exchange resins. As finally assembled, the early Cohn centrifuges were mounted within a portable refrigerated cabinet that housed the motor and controls for rotating the bowl.

A Conference on "New Mechanized Equipment for Collection and Processing of Human Blood" was held at the Blood Characterization and Preservation Laboratory on June 14 and 15, 1951.[2] Among the institutions sending observers to this meeting were the Dow Chemical Company, Rohm and Haas Inc., Abbott Laboratories, Ionics Inc, The Permutit Co., American Optical Co., International Equipment Co., The Sharples Corporation, Mundet Cork Co., Harris Refrigeration, Artisan Metal Products, and Emerson and Cuming. The Pharmaceutical Industry was represented by individuals from Lederle Laboratories, Sharp and Dohme, E.R.Squibb & Sons, Upjohn, Cutter Laboratories, and Armour & Co. The Plastic Equipment industry was represented by Abbott Laboratories and Fenwal Laboratories, American Hospital Supply Corporation, and Baxter Laboratories, Massachusetts State Biologics Laboratories, Mallinckrodt Chemical Laboratory of Harvard University, the NIH Division of Biologics Control, American Red Cross and associates of Edwin Cohn.

In opening this first of many meetings devoted to the design and use of the new centrifuge, Cohn explained that although research on red cell preservation had been progressing satisfactorily in several laboratories in 1945, it came to an abrupt halt when the U.S. Office of Scientific Research and Development went out of existence in 1946. He said "I consider centrifuges to be the core of a modern blood program." Only a few moments elapse, he continued, before the optimal conditions for blood preservation have been dissipated. This conclusion suggested the need to develop processing equipment for immediate use following blood collection at blood centers. "Two years ago, I came to the conclusion that a proper blood program could not be developed unless two conditions were satisfied: centrifuges must be developed to operate in a closed system; and centrifuges must be refrigerated so that low temperatures could be maintained throughout processing."

Noting the large number of semi-public and governmental agencies that were involved in a blood program, Cohn concluded there was enough disagreement among those responsible for various programs to indicate that changes will be introduced more rapidly in some cases than in others. Continuing, he observed that the new process eliminates the blood collecting bottle. The blood flows from the arm of the donor through the vessel containing the anti-coagulant and the heat exchange unit into the new light centrifuge. The separation of plasma takes place at the time and at the rate at which the blood is drawn from the donor. Processing an aliquot of plasma under the best conditions should yield an estimate of the concentration of any enzyme for which there is a rapid analytical method.

In a heat exchange/ion exchange resin unit, the upper vessel contains the exchange resin. Once the calcium has been removed, the danger of clotting is averted. When the temperature is reduced, the whole system is more stable. Whether the aim is the collection of whole blood, blood cells, plasma or plasma fractions, he believed that improvements in processing were possible. "Substituting an ACD mixing column for the resin column took Mr. Ellis about a week to achieve. How long it will take those responsible for blood collection to adopt it is another matter." Conferences to consider this fundamental problem led to the realization that adequate evidence had accumulated indicating that if blood was to be collected in its state in nature, one should not use glass; one should not use rubber; and one should not use citrate. A great difference thus existed between knowledge about blood and the recognized practice in blood banks the world over. Cohn told that when considering the problems of blood collection in order to determine where further research was needed, the tendencies which led to our present difficulties became only too clear. The question of apparatus raises considerations different from those concerning reagents. Thus in apparatus of most kinds there are disposable parts and parts that can be used again and again.

He admitted that every accretion to knowledge brings about changes in processing, which are foreseeable far in advance. The equilibrium that will be reached following any scientific advance is thus readily predictable. The kinetics of the changing operation is, however, more difficult to predict for the interactions of personalities become involved as well as other forces, and knowledge of the resistance to change is necessary before prophesying how long it will take for an advance to be understood and accepted. A predictable element of resistance derives from the prejudices of the medical profession. These prejudices are often well grounded in past experience, he believed, "and for the most part I have always been prepared to respect them." They are born of concern regarding mistakes that clinicians have had to cope with as a result of over-enthusiasm on the part of investigators. "It behooves us to take precautions in a program of this kind so that no unfortunate mistake is made." Such precautions demand a great deal of research and a great deal of painstaking developmental work at the pilot plant level, which a single industry is rarely in a position to undertake. He mentioned that those assembled for this conference constituted a heterogeneous group. The meeting included experts upon each of the reagents that is involved in the new methods, experts regarding plastics, resins, the coating of glass and the toxicity of metals, and experts on the manufacture and use of various kinds of centrifuges. All phases of biologics, as well as blood programs, should benefit from technical conferences of this kind.

The development of measures to reduce the risk of transmitting viral diseases remained a high priority objective. However, those attending the June meeting heard little encouraging information about this problem. Extensive efforts to adapt ultraviolet irradiation for use in sterilizing plasma suffered from the limitation that the irradiation had to be applied directly to a thin film of plasma, without interference by glass or plastics. In almost every instance tested, the harm done to viruses by irradiation was accompanied by unacceptable degradation of the proteins as well. Studies by John Trump and Kenneth Wright at MIT had shown that cathode ray irradiation penetrated plastic vessels, but the strength of irradiation needed to destroy the hepatitis virus had not yet been determined.

An Exhibition on the Development of Knowledge of Blood

In collaboration with the Harvard University Library, and the Committee on Higher Degrees in the History of Science and Learning, Edwin Cohn had helped to arrange an exhibit at the Widener Library on the development of knowledge of blood as represented by manuscripts and selected books published between 1490 and the 19th Century. The exhibit was open to the public for several weeks in the spring of 1951. In his Foreword to the catalogue of the exhibit, Cohn explained that universities perform many functions, among them the function of acquiring new knowledge and preserving the accumulated learning of the past. "Reinterpreting the past is the special privilege of each generation, shared by the trained historian and the amateur alike." At the end of World War II, he revealed, it seemed important to him, and to others, to consider the direction of future scientific research by reexamining the record of past accomplishments from the vantage point of the present. Considering the contributions of the investigator in terms of his times, his environment and his genealogy, especially his scientific genealogy, proved illuminating, and arranging the names of those who contributed to a given field, in the manner of a family tree, created a pleasant illusion of objectivity.

The stimulus of Cohn's thinking about these excursions into the past had led him to search for the origins of knowledge in every field he was investigating. The very titles of many of the early books made him wish to see the first editions, on the assumption that from their form, he might learn more of the spirit of the investigator, unless, of course, the form characteristic of the author had been obscured by the publisher. In compensation, there were volumes, such as those of Thomas Willis, which acquired two-fold significance from the illustrations of an understanding colleague, who in this case was Christopher Wren. Cohn's desire to share his experience with others led to the suggestion and to the immediate acceptance by his colleagues to arrange for this exhibition.[3]

IV International Congress of Blood Transfusion, 1951

Cohn attended the IV International Congress of Blood Transfusion in Lisbon on July 23–29, 1951. In a brief address to the Congress as part of the opening ceremonies, he stated that while the blood bank had long been considered as an aid to the surgeon, its great value had in recent years become apparent in the treatment of shock and burns and even in atomic injury. Turning to the future, he noted the ability to routinely and automatically analyze a small aliquot of the blood from a normal donor, as an integral part of blood collection, has only recently become apparent. It would now be possible, he continued, to supplement measurement of empirical iron levels, hematocrit readings and albumin/globulin ratios of hospital laboratories with increasingly accurate knowledge of the concentration of specific antibodies, type of the blood, specific enzymes, as well as steroid hormones, lipid binding globulins, the iron binding globulin and even the binding of drugs by albumin. In those terms, he predicted that transfusion services would become of great significance to communities as a powerful tool of preventive medicine for use in the interest of the public health.[4]

Cohn had arranged for James Tullis to accompany him to Portugal and demonstrate the new blood centrifuge at the Lisbon Congress. In preparation for the journey, the only existing centrifuge in the Blood Characterization and Preservation Laboratory had been thoroughly cleaned, and its surfaces treated with silicone. It had been fitted with a fresh ion exchange column and new plastic tubing and blood bags. Tullis and the centrifuge traveled by ship directly from Boston Harbor to Estoril, Portugal. The demonstration took place in an amphitheater before an audience of blood experts. Cohn arrived separately and delivered a brief lecture while Tullis readied the apparatus. Then, still talking to the audience, Cohn lay down on the donor cot, the venipuncture was performed, a stopwatch was started and Tullis bent over to observe the blood as it entered the spinning bowl. The audience leaned forward in awe. At the Blood Characterization and Preservation Laboratory, the first overflow of yellow plasma always occurred in exactly four minutes, but even though four minutes passed, there was no overflow of plasma. Cohn calmly continued to lecture. Suddenly, at five minutes, pandemonium erupted among the audience. Those in the front row clambered over the backs of their seats in an attempt to reach higher ground. A fine spray of bright red Cohn blood drew a horizontal line across the row of seats. Cohn maintained his aplomb and resumed his lecture. It later emerged that an enthusiastic laboratory technician in Boston had thoroughly cleaned all parts of the apparatus, but had used a new type of gaseous silicone vapor to render the internal surfaces non-wettable, thus blocking the outflow of plasma.

A Meeting of the American Association of Blood Banks in Minneapolis

Edwin Cohn was a featured speaker at the October 1951 meeting of the American Association of Blood Banks in Minneapolis. He chose that occasion to describe his new Biomechanical Equipment for Blood Collection, presenting a slide entitled "Principles of the new Blood Centrifuge." He began by stating that it became necessary to develop a simple centrifuge in which plasma could be separated from blood in a completely closed system thus obviating the problem of drawing off plasma under a bell or a hood as well as being economical in blood and in cost. All surfaces of this centrifuge had to be non-wettable and readily sterilizable. In the new blood centrifuge, the separation of plasma from the cells is a continuous process. The volume in the centrifuge vessel, at any one time, is therefore always less than the volume of blood to be processed. As a result, all parts of the apparatus may be of very light construction. The shape of the centrifuge vessel is critical for the separation of red blood cells from white cells and platelets. It should vary with the centrifugal force employed and with the specific gravity of the blood cells, plasma fractions or solutions employed in processing. The forces which maintain blood cells in a centrifuge rotating around a vertical axis depend upon the diameter and revolutions per minute of the centrifuge, the difference in specific gravity of the cells, and the medium. The vertical force of gravity upon the sedimentation of red blood cells in a centrifuge operating with a force of no greater than sixty times gravity cannot be considered negligible. In a vertical, cylindrical centrifuge, the red cells arriving at the outer layer accumulate at the base of the centrifuge vessel and prevent all of the separated plasma from leaving the centrifuge. The shape of the new centrifuge vessel overcomes this effect.

The path of the various formed elements varies with the dimensions of the vessel. In an empty vessel, red cells pass down the sloping sides of the vessel more rapidly than do the white cells, which, as in other forms of centrifugation, are forced to the center. In the new centrifuge red cells pass from the first collecting bowl into a lower bowl. If a part of the retaining volume of the centrifuge is filled with a high-density solution, the separations are even more satisfactory. The higher the specific gravity of the solution, and the closer it approximates to the specific gravity of the white cells, the more the separated red cells are free of white cells. The theoretical conditions for the quantitative separation of red cells from white cells, platelets and plasma can be achieved by interposing a layer of solution in the centrifuge of such specific gravity that it is rapidly traversed only by the red cells. Under these conditions, red cells are freed both of the other formed elements and of the plasma. Moreover, the recovery of plasma from blood is essentially quantitative.

In concept, this simple technology presented almost unlimited opportunities for collecting, processing, and separating the formed elements and plasma products from the blood of individual donors. The retention volume of the bowl and its speed of rotation could be varied to deliver a desired volume of plasma, leaving the packed blood cells in sufficient plasma to flow out as a thick suspension when stopped. In the simple version, the centrifuge delivered a unit of packed red blood cells and a unit of plasma. Relatively simple design changes permitted separating the white cells and the platelets into their own plastic bags ready for infusion. Alternatively, modifications of the bowl made it possible to wash the red cells free of plasma in countercurrent fashion. Still another procedure, termed plasmapheresis, enabled the separation of a unit of fresh plasma, while returning the red cells to the donor.[5]

In December 1951, President Truman issued an Executive Order designating the Director of the Office of Defense Mobilization to provide, within his office, "a mechanism for the authoritative coordination of an integrated and effective program to meet the nation's requirements for blood, blood derivatives and related substances." This order noted that a subcommittee on blood had been appointed within the Health Resources Advisory Committee to develop a National Blood Program. "It was the President's desire that the activities of all departments and agencies in the field be coordinated through this mechanism."[6]

Policies of the New University Laboratory

As the 1950–51 academic year drew to a close, Cohn prepared a new pamphlet entitled *History of the Development of the Scientific Policies of the University Laboratory of Physical Chemistry Related to Medicine and Public Health, Harvard University.* It was his expectation that the research to be carried out in the new laboratory would represent a continuation, without interruption, of the research that was being carried out in the old Department of Physical Chemistry.

In this new pamphlet, he suggested that the most satisfactory method for evaluating the incisiveness and direction of research in the new laboratory should stem from projection from the published records of the Department of Physical Chemistry. Indeed, he assigned a new subtitle for this section, calling it *A Memorandum on the Unwisdom of Projects and Reports.* Edwin Cohn often described himself as a conscientious objector to announcing projects in advance of publication or reporting them in any but the established methods of universities, scientific academies, or learned societies.

He advocated what the credo for the new University Laboratory of Physical Chemistry Related to Medicine and Public Health should be:

> The record of past accomplishment, rather than a project, should be made available by the University Laboratory in requesting collaboration or assistance of any kind.
>
> The record of the work accomplished should always be made available, as promptly as possible, in published form or—when desirable to aid others working in the same or a related field—by tape recording, in mimeographed sheets or in privately published pamphlets.
>
> By the same token, a statement should not be made in advance as to the probable research that a pre-doctoral or post-doctoral fellow would pursue while in training.

For some years, the following sentence was routinely inserted in correspondence with those who wished to come to the laboratory for training:

> In order to study proteins effectively we have also found it necessary to carry out all work in this department as group investigations and you would therefore participate in one of another of the programs that are actively being pursued at the time you are here.[7]

Commission on Plasma Fractionation

The Commission on Plasma Fractionation and Related Processes met in December 1951 for a brief meeting. The principal item of business was a discussion concerning the production of albumin and gamma globulin. Unfortunately, the National Blood Program, having made arrangements to produce human serum albumin from the large supply of surplus World War II dried plasma, had failed to take corresponding steps to recover the gamma globulin from Fraction II+III. Norman Topping and John Enders, members of the Commission as well as the Committee on Immunization of the National Foundation for Infantile Paralysis, expressed concerns that the demand for gamma globulin for treating poliomyelitis would far exceed the supply. The Commission also heard a report that the costs of testing and release of products produced under the Cohn patents, expressed in terms of the cost per vial of the products, amounted to between 0.6% and 0.15% of the cost per vial of human serum albumin and gamma globulin, respectively.

At the end of 1951, the Commission was administering five active U.S. Patents, including the main Cohn patent for the ethanol-water process for fractionating human plasma, (U.S. 2,390,074), a Cohn-Ferry plastic composition patent, a Ferry patent on plastic materials, a Cohn patent on selective extraction, and a Ferry-Morrison fibrin products patent. At this same meeting, the Commission reviewed the status of foreign patent applications and patents. This revealed the following active patents or applications that covered the ethanol-water process for fractionating plasma:[8]

Country	Status	Expiration date
Australia	active	1961
Belgium	active	1966
Canada	active	1965
Great Britain	allowance "probable" 1952	
South Africa	active 1959	
Sweden	amendment filed 1951	
Switzerland	new claim filed 1951	

Edwin J. Cohn Jr. and Alfred B. Cohn

Almost five years had elapsed since Ed's return from Europe. In the intervening years Ed had encountered his father on various occasions, sometimes in New York, sometimes in Cohasset and almost every

Christmas in Cambridge. Ed had gained a better understanding of himself and his family environment in the years since his return from Europe. He was no longer intimidated by his father; indeed for almost the first time in his life, he was able to stand up to him. A modus vivendi of sorts had thus developed. For his part, Edwin Cohn made an effort to avoid intruding in Ed's life and had regained the respect of his son. In 1951, Ed Cohn brought his fiancee, Katherine Sloss, to Cohasset to meet Edwin and Rebekah Cohn. Katherine and Ed had met in New York where she was working in the publishing field following her graduation from Sarah Lawrence. During the visit to Cohasset, there were moments when Edwin and Rebecca Cohn seemed ill at ease. Kath obviously knew about Edwin's difficulty in dealing with his Jewishness. She was therefore amused when Edwin said at one point when they were alone, "I'm glad you're Jewish. It bothered Ed's mother and I know it bothers Ed, but it has never bothered me."

Ed and Katherine were married at her family home in San Francisco in June 1951 while Edwin Cohn was in Portugal. Fred and Barbara Cohn represented the Cohn family, with Fred doing the honors as best man. Ed Cohn completed his doctoral dissertation and received the Ph.D. in economics from Columbia in the spring of 1952. Ed and Kath left shortly thereafter for Turkey where Ed had taken a post as economist with the Mutual Security Agency administering the U.S. Foreign Aid Program there. From Turkey, they moved on to Afghanistan where Ed served as an adviser on economic planning under the United Nations Technical Assistance Program.[9]

Fred Cohn graduated from the Harvard Law School in 1952 and became involved in the Cambridge Civic Association, serving as its president. He designed and built row house projects in Cambridge using prefabricated units that were made in a shop in North Cambridge. Barbara Norfleet Cohn received her Ph.D. degree from Harvard in 1951. Having enjoyed a warm relationship with her own father, Barbara tried to establish a good relationship with her new father-in-law, but without success. Edwin Cohn was unable to tolerate having a bright young daughter-in-law who was accustomed to acting independently and expressing her own ideas. After a few contretemps, Fred and Barbara found it best to avoid expressing their opinions when they were guests at 183 Brattle Street. It was best to be seen but not heard. Barbara Norfleet Cohn taught at Harvard between 1960 and 1996. She founded the Photography Collection at Harvard, and served as its first curator.[10]

18 IMPLICATIONS OF NEW KNOWLEDGE

"IMPLICATIONS OF New Knowledge About Proteins, Protein Enzymes and Cells" was the title of a two-day conference that was held on January 7 and 8, 1952. Its purpose was to communicate the scientific program of the University Laboratory of Physical Chemistry Related to Medicine and Public Health to its broader constituencies within Harvard University. The sessions were held at three sites—at the Harvard Medical School, at the Blood Characterization and Preservation Laboratory in Jamaica Plain, and at Harvard Hall in Cambridge.

In welcoming those attending the conference with remarks that reflected the serious world situation at the time, President Conant confessed that he could not help thinking that this conference was one at which a number of different streams of activity and thought had met. Not only the fact that advances in knowledge about proteins, enzymes and components of blood represented the coming together in this mid-twentieth century of the vast amount of research in physics, chemistry and biology, but also the fact that this work was so directly related to medicine and public health meant that this was a coming together of pure and applied research. Because of the nature of problems connected with blood preservation and all the social, political and military implications of the need for that new knowledge, he sensed that one could feel that this was perhaps one of the occasions in a very grim time when one could both survey advances in applied science essential for the peacetime well-being of a community, and also rejoice in the opportunity to look at advances in knowledge for their own sake—

"pure research, if you will"—and find that these are related to the anxious times in which we live.

He supposed that every one must wonder at times whether the work in which he is engaged is the best thing for the welfare of all concerned. In a total war, like World War II, almost everyone could ask, is the work I am doing in applied science the best contribution I can make to the winning of the war? He pointed out that we are in a strange period, a shooting war, with casualty lists coming in every day from Korea, yet not in a total war. Men are called back to service, often at great sacrifice to their families. With the world still desperately divided and showing no signs of being less divided, one must wonder where does the path of duty lie for any individual connected with science and its application? Conant observed that no one can say what may lie ahead in the next half dozen years. Some of the work has surely been stimulated by the need to be prepared for a calamity on a vast scale which would result in World War III and the use of the atomic bomb. This research is connected with the emergency and with national defense. Nonetheless, he reflected, "If, as I believe, we can get through without a global war, the work will have vast implications for the health and well being of civilian populations in time of peace."

This symposium is an example, Conant opined, of the importance at the most applied level of the quest of human knowledge by the man whom he ventured to call the uncommitted investigator. "Indeed, Dr. Cohn's laboratory itself could be written about in terms of the need for people to come together, as you have, to learn about the implications of new knowledge for practical results. No one is wise enough to foresee what particular line of investigation is going to prove fruitful within the lifetime of an individual." He concluded by pointing out that, in science, you do not have to wait for another generation, as you do in forestry, to see the results of one's labors. In recent times the organic chemist is coming in again; the whole complicated structure of the protein molecule in terms of amino acids is being unraveled in other laboratories. "The frontier is still open and what lies ahead will make even these advances seem feeble steps."[1]

The Conference dealt with four main topics, of which the first was concerned with the interactions of proteins with each other, with cells, with metals, specific polysaccharides and other molecules, including presentations about the stroma of red blood cells, about interactions of plasma proteins with zinc, mercury and lead, and about plasma enzymes, proteins of blood and liver and the structure of the fibrin clot. Much of the scientific work in the University Laboratory in the post World War II period focused on interactions of proteins with metal ions, particularly zinc, mercury, calcium and barium. Hughes performed the earliest study of this type in 1947 when he found an albumin fraction isolated from human plasma as a crystalline mercuric salt.[2] The ratio of mercury to albumin in this complex of mercury and human serum albumin proved to be one to two, suggesting that the formula of the product was Alb-Hg-Alb. Consistent with that, the crystals were found in the ultracentrifuge to have a molecular weight twice that of albumin. Hughes also showed that the new complex was easily dissociated by treatment with a solution of cysteine, the amino acid with a sulfhydryl group. In the presence of excess cysteine, the mercury was complexed with the cysteine, thus destroying the complex with albumin and yielding albumin of the "native" molecular weight. Consistent with these findings, Hughes showed that the formula of the mercury complex was Alb-S-Hg-S-Alb, proving that the mercury ion reacted with a free sulfhydryl group of the amino acid cysteine in peptide linkage in each albumin molecule. This picture was further clarified by showing that there was only one cysteine residue per molecule of albumin.

In a similar study of the interactions of zinc ions and human albumin, Gurd and Goodman found that up to sixteen atoms of zinc could be bound per molecule of albumin. On investigating what groups on the albumin molecule could have bound the zinc, they demonstrated that the zinc binding involved imidazole groups of the amino acid histidine.[3] This was a particularly satisfying result since Brand, Kassell and Saidel, in their study of the amino acid composition of the Cohn fractions at Columbia University during the war, had found that there were sixteen residues of the amino acid histidine in each molecule of human serum albumin. Taking these observations together, this confirmed the hypothesis that all the histidines in albumin, of which there were sixteen, existed in polypeptide linkages. To add perspective, the amino acid analyses of albumin by Brand et al revealed that there were 576 amino acid residues in albumin. Moreover, the Gurd and

Goodman study indicated that the zinc reaction was quite specific, involving only histidine residues and no others. Thus zinc binding was a marker for histidine side residues which comprised less than three percent of the amino acid residues in the huge albumin molecule.

The other topics which made up the "Implications" program were:

> Physical Properties and Chemical Reactions of Various Purified Proteins. This topic provided reports concerning the nature of the chemical groups of plasma proteins, light scattering studies of protein interactions, X-ray diffraction studies of insulin, and ultracentrifugal diffusion studies of various plasma proteins.
>
> Fundamental Studies of the Formed Elements of the Blood. This dealt with studies on blood platelets, the role of platelets in blood clotting, a group of reports on the properties of leucocytes and new developments in the use of biomechanical equipment.
>
> Implications of new Knowledge of Blood and Tissues for the Public Health. This included observations on the state of knowledge concerning the separated formed elements of blood, on the development of biomechanical apparatus for the investigation of blood, and on the clinical and immunological testing and control of biologic products prepared from blood.

In his presentation on "Clinical and Immunological Control of Biologic Products," which was included in the final part of the program, Janeway dealt with one of the major public health questions at the time, the risk of transmission of viral infection by blood transfusion. Stating that biologic products prepared from living sources can transmit disease, he pointed out that "the transfusion of blood is an example of serial passage, whereby biologic agents from one person are transmitted to others." As the frequency of the use of blood and blood products increases, he commented, the risk of spreading disease is also increased. The regulation or control of the quality and safety of biologics prepared from blood thus had two aspects. There was the governmental regulatory aspect that operated by requiring adherence to a set of minimum standards. And there was the broad public interest in the quality and potency of a product beyond that required in minimum standard terms.[4]

The Commission on Plasma Fractionation and Related Processes

The Commission met in January 1952 with two new members: Gilbert Dalldorf, the Director of the Division of Laboratories and Research of the New York State Department of Health, and T. Duckett Jones, a rheumatologist and now Medical Director of the Helen Hay Whitney Foundation in New York. When the Commission was founded in 1944, the phrase "and related processes" had been included in its name in the hope that royalties might flow from licensing of the Cohn alcohol-water protein fractionation methods for industrial uses outside the therapeutic field. Prior to the end of the war, Cohn had initiated discussions with firms in the corn products industry about the development of new commercial protein products from corn, based on zein, the predominant protein constituent of corn. In response, the American Maize Products Company, the Corn Products Refining Company, and A.E. Staley Manufacturing Company had provided funds to create training fellowships at the Department of Physical Chemistry, each sponsoring a fellow for one year. However, after a year at Harvard, the trainees had returned to their respective firms, but had moved on to other careers within a short period of time.

In the period since its establishment during the war, the Commission on Plasma Fractionation had established an excellent reputation for its role in the production and release of royalty-free therapeutic products from human plasma. In 1951 alone, a total of 120,000 vials of human serum albumin and 700,000 vials of immune serum globulin, produced by three firms, had been released by the Commission. Cohn was confident that new products and new uses would arise from the applications of the new biomechanical equipment then being developed. Indeed, a memorandum recommending the organization of a new company for the further development, production and control of his new biomechanical equipment was being discussed at the time.

Another important function of the Commission was the fostering of special investigations of interest to the general public as well as to the pharmaceutical industry. One example was the 1951 study by R. S. Paine and Charles Janeway aimed at confirming that homologous serum jaundice was not being transmitted by human serum albumin. Two hundred and thirty-seven patients in four Boston hospitals who had received albumin during an eighteen month period between 1950 and 1951 were enrolled. After excluding those cases in which the patients had died within five months of hospital admission, or who were diagnosed as having liver disease, the remaining patients were sorted into groups depending on whether they received albumin alone or albumin in combination with plasma or other products from blood. Among thirty-three patients who received human serum albumin alone, none of these were found to have developed hepatitis when followed up. In contrast, among ninety-three cases who received blood or plasma in addition to albumin, there were two cases of probable homologous serum jaundice, but in both those cases there were probable explanations unrelated to the use of albumin. Taken together, these results were interpreted as confirming that heat treated human serum albumin did not carry any risk of transmitting homologous serum jaundice.[5]

Scientific Meetings in Europe

The Official Announcement of the Second International Congress of Biochemistry to be held in Paris in July 1952, together with notices of satellite meetings before or after the Congress itself, stimulated considerable interest within the University Laboratory. The Faraday Society in England announced a General Discussion on the Physical Chemistry of Proteins, to be held in Cambridge two weeks after the Paris Congress, to which several members of the University Laboratory were invited. Edwin Cohn decided to demonstrate the new biomechanical equipment in France and Switzerland. Before long, it became evident that a large delegation from the laboratory would attend the meetings in Europe. The venue of the meetings in Paris meant opportunities for reunions with alumni of the laboratory, young and old. Most important, the Congress provided important opportunities for the new postwar generation of students in Europe to meet their peers in an international meeting for the first time.

Arrangements were made with U.S. Lines for the author and his spouse to bring a Cohn Centrifuge to Europe in a station wagon, departing from New York just before the Congress. Edwin Cohn liked nothing better than planning European trips for his friends. At every opportunity on our itinerary on the continent and in England Cohn arranged accommodations and identified the places to be visited. This began with a leisurely tour up the Seine to Paris, pausing for a night at an inn on the riverbank at Le Petit Andely and a visit to the chateau Gaillard, then on to Chartres before proceeding into Paris. The Cohns, accompanied by Mrs. Cohn's daughter, Susan Higginson, arrived in Paris a day before the formal opening of the Congress, having driven over by car from their villa in Switzerland. With assistance from Alex von Muralt, the Cohns had taken a villa in Switzerland that Edwin and Rebekah could use as their European base for several weeks. They had leased a Mercedes cabriolet with a unique feature, a velvet lined tool compartment that Cohn liked to show to his friends.

Several thousand scientists attended the Congress, at which over a thousand scientific papers were presented, grouped into thirty topical sections. The sessions were held in lecture halls and classrooms in the Sorbonne. Under favorable summer skies, attendees could stroll out into adjacent courtyards for conversations and meetings with colleagues. The Congress program included a variety of social events, tours and special programs for attendees and their spouses. On the Tuesday afternoon after the formal opening of the Congress, Edwin Cohn gave one of a series of formal lectures listed in the Congress Program. At the appointed hour, the large lecture hall was filled with dignitaries, scientists, students and spouses, all anxious to see and hear one of the most famous scientists of the period. Alex von Muralt introduced Cohn in warm terms that only a close personal friend and associate could use. Addressing his remarks particularly to the new young generation of European scientists present, he offered a brief review of the enormous contributions that had been made by Cohn and his associates during the war, stressing the implications of his work for the future of protein science. Edwin Cohn lived up to expectations. As always in that kind of setting, he began with some formal remarks, and then proceeded to present some of the newer scientific findings from studies in process in Boston. Following his lecture, he was surrounded by a host of old friends as well as members of the new generation attending their first international congress.

One morning during the Congress, Cohn demonstrated his new centrifuge at the French Establissement Central de Transfusion et Reanimation de l'Armee in Clamart, a Paris suburb. There, he was ceremoniously welcomed by Col. Jean Julliard, the Chief of the Establissement Central and a large group of military and civilian dignitaries, including a number of ladies. After introductions, the visitors were ushered into an amphitheater where the new centrifuge with its refrigeration unit running quietly, stood in front of a long table. Col. Julliard called the meeting to order and introduced Cohn who acknowledged his pleasure at being able to demonstrate the new machine in France. A blood donor lay down on a donor cot, the phlebotomy was started by an Army physician, and blood from the donor could be seen to flow from the donor's arm through clear plastic tubing into the machine. At first, while the spinning bowl slowly filled with blood, there was little to be seen. After about four minutes, clear yellow plasma could be seen to emerge from a transparent plastic tube into a clear plastic bag hanging below. The phlebotomy continued. At a signal, the physician withdrew the phlebotomy needle from the donor's arm. By that time, the bag of plasma was full. The machine was then turned off, and the thick suspension of red blood cells drained into a second plastic bag. After the blood donor was helped off the table, Col. Julliard came forward and paid a brief formal tribute to Edwin Cohn, in French and without translation. Enthusiastic applause followed, after which it became apparent that all attendees had been invited outdoors into a courtyard. There, in the warm sunshine of a beautiful Paris day, a warm celebration took place with champagne corks popping in the background. It had been a perfect demonstration. Among those present was the Baroness de Waldner, a patroness of the transfusion center, who was instrumental in acquiring one of the first Cohn centrifuges for the Establissement Central de Transfusion. That afternoon, a large photograph of Edwin Cohn standing in front of his Blood Centrifuge appeared on the front page of one of the Paris evening papers under the headline, "Le Roi des Proteins."

The hospitality extended by the French hosts throughout the week in Paris lent a special element of warmth and graciousness on many occasions. The Harvard visitors were swept into a whirl of meetings and social events with former associates and their spouses. One evening there was an official reception at the Louvre. On another evening, the Harvard group were the guests at a dinner party hosted by Professor D.M. Macheboeuf at a restaurant in the Forest of Vincennes. After aperitifs, dinner was served at a long narrow table under the trees. Following French custom there were no place cards. Macheboeuf waited until everyone else had been seated and then took the last empty place. There were no other formalities and no speeches. It was a memorable last evening in Paris for the Harvard guests.

The following day, we picked up Margaret Hunter and Henry Isliker at their hotels and departed for Switzerland. Following Cohn's suggested itinerary, we stopped briefly at Versailles and later in the day visited the Romanesque Benedictine abbey in Vezelay, the latter a particular favorite of Edwin Cohn. After spending the night in Beaune, we arrived in Rolle the next afternoon and checked into our inn, the Gay Rivage on the shore of the Lake of Geneva. Almost every day we were in Switzerland, scientific matters were interspersed with sightseeing activities.

The Cohn villa was a spacious mansion set back about half a mile from the edge of the Lake of Geneva with an unobstructed view across the lake to Mont Blanc on the far shore. In Rolle, Cohn spent his days much as he did in Cohasset, working out of his briefcase in the morning. Luncheons were served out of doors under tall trees. Most afternoons, Cohn, with a bathrobe over his bathing suit, led any interested guests down through a pasture for a swim. On our second day, we unloaded the Cohn machine at Professor von Muralt's laboratory in Bern and spent the night in Interlaken. On the following day, as von Muralt's guests, we rode the train up the Jungfrau and visited the Swiss High Altitude Physiological Research Station. The following day, we demonstrated the Cohn centrifuge at the University of Bern, after which we returned to Rolle, arriving in time for a swim with Cohn. On our last day in Switzerland, the Cohns gave a luncheon at their villa for Professor von Muralt and guests from the University of Bern and the University of Geneva and several scientists who had trained under Cohn in Boston. For that occasion, Rebekah Cohn had arranged a pleasant luncheon, served with Swiss elegance, at round tables, with all the windows and doors of the villa open to the outdoors on a beautiful summer day.

The Faraday Society

A General Discussion of the Faraday Society on "The Physical Chemistry of Proteins" was held in the Zoology Department of Cambridge University on August 6, 7, and 8, 1952. This meeting had an unusual format. It was intended to foster communication among an international group of scientists about five scientific topics deemed by the Society to have reached a stage where a general discussion among experts was expected to be fruitful. In a Section on "Experimental Techniques," papers were devoted to "Zone Electrophoresis in Filter Paper," to "Polarization of the Fluorescence of Labelled Proteins," and to the application of solubility measurements to the study of complex protein solutions and isolation of individual proteins. A second section was devoted to "Low Molecular Weight Proteins." It included papers on "Thermodynamics of the Association of Insulin Molecules," "Molecular-kinetic Properties of Trypsin and Related Proteins," "Electrophoretic Studies of Enzymatically Modified Ovalbumin and Casein," and "Globular-fibrous Protein Transformation." The third section, devoted to "High Molecular Weight Systems," included seven papers. The fourth section dealt with "Protein Interactions." It included the Cohn paper and six others. Five papers made up the last section which was devoted to "Conjugated Proteins."

To introduce the general topic of the meeting, John Edsall, then finishing a sabbatical year as a Fulbright Visiting Scholar at Cambridge, delivered the Sixth Spiers Memorial Lecture on "The Molecular Shapes of Certain Proteins and Some of their Interactions with Other Substances." This was Edsall at his scholarly best. For his major topics, he chose to discuss the "Size and Shape of Fibrinogen;" the "Conversion of Fibrinogen to Fibrin;" the "Size and Shape of Serum Albumins;" the "Possible Flexibility of Protein Molecules;" and the "Interactions of Serum Albumins with certain cations."

The Harvard paper bore a characteristically long title: "The Interaction of Plasma Proteins with Heavy Metals and with Alkaline Earths, with Specific Anions and Specific Steroids, with Specific Polysaccharides and with the Formed Elements of the Blood." Its authors were Edwin J. Cohn, Douglas M. Surgenor, Karl Schmid, W. H. Batchelor, Henry C. Isliker and Eva H. Alameri. Since Cohn had not come to

England, the paper was presented by Surgenor. It stimulated considerable discussion from Professor E. Schoberl of the Tierarztliche Hochschule in Hanover, Germany, Dr. B. Robert of the Institut Pasteur in Paris, Dr. L. Robert of the Faculté de Medecine de Paris, and from Professor Bo G. Malmstrom of the Biochemical Institute, Uppsala, Sweden.[6]

On the first evening of the Faraday Society meeting, the overseas members and visitors had dinner at Trinity College in Cambridge. There the Surgenors were seated at the high table with Professor Melwyn-Hughes, a charming Welshman and a physical chemist at the University of Cambridge. Melwyn-Hughes, who resembled Abraham Lincoln, added greatly to our pleasure with a steady stream of comments about the formalities and personalities at the long high table. Following dinner we drove him home and met Mrs. Melwyn-Hughes, a charming lady who explained that she did not ordinarily attend high table. We arrived back at our hotel only to find that the doors had been locked! On a subsequent afternoon, we punted on the Cam with Ephraim Katchalski, a theoretical chemist who had completed a year in Cohn's laboratory and was on his way back to Israel. (He later served as the President of Israel.)

Impending Change in Administration of the Cohn Patents

In the spring of 1952, the Research Corporation had requested to be relieved of the responsibility for administering the Cohn patents, as it became clear that prospects for earning royalties from their application to non-public health uses were dim. As was his custom, Cohn sent two responses to President Conant. In a formal response, Cohn wrote, "I request that the Corporation itself appoint Trustees, or let me do so, in whom our patents would be vested and who would be responsible for their administration." Although vesting the patents in Trustees would require the permission of the Corporation, Cohn was confident that if Chester Barnard were willing to undertake executive responsibility, there would be enthusiasm on Conant's part as well as his own, for this solution.[7] In an informal second letter to President Conant, Cohn wrote, "I am satisfied that neither Research Corporation nor any other existing organization of which I know, can satisfactorily assume this responsibility. It seems best therefore to return to the idea of vesting the patents in a Board of Trustees and to that end I enclose a formal letter making that suggestion, together with the names of some who, in my opinion, would be useful members of that board."[8]

President Conant responded a few days later: "The Corporation agrees that you may assign your patents, present and future, to a Board of Trustees that you yourself will select and persuade to serve. It would be understood that the Trustees would administer these patents according to the University policy, and the University assumes no financial or other responsibility in connection with the operation of this Board."[9] Cohn immediately contacted Chester Barnard, and by June 9, 1952, had a letter in which Mr. Barnard expressed his willingness "to serve on such a board and even be its unpaid chairman if it wanted me to be."[10]

By December 1952, Edwin Cohn had identified seven men who had agreed to serve as Trustees of a new corporation, which Cohn wanted to call "Protein Foundation," to provide the services previously supplied by the Research Corporation to the Commission on Plasma Fractionation and Related Processes. R. Ammi Cutter, a Boston lawyer, had been asked to develop a set of by-laws and draft a charter for the new corporation to replace the services that had been previously been provided by the Research Corporation. The new trustees included James S. Adams, general partner, Lazard Freres & Co, New York; Chester I. Barnard of New York, Former President of New Jersey Bell, Former President of the Rockefeller Foundation, and former Chairman of the National Science Foundation; Detlev W. Bronk, President of Johns Hopkins University, and President of the National Academy of Sciences; Paul F. Clark, President of John Hancock Mutual Life Insurance Co, Boston, and Chairman of the National Advisory Committee, American Heart Association; Wilbur K. Jordan, President of Radcliffe College and Harvard Professor of History; William M. Rand, Lincoln, Massachusetts, former President of the Monsanto Corporation, and Deputy Director, National Security Agency; James N. White, Partner, Scudder Stevens and Clark, Boston.

Early in January 1953, President Conant wrote to each of these men setting forth the interest of Harvard University in the matter and enclosing a statement of the Harvard patent policy. In his letter, he said that while the University ordinarily leaves professors entirely free to take out and handle patents on their discoveries in any manner they desire, an exception is made regarding discoveries in the field of therapeutics or public health, in which case the University requires that such patents be dedicated to the public. The reason for this policy is that the University believes that neither its professors nor itself should profit at the expense of the public's health. Conant then told that Cohn and some of his associates in the University Laboratory had applied for permission to seek patents on some of their discoveries on the grounds that the discoveries are of such a nature as to require carefully designed controls to prevent great danger to public health. He explained that the form of control Dr. Cohn desired will frequently produce highly complicated and far reaching legal and policy problems, the resolution of which will call for the judgement of men of specialized knowledge and general wisdom. Pointing out that the trustees had been selected to meet this need and that their services would be a real help to the University, he expressed the hope each person chosen by Cohn would consent to serve.[11]

The new corporation was chartered by the Commonwealth of Massachusetts on February 5, 1953. Governor Christian Herter presented the charter of the Protein Foundation to Chester Barnard on March 25, 1953 in his executive office in the State House in the presence of Roger I. Lee, W. K. Jordan and Edwin Cohn. At the first meeting of the new Trustees, Barnard was elected Chairman. In one of his first acts as Chairman, Barnard wrote to Joseph Barker at Research Corporation in New York notifying him "of the formation of the Protein Foundation . . . with the principal purpose of administering, in the public interest, patents in the fields of therapeutics and public health" and informing him of the Foundation's plan to take over the responsibility for administering the Cohn patents, effective on July 1, 1953.[12] Through an arrangement with Harvard, a house at 78 Mt. Auburn Street in Cambridge became available to the Protein Foundation, providing space for offices and meeting rooms, including the office of the Commission on Plasma Fractionation.

Further Development and Use of the Cohn Centrifuge

The successful demonstration of the new biomechanical equipment in Paris, and the interest it stimulated among Col. Julliard's associates to acquire a Cohn blood centrifuge, marked the end of the first phase of the development of the Cohn centrifuge. During that stage, it had been implicit in Cohn's thinking that the new equipment be capable of supporting a variety of blood and plasma related operations, including the decalcification of blood and plasma; of cooling blood to a specified temperature; of producing separate specific formed elements, including pathogenic constituents; and of processing plasma to yield stable plasma protein solution. On his return from Europe, Cohn conceded that the glass used in the early centrifuge bowl, so important visually during the early stages, should now be replaced by more durable material. Crude provisions for preventing bacterial contamination of the blood should also be replaced, and ease of cleaning and reuse of all parts coming into contact with blood should be addressed. Each modification would require thorough testing before incorporation into the units.

The machine itself, about the size of a small refrigerator, was made of stainless steel, glass and plastic. It took fresh blood directly from a blood donor and quickly processed it into important constituent parts. In its simplest form, it separated clear yellow plasma and red blood cells into separate plastic bags ready for transfusion. In developing the machine, great care was taken to reduce the temperature of the blood in order to remove the calcium to prevent clotting and to do so with minimum damage to the white blood cells, platelets, and plasma proteins. The parts of the machine that came into contact with the blood were contained in small sterile cartridges that included the centrifuge bowl, needles, tubing and plastic bags to hold the separated parts of the blood. A new cartridge, one for each blood donor, was inserted into the machine before use. Among Cohn's associates at Harvard who collaborated with him in developing this machine were J. Lawrence Oncley, James L. Tullis, John G. Gibson, 2nd, Henry C. Isliker, William H. Batchelor and the author. The key mechanical experts were Charles Gordon, Fred Gilchrist and Robert Tinch. Once the basic design and functions had been established, it quickly became apparent that the technology could support several different blood processing tasks in addition to simple separation of formed elements and plasma. These included, for example, the washing of blood cells and the fractionation and separation of plasma proteins. As diverse new uses of the equip-

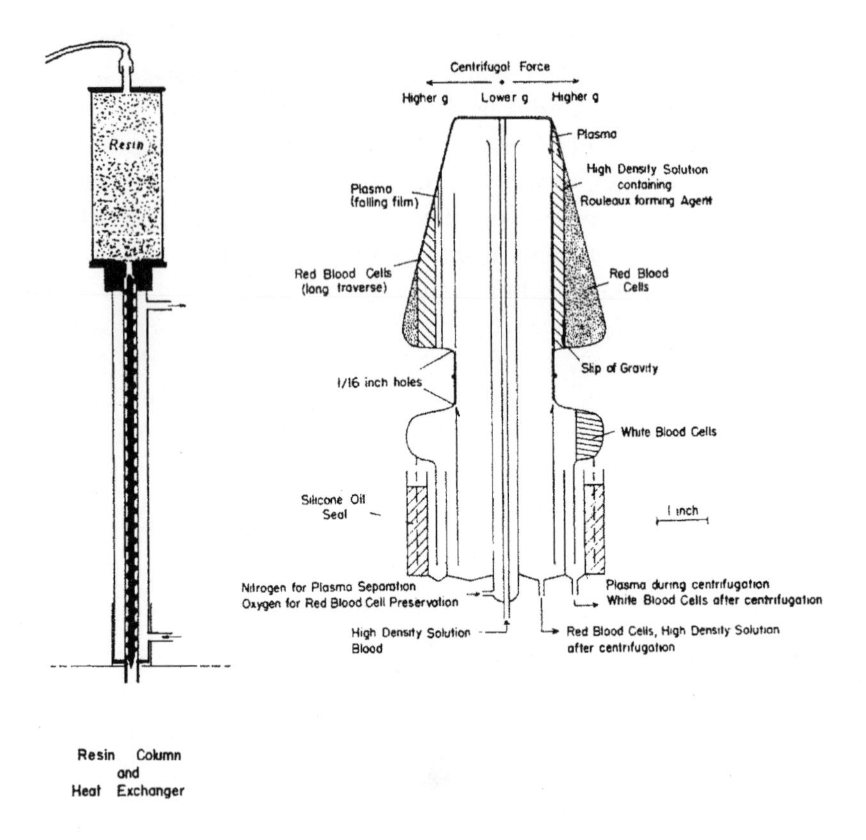

Fig. 9. Principles of the Cohn centrifuge.

ment suggested themselves, further refinement of the mechanism took place.

After a problem developed in the machine in 1952, Cohn asked his friend Earl Stevenson, President of the Arthur D. Little Company in Cambridge, for some engineering assistance and, in response, Allen D. Latham, an M.I.T. graduate in mechanical engineering and a Vice President at Arthur D. Little, came over to see him. Latham immediately recognized that the problem was caused by a flaw in the mechanical drive of the machine. Latham had one of his mechanical engineers design a reliable drive mechanism. In their conversations about the machine, Cohn suggested to Latham that a single use drive and disposable bowls would eventually be needed.[13] This was the time, Cohn felt, to develop agreements for the manufacture, distribution and servicing of new equipment. Preassembled sterile cartridges should be developed and made available to users as well. To that end, he recommended that Chester Barnard follow the custom of the Commission on Plasma Fractionation and appoint three experienced individuals as engineering consultants to the Protein Foundation. Cohn recommended the consideration of Victor Conquest, Vice President of Armour's Research Division, John Perkins, Research Director, American Sterilizer Co, and George Ryan of Abbott Laboratories.

Work was well underway on the development of new stainless steel bowls, using stainless steel parts that could be easily cleaned, assembled and steam sterilized, much like the old cream separators then still in use in dairies. The problem of maintaining sterility during operation was also resolved. In conversation with his consultants, Cohn learned of a rotary seal that was then being manufactured by the Sealol Corporation in Providence. Their seal was being used in the "fluid drive" of Chrysler automobiles. The critical functional parts of this seal were two rings, one machined of graphite, and the other of stainless steel. The graphite ring rotated against the rigid surface of the stainless steel ring. Whether static or spinning, an effective seal was maintained, as confirmed by bacteriologic testing. Cohn developed a plan by which the Arthur D. Little Company would fabricate ten Cohn centrifuges, each estimated to cost about ten thousand dollars. In this way, investigators in other institutions could acquire machines for their own use, the quid pro quo being that they would share experiences and thus provide feedback for use in further improvement and production of the machines.

The first training program for users of the new technology consisted of a three day program conducted at the Blood Characterization and Preservation Laboratory on May 6, 7 and 8, 1952. This provided hands-on experience with all aspects of use of the machine in blood collection. In three days, the training program gave the trainees experience in operating the centrifuge, including the washing and siliconing of the glass surfaces, assembly of the cartridges, and working with a nurse and blood donors in producing acceptable blood products such as platelets and red blood cell concentrates.

Gamma Globulin

A small field trial of gamma globulin in poliomyelitis had been proposed by Sidney Kramer of the Michigan Department of Health in 1944, and an application for grant support had been submitted to the National Foundation for Poliomyelitis. However, it had not been approved since "it was felt that with the war on, this was not the time to be conducting such trials."[14] As a result, the first field trials of gamma globulin were delayed for several years. Funding by the National Foundation for Infantile Paralysis became available in September 1951, with the first report being made public in 1952,[15] and with larger trials being conducted in Houston, Texas and in Sioux City, Iowa in 1953.[16] The results indicated that gamma globulin greatly reduced the incidence of the disease. They also revealed that the content of antibodies against the three types of poliomyelitis virus varied widely. This led to the realization that testing of gamma globulin for antibodies against polio prior to release would be needed to control the potency of the product in treatment of poliomyelitis. On October 26, 1952, the *New York Times* carried the news item: "Gamma globulin gives hope that polio can be checked." Reporting on a meeting of the American Public Health Association, it told that 54,772 children in Utah, Texas, Iowa and Nebraska were involved in the largest field trials in medical history, supported by a $1,000,000 grant from the National Foundation on Infantile Paralysis. The gamma globulin used was prepared from pools representing between 50,000 to 70,000 blood donors. There were no reported complications or increases in paralysis.[17]

In response, a conference was convened at the National Research Council in August 1952 for the purpose of evaluating the availability of stockpiles of gamma globulin to meet the needs in the summer of 1953. At the time, the Red Cross was producing about 70% of the national supply of gamma globulin, the balance being supplied by pharmaceutical firms processing plasma that had been procured privately. The U.S. plasma supply was diverted entirely to the production of human serum albumin and gamma globulin in 1953.[18] By that time, the involvement of the National Research Council and the Armed Services with respect to the national blood supply had ceased. Much later, in 1966, Joseph Stokes wrote: "Had the urgency of the theme of control affected the authorities in the National Foundation for Infantile Paralysis as forcefully as it did Dr. Sidney Kramer and myself in the early 1940s, again nature might well have been brought into balance at a considerably earlier date than actually obtained."[19]

Isolation of Gamma Globulin by Method 12 of Plasma Fractionation

As initially introduced late in 1950, Method 12 made it possible to obtain stable plasma protein solution and plasma globulin precipitate from plasma in an aqueous system at neutral pH and without relying on the use of ethanol-water mixtures. However, it was not until late in 1952 that important modifications were introduced in Method 12, making it possible to selectively extract gamma globulin from the plasma globulin precipitate. This depended on replacing the zinc acetate reagent with zinc glycinate in the original zinc precipitation step, thus providing a hydrogen ion buffer as well as a zinc buffer for the system. After separating and washing the plasma globulin precipitate so obtained, the gamma globulin was extracted with a dilute zinc acetate solution saturated with carbon dioxide. Once extracted, the gamma globulin could be precipitated in highly purified form simply by removing the carbon dioxide. Cohn was optimistic that this new gamma globulin would prove to be safe for intravenous use, while at the same time being free of the risk of transmitting the serum hepatitis virus. Nevertheless, he warned that it would be necessary to validate the safety of the new product with respect to its freedom from the virus of serum hepatitis before clinical testing in man could proceed.[20]

Immunopheresis

Continuing the custom of holding a January Conference, the fifth conference on "Implications of New Knowledge about Proteins, Protein Enzymes and Cells" was held on January 15, 1953. In a report at that meeting, Stokes and Smolens of the Children's Hospital of Philadelphia, citing the successful role of gamma globulin in the control of paralytic poliomyelitis, including the attenuation of polio when given in the latter part of the incubation period, called attention to the resulting increased need for gamma globulin of high antibody content with respect to the three types of polio. They described a new procedure, called immunopheresis, that could be carried out by using the Cohn Centrifuge. This involved drawing a unit of blood from a donor, separating and retaining the plasma, and immediately returning the formed elements to the donor. By repeating this process, even on a weekly basis, immunopheresis offered the potential of markedly increasing the amount of antibodies that could be donated by individual donors. Immunopheresis assumed added importance because the success of the field trials in poliomyelitis posed a significant added demand on the national supply of gamma globulin. This would offer cost effective ways of supplying needed antibodies for public use.[21]

At this same meeting, Ivan Brown, a Duke University surgeon, reported on the adaptation of the Cohn centrifuge for the separation and preservation of red cells in the frozen state at sub-zero temperatures. Brown's talk was followed by a report by Henry Isliker on interactions of antigens with antibodies and red blood cells. Isliker used a preparation of the stroma of outdated red cells linked chemically to an ion exchange resin to absorb isoagglutinins, which were then eluted with a sugar, yielding a high titer blood typing reagent. In another report, Cohn and Frederic Richards, a graduate student in the University Laboratory, reported on the use of the new zinc reagents to fractionate the proteins of milk. The final report was a paper by Monroe Eaton and Stephen Chapman of the Bacteriology and Immunology Department of the Harvard Medical School dealing with interactions of animal viruses with zinc and specific anions.

Issues Facing Protein Foundation

Following the selection of the Trustees and completion of the process of incorporating the Protein Foundation, Edwin Cohn wrote a series of letters to familiarize Chester Barnard, the chairman of the new foundation, with several issues that lay ahead. Concerning the introduction of biomechanical equipment into general use, he pointed out that it would require a type of regulatory control for which there was no precedent. The Public Health Service, responsible for minimal requirements, was not yet ready nor qualified to replace the Commission on Plasma Fractionation. Cohn stated, "The interpretation of the Harvard Patent Policy to which I am committed, demands maximal protection of the public by the control program, at minimal expense to the industrial firms licensed for production lest the charge be made that the costs of control were in fact a form of royalty. This charge must ever be considered latent and must determine policy." It would be a violation of this interpretation to increase charges for the control program to permit the financing of new developments. Public health studies made as a by-product of the control studies should not be directly financed by charges which would increase the cost of the biological product, he declared, adding: "It is my personal belief, frequently restated, that unlike the other parts of blood, antibodies should, in general, be used in the area in which they are collected. Thus the greatest value to our people of antibodies collected abroad is clearly to give temporary protection by passive immunization to civilians or military forces entering the area in which the gamma globulin antibodies were collected." With regard to funding the control of biologic products, Cohn commented that "the financing of advances in the development of biologic products can never be put on a strictly business basis . . . Product and process patents that Protein Foundation will hold after July 1, 1953 should never lead to exclusive rights."

Vis-à-vis meeting the expected demands for gamma globulin in 1953, Cohn reminded Barnard that the heavy use of gamma globulin for poliomyelitis required that a new specification be drawn for antibodies to the three prevalent strains of the poliomyelitis virus. The public health would be greatly in jeopardy in a largely expanded blood program if proper steps were not taken to assure that the gamma globulin for treating poliomyelitis actually contained the antibodies to the three strains of poliomyelitis. In the case of measles, the Commission had to rely on the routine assay for certain other common antibodies as surrogates of the measles antigen. The variation in antibody content of gamma globulin from a segment of the population should always be considered before using a particular antibody in prophylaxis and/or therapy. The supply of the reference standard gamma globulin that had been in use since the war was low. Funds should be procured to provide for an adequate supply of a new reference standard as soon as possible. The use of the term gamma globulin, which characterizes a complex group of antibodies, brought into focus the need for the Protein Foundation to adopt policies which would take cognizance of variations in antibody distribution.

Concerning the licensing of existing products in foreign countries and the need for caution in introducing new fractionation methods and biomechanical equipment, Cohn believed that new products and new procedures should be recommended after adequate experience has been gained. The responsibility for appraising the results of such experience rests with the Commission on Plasma Fractionation and Related Processes. The Commission members recognized that the present challenge exceeded any that had been faced since World War II, when the entire responsibility for training the personnel to introduce the new processes of plasma fractionation was vested in Cohn by the Navy and largely financed by the Committee on Medical Research of the Office of Scientific Research and Development.

Concerning the establishment of satisfactory public relations, Cohn advised Barnard that it had been his practice not to permit direct quotations. He also insisted on seeing articles in advance of publication so that care could be taken in making suggested corrections. Cohn said "a newspaper man has no obligation to accept these suggestions, and it is quite often impossible to explain to him the reasons, from the public relations viewpoint, for having a completely factual statement appear. The right of deletion, therefore, I have always claimed."[22]

Related Scientific Events

In January 1952, the National Institutes of Health Division of Biologics Control, in the interest of reducing the transmission of serum hepatitis, substantially tightened its minimum requirements for dried plasma by reducing the maximum permitted size of the pools of plasma for drying from four hundred donors to fifty donors in each pool. However, this change could not be put into effect immediately at licensed pharmaceutical firms. Late in October, 1952, an NRC Committee on Sterilization of Blood was recommending that dried plasma should be used only in an emergency. When active combat in Korea terminated in March 1953 (although a formal treaty had not been concluded), final action was taken by the Department of Defense to halt further production of dried plasma because of its risk of transmitting hepatitis.[23]

During this period, several landmark developments were recorded in the world scientific literature. In 1953, Sanger and Thompson completed the amino acid sequencing of the A and B chains of insulin.[24] At almost the same time, Watson and Crick postulated the double-helix model for the structure of DNA.[25] The first three cases of agammaglobulinemia were reported from Boston in 1952 by Bruton, Apt, Gitlin and Janeway.[26] Within a year, nine additional cases had been identified; six were seen at Children's Hospital in Boston, while the other three cases were reported by alert pediatricians in other institutions. These patients, all of whom were children, contained little or no circulating gamma globulin in their plasma. Monthly injection of gamma globulin reportedly gave them good protection.[27]

President Conant resigned as President of Harvard University at the end of the 1952-53 academic year in order to accept appointment as U.S. High Commissioner for Germany. In 1955 he was named U.S. Ambassador to the Federal Republic of Germany, a post that he held until 1957.

The Fine Structure of Proteins

All protein chemists in Cohn's laboratory and elsewhere engaged in speculation about aspects of the structure of the proteins they were studying. Cohn periodically expressed his views on such matters. In April 1952, he had occasion to reexamine the data in the 1944 paper of Brand, Kassel and Saidel in the *Journal of Clinical Investigation* on the amino acid composition of plasma proteins.[28] From that review, he was struck by the realization that a single molecule of human serum albumin was comprised of about 576 amino acid residues in peptide linkage. The most frequent amino acids in albumin were leucine, lysine, and glutamine, each with fifty-nine residues. The least frequent amino acid was cysteine, with only one residue. For another group of amino acids, Cohn was impressed with the realization that the number of residues was either sixteen, thirty-two, or forty-eight. Intrigued by this observation, this became the point of departure for Cohn into a series of cryptographic explorations seeking to fathom the relationships between the numbers of amino acids and the structure of the proteins involved. Although he had admitted in his 1948 Richards Medal Lecture that "the error of assuming uniformity, where more detailed observation of natural products has disclosed diversity and specificity, has been repeatedly made in the field of protein chemistry," he nonetheless forged ahead. Indeed, he became so impressed with his findings that he discussed them with his colleagues in the laboratory. In May 1952, he told President Conant, "in my mind there is no longer any possibility of the new concepts (in this work) being correct as a working hypothesis, except in detail, which can only be determined by experiments carried out on the basis of the hypothesis."[29]

Taking advantage of his privilege as a member of the American Philosophical Society in Philadelphia, Cohn submitted an abstract of a paper entitled "The Fine Structure of Proteins: A Contribution to Order in Biological Systems." In that presentation, he postulated the existence of certain important amino acid sequences in the peptide chains of certain proteins, concluding that "extensive examination of a large number of proteins suggests that all parts of proteins are not uniform."[30] Cohn never published anything more about his studies on the fine structure of proteins. Much later, Fred Cohn likened his father's studies on the fine structure of proteins to that of Mendeleyev who, in 1869, devised the periodic table of the chemical elements and successfully predicted the existence of three new chemical elements.

The Demise of Edwin J. Cohn

Late in the morning of October 1, 1953, while talking on the telephone with George Scatchard from his desk in the Protein Foundation office on Mount Auburn Street in Cambridge, Edwin Cohn suffered a massive cerebral hemorrhage. His last words were: "Georgie, I can't hear you." The telephone handset had fallen to the floor.[31] He died a few hours later in the Peter Bent Brigham Hospital. Post mortem studies revealed that Cohn had a pheochromocytoma, a rare adrenal tumor associated with sustained hypertension. Reportedly, on at least two earlier occasions, he had been persuaded by his physician to enter the hospital for a work-up. However, before all the tests could be completed, and against advice, he had signed himself out of the hospital, indicating that he was too busy to remain inactive for another day.[32]

A Memorial Service was held in the Memorial Church in Harvard Yard, and on November 12, 1953, the laudatory editorial quoted in the Foreword appeared in the *New England Journal of Medicine.* The Trustees of Protein Foundation adopted the following Memorial Minute prepared by W.K. Jordan at its December 11, 1953 meeting.

> Edwin Cohn was perhaps a great scientist for the very reason that he pressed on eagerly and almost impatiently to the next dim frontier of knowledge. Indeed, as one views his life, it is evident that there was in the career a metamorphosis from an exacting and rigorous protein chemist to what can only be described as a natural philosopher who came at the end to view knowledge broadly and synoptically. Edwin Cohn was a complex human being, a man at once a gifted and most exact scientist and an imaginative searching philosopher. He was a remarkable amalgam of the qualities of his own age and those which were more characteristic of the eighteenth century. His manners, his great eloquence, his concern for first principles, his scorn for the immediate—all these attributes are those of a century now too little remembered for its contribution to the graces as well as to the wisdom of mankind.[33]

19 AFTEREVENTS: THE LEGACY OF EDWIN J. COHN

SHORTLY BEFORE HIS DEATH, Edwin Cohn was notified that the University Laboratory of Physical Chemistry Related to Medicine and Public Health, Harvard University had been awarded a 1953 American Public Health Association Lasker Group Award. The citation read:

> For more than 30 years, Dr. Edwin Joseph Cohn and his associates have made brilliant observations on amino acids, peptides and proteins. These include fundamental contributions on the isolation, characterization, chemical differentiation and biological functions of these substances.
>
> Exceptionally meritorious has been the fractionation of human blood plasma into constituent purified substances of preventive and therapeutic value. These researchers have already produced biological preparations extensively used to alleviate suffering and disease: albumin for combating shock; isohemagglutinins for blood grouping; fibrinogen and thrombin for homeostasis; gamma globulin for passive immunization against epidemic disease. Using new chemistries and biomechanical equipment, these scientists have been able for the first time to separate blood and other biological materials into constituent formed elements and chemical substances in a closed system and a continuous process. Using methods closely approximating chemical interactions used by nature in the process of continuous differentiation to meet complex and changing functions and equilibrium needs of biological systems, new vistas are opened for the preparation, study and utilization of biological substances . . . These con-

> tributions are of untold benefit to man and form an important component of an emerging natural philosophy.

On behalf of Cohn's colleagues, the award was accepted by the author at a ceremony in New York City in October 1953.[1]

Following Edwin Cohn's death, Nathan Pusey, Harvard's new president, asked A. Baird Hastings to chair a Medical School committee to make recommendations regarding the academic reassignments of Cohn's former associates. As a result, the faculty appointments of Oncley, Hughes, Surgenor, Hunter and Low were transferred to the Department of Biological Chemistry in the Medical School, where they retained their offices and laboratories on the fourth floor of Building C-1. In less than ten years, all five had moved on to posts at other institutions—Oncley and Hunter to the University of Michigan, Hughes to the Brookhaven National Laboratory, Surgenor to the University of Buffalo, and Low to Columbia University. John Edsall had previously moved to the Harvard Biological Laboratories in Cambridge.

At the same time, the Trustees of the Protein Foundation submitted a proposal to the President and Fellows of Harvard University confirming their interest and willingness to assume the responsibility for Cohn's ongoing scientific program in blood characterization and preservation. Their proposal was accepted by Harvard, which leased to the Foundation the space in Jamaica Plain that had been used by the Blood Characterization and Preservation Laboratory. Although these arrangements did not immediately alter the long standing collaborative relationships in research between the scientists who worked at Protein Foundation and those who continued to work at the Medical School, the lack of a strong leader soon became a serious problem.

One of the first decisions of the Protein Foundation's Board was to continue the series of annual two-day November scientific meetings begun by Edwin Cohn. The first, a Conference on the Cellular Elements and Plasma Proteins of Blood, was held in Cambridge on November 15-16, 1954. The program of that conference told that the conference had been convened by the Protein Foundation, Incorporated, a nonprofit organization dedicated to research on blood and related biological problems and to the qualitative control of biological products in the public interest. In opening that 1954 Conference, Chester I.

Barnard stated that the therapeutic uses of blood had come a long way since the days when it was used for crude replacement purposes. Barnard characterized Cohn's discoveries as the "opening of a new scientific subcontinent." He also termed "spectacular" the development of gamma globulin as a prophylactic agent against measles, poliomyelitis and infectious hepatitis. He announced that the American Sterilizer Corporation had completed a series of sterility tests on the plasma from more than seventy blood collections done with the Cohn centrifuge. Every single unit of plasma had proved to be sterile and free of bacterial contamination. This represented an important milestone on the road to further development of the Cohn centrifuge technology.

Human Serum Albumin

Late in 1953, Roderick Murray of the NIH Division of Biologics Control reported to the Commission on Plasma Fractionation that human serum albumin fractionated from plasma by use of Cohn's new Method 12 had transmitted hepatitis. As a consequence, Method 12 was peremptorily abandoned. Today, all human serum albumin is prepared by using Cohn's Method 6 and is heated for ten hours at sixty degrees centigrade in the final containers to inactivate viruses.[2] In 1999, E. Tabor, at the U.S. Office of Blood Research and Review of the Food and Drug Administration, praised albumin for its excellent record of safety with respect to virus transmission and concluded that "albumin has the longest record of safety of any plasma derivative."[3]

Albumin is widely used today, although its value in fluid resuscitation to restore adequate circulation and perfuse critical organs such as brains, kidneys, liver and lungs is under some debate. In a recent review, J. J. Skillman, writing from the Beth Israel Deaconess Medical Center, stated that the significant cost and the limited supply of albumin, as well as its possibly adverse effects on the functions of critical organs, were factors in the debate. Conceding that the use of albumin for fluid resuscitation of certain critically ill patients is not supported by the available data, he nonetheless believed that the alternative of never using albumin in caring for patients was unreasonable. He concluded, "Albumin, the bell may be clanging softly somewhere in the distance, but, perhaps, it does not yet toll for thee."[4]

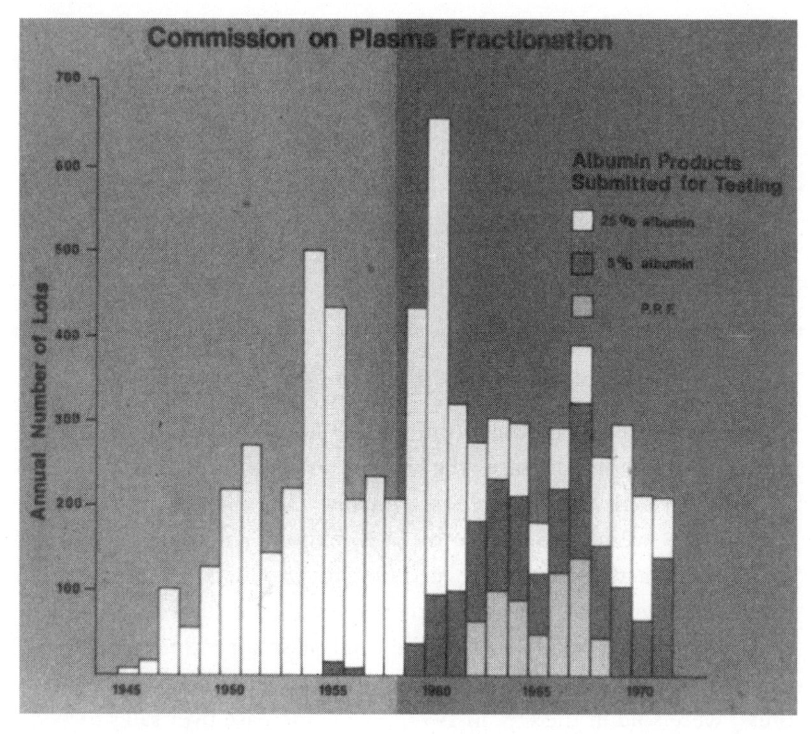

Fig. 10. Post war production of human serum albumin products prior to the expiration of the Cohn patents in 1971.

Gamma Globulin

Fraction II+III, prepared by the Cohn Method 6 and further purified by use of Method 9 of Oncley et al, is still the starting material for the isolation of all intramuscularly administered immune globulin (IGIM) and for the preparation of intravenous immune globulin (IVIG).[5] During the 1970s, Barundun, Hässig and Isliker at the Swiss Red Cross showed that removal of molecular aggregates from Fraction II rendered it safe for intravenous infusion. Today, nine different IVIG products have been approved by the U.S. Food and Drug Administration for clinical use in the United States.[6] Neither IGIM nor IVIG transmit hepatitis viruses or HIV.[7] Both IGIM and IVIG are prepared from the plasma of thousands of donors. IVIG has therapeutic value in treating antibody deficient syndromes such as neonatal sepsis and possibly

Kawasaki syndrome. It has been found to have beneficial actions in several autoimmune and inflammatory diseases. Half a century after Cohn's death, the Cohn methods for the preparation of immune globulin (IGIM and IVIG) continue to be used around the world. Although the Cohn patents covering plasma fractionation expired in the 1970s, the safety of the Cohn products in the United States is assured by the Minimum Requirements of the FDA which continue to specify that the Cohn methods be used in the preparation of gamma globulin.

The Market for Plasma Derivatives

A 1997 estimate placed the world market for all plasma derivatives at $4.2 to $5.0 billion. Of these, the market for albumin was estimated to lie between $2.1 and $2.4 billion annually.[8] U.S. sales of plasma products in 1999 were estimated to be $1.63 billion. The total U.S. market for plasma fractions including recombinant products and monoclonal respiratory syncytial virus antibody was estimated to be $2.57 billion. IVIG represented nearly half of the total plasma product sales in the US in 1999. About 16.7 metric tons of IVIG (6.68 million 2.5 gram units) were sold in the U.S. in 1999, a 7.4% increase over sales in 1998. Between 1998 and 1999, the cost of albumin fell from $4.00 to $3.30 per gram according to the Market Research Bureau.[9]

While the above figures are impressive from the viewpoint of a Wall Street financial analyst, their citation here is intended to demonstrate, through the considerable sales generated, how important blood and its derivatives have become in the delivery of modern medicine. Just as the surgeon has come over the last century and a half to rely on anesthesia, antisepsis, X-ray and other technologies that make the medicine of today seem near miraculous, he would not undertake his work without the assurance of an adequate supply of blood or its constituents and derivatives. Their use also falls within the sphere of the au courant internist, immunologist and other medical specialists in the day-to-day carrying out of their work. Many of the blood products that have come into standard use today had their beginnings in the work of Edwin J. Cohn or spun out of innovations he pioneered.

The Cohn Centrifuge

In 1956, Tullis, Surgenor, Tinch, D'Hont, Gilchrist, Driscoll and Batchelor submitted to the journal *Science* the first scientific paper describing the Cohn centrifuge.[10] Its authors cited the results of eighty consecutive trial runs involving collections of blood from donors in Boston, separation of red cells and plasma, and shipment of the red cells and plasma by air to the American Sterilizer Company in Erie, Pennsylvania. When tested there, all the bags except one, a bag of plasma that had ruptured during shipment, proved to be sterile. In a separate, more elaborate study, the adequacy of the mechanical seals in preventing contamination was proved in tests on more than 500 samples of blood products.

The use of the Cohn centrifuge for the long term preservation of blood cells stemmed from a 1951 discovery by Audrey U. Smith, an English veterinarian, that the damage to red blood cells caused by freezing was greatly reduced if the cells were first equilibrated with a glycerol solution.[11] The centrifuge was ideally suited for use in the glycerol process to preserve human blood cells. It made possible the deglycerolization of glycerolized red cells in a closed system without limiting the volume of wash solutions. Efforts in this direction were first described in a 1958 Protein Foundation paper entitled "Use of Biomechanical Equipment for the Long-Term Preservation of Erythrocytes." Based on 175 experimental runs, it confirmed the work of Smith by showing that human red cells can be stored for three months or longer at temperatures below −80° C when mixed with glycerol and confirmed that sterility had been maintained. Loss by hemolysis did not exceed 20% and the time required for removal of the glycerol had been substantially reduced.[12]

These results led directly to a more ambitious long-term test of the feasibility of establishing and operating a hospital blood bank based on the use of frozen glycerolized red cells. The test was carried out at the United States Naval Hospital in Chelsea, Massachusetts between 1958 and 1963. Using Cohn centrifuges, bloods from volunteer naval personnel were collected in glycerol and stored in the frozen state in the hospital blood bank. When blood transfusions were needed by patients in the hospital, frozen red cells were deglycerolized and transfused. The use of 1,014 units of frozen glycerolized blood and their administration to patients with good clinical results and a low reaction rate were reported in the *Journal of the American Medical Association* in 1960.[13] In this way, the Chelsea Naval Hospital met all its blood needs from frozen blood for a period of twenty-eight months.

The Next Generation of Centrifuges

From the time he had been called on by Cohn to help re-engineer the original mechanical seals on the Cohn centrifuge, Allen Latham, Jr. had been imbued by the importance of Cohn's work. Latham later became Chairman and CEO at Cryogenics Technologies, Inc., an Arthur D. Little (ADL) start-up company that had been formed to commercialize ADL's research in cryogenics. There his greatest achievement had been the design and building of a refrigeration system capable of maintaining a temperature of −450°F—near absolute zero—for the preservation of large quantities of liquid deuterium. After Cohn's death, Latham resolved to do everything possible to see that Cohn's vision for the blood centrifuge was carried forward. During this period, Latham frequently dropped in at the Blood Characterization and Preservation Laboratory to see Robert Tinch, who had been Cohn's technician. Latham presented papers at most of the annual Protein Foundation Conferences. One of the first of these was entitled "Simple Fractionators for Special Use." Another paper offered a detailed analysis of the hypothetical costs involved in using Cohn centrifuges in large blood centers. At another time, he reported the results of tests of the rate of glycerol transfer during deglycerolization of frozen red blood cells. Once, he devoted a month-long "sabbatical leave" to the development of an expendable plastic liner and automatic solution system for deglycerolization of frozen red blood cells. One of his most important inventions was the design of a new one-inch mechanical seal to replace the four-inch seal that Cohn had used. This new seal became the key to Latham's design of a completely new centrifuge. Latham never forgot Cohn's suggestion that a single use disposable centrifuge bowl would eventually be needed. His own studies had convinced him that reusing the same bowl would be time consuming and, unless circumvented, would be a barrier to commercial success. This strongly motivated him to devise a disposable bowl. At lunch one day with a friend, Latham learned about a new thermoplastic called polycarbonate. Within a month, he had fabricated the prototype of the disposable bowl.

In 1968, realizing that outside resources would be needed to commercialize the blood centrifuge technology, Latham began to explore the possibility of a strategic partnership between Arthur D. Little and a company with strong manufacturing and marketing capabilities in the plastics field. This led to an agreement between ADL and Abbott Laboratories under which Abbott would manufacture and market sterile cartridges designed by Latham, while ADL would supply the mechanical systems. However, in the next three years it became apparent that Abbott was having difficulties living up to the agreement. Latham's final decision to pull out came in 1970 when Abbott encountered an unrelated FDA action that threatened one of its major product lines. The continued development of the disposable bowls and cartridges was in jeopardy.

Latham resigned from ADL and turned to his last alternative, which was to start a new company on his own. He talked with Gordon Kingsley, a friend and experienced engineer who at the time was looking for an interesting business venture. Kingsley was intrigued by the blood separation technology and the opportunity to help shape a new company in a fundamental way. After considerable thought, they articulated three founding principles. These were, first, a commitment to quality, since quality of their products would be crucial to their success, given the medical nature of the company's activity; second, the recognition of the importance of bringing into the organization a group of outstanding people; and, third, a commitment to offer outstanding products and services, with an awareness of the importance of listening to the customer. What really mattered to Latham was making quality blood components available to people who need them.[14]

Haemonetics

Haemonetics was founded in 1972 with the help of the venture firm, Breck, McNeish and Nagel. Within a year, a disposable bowl, now known as the Latham Bowl, was introduced, together with a machine called the Model 10 Cell Washer, to drive the bowl. This combination enabled users to perform several blood related procedures. The commercial availability of the Cohn centrifuges led to further differentiation of the centrifuge technology. In 1975, a cell saver machine was introduced for use in salvaging spent patient blood in hospital operating rooms. In 1984, a new plasma collection system followed for use in commercial plasma collection centers. In 1991, an Initial Public Offering was completed and the company moved into a new plant in Braintree, Massachusetts. In 1994, a new manufacturing plant was completed in Scotland. In fiscal 1998, Haemonetics reported sales to blood banks of $121 million out of an estimated market of $375 million. For the commercial plasma sector, sales were $85 million out of an estimated $185 million market. For the surgical sector, the sales were $62 million out of $125 million. The company employed 1,400 people and had offices in fourteen countries. Its revenues in 1999 amounted to $285 million.[15]

In 1991, Allen Latham, then eighty-three years old, addressed the student body at the Roxbury Latin School that he had attended as a boy. He spoke about "The Rewards of Being a Mechanical Engineer." As he finished his talk, he said:

> During the recent Persian Gulf hostilities, the field hospitals and military hospitals in Germany were well stocked with frozen red blood cells along with the supplies and equipment for their deglycerolization. Today, every major cancer treatment center is equipped to collect single donor platelets for support of its patients. Not all of them use Haemonetics equipment. There is strong competition, but we lead the way. I hardly need to say that I experience an abundant flow of psychic reward whenever these thoughts come to mind.[16]

The Center for Blood Research

In 1965, finding it difficult to raise funds on behalf of an institution called The Protein Foundation, the trustees changed its name to the Blood Research Institute. In 1970, after a successful fund raising drive, the present building at 800 Huntington Avenue was acquired and its rehabilitation was begun. In the first of several construction phases that followed, after 12,000 square feet of laboratory space was equipped at the western end of the second floor, the scientific staff of the former Protein Foundation were moved in, thus terminating all operations in Jamaica Plain in favor of renewed presence in the Longwood area. In a second stage in 1972, the Blood Grouping Laboratory of the Childrens Hospital, founded by Louis K. Diamond, was merged with the Blood Research Institute. Under Diamond, the Blood Grouping Laboratory had discovered several new blood groups and had developed the successful exchange transfusion treatment of Rh disease of the newborn.

The new organization was named The Center for Blood Research; Douglas Surgenor was elected its President, and Chester Alper, an expert on the molecular basis of the nonresponder state, was named Medical Director. Expansion of the scientific staff and rehabilitation of the space at 800 Huntington Avenue to provide additional laboratories was completed in subsequent steps. The final project saw the exterior of the building completed as it looks today. Informal discussions between Harvard Medical School Dean Tosteson and Surgenor led, in 1986, to an affiliation agreement under which The Center for Blood Research was designated a Harvard Medical School Affiliated Institution.

Fred S. Rosen, the James L. Gamble Professor of Pediatrics at the Harvard Medical School and for many years Chairman of the Committee on Primary Immunodeficiency of the World Health Organization, succeeded Surgenor as President of the Center for Blood Research in 1987. Under Rosen's brilliant leadership, the focus of research at the Center has shifted to the study of the immune response, the body's response to antigens. Today, CBR enjoys worldwide stature in its field. It now supports twenty-eight doctoral level investigators, all of whom hold academic appointments at the Harvard Medical School. In 1999, CBR ranked eleventh in total funding among the seventy-one of the Association of Independent Research Institutions which received NIH support during 1999. Thus, the legacy that Edwin J. Cohn initiated rigorously lives on.

Afterword

In early June of 2000, Hans Jörnval, Secretary of the Nobel Assembly in Stockholm, responded to an inquiry made by the author. After noting that the Nobel archives are kept strictly confidential, Jörnval informed him that Edwin Joseph Cohn had been nominated for a Nobel Prize in Physiology or Medicine by one nominator in 1929, by two more nominators in 1946, by five nominators in 1948, and by six nominators in 1949. However, no further information was forthcoming, and no names were mentioned.[17]

Notes

Foreword:

1. "Editorial. Edwin J. Cohn," *The New England Journal of Medicine* 249 (1953): 824–825. Reprinted in part with permission of *The New England Journal of Medicine.*

Chapter 1:

1. Edwin J. Cohn, Jr., interview with the author, Washington, D.C., April 19, 1993, author's files.
2. Alfred E. Cohn, *No Retreat from Reason* (New York: Harcourt, Brace, 1948).
3. E. Albert Berol, interview with the author, Bedford Hills, New York, April 10, 1992, author's files.
4. Ibid.
5. Ibid.
6. Edwin J. Cohn, Jr., interview with the author, Washington, D.C., April 19, 1993.
7. G. W. Gray, "Confidential Monthly Report for the Information of the Trustees of the Rockefeller Foundation, No. 122, June 1, 1950," pp. 2–19, of which a copy is in the E. J. Cohn File in the Archives of the Center for Blood Research, cited hereafter as ACBR.
8. J. T. Edsall and W. H. Stockmeyer, "George Scatchard, 1892–1973," *Biographical Memoirs of the National Academy of Sciences* 52 (1980): 334–377.
9. George Scatchard, "Edwin J. Cohn and Protein Chemistry," *Vox Sanguinis* 19 (1970): 37–44.

10. G. W. Gray, "Confidential Monthly Report for the Information of the Trustees of the Rockefeller Foundation, No. 122, June 1, 1950."
11. D. W. Lovelace, "Who's Who in the News," *Newark News*, June 7, 1944, in Marianne Brettauer Cohn Scrapbook, ACBR.
12. G. W. Gray, "Confidential Monthly Report for the Information of the Trustees of the Rockefeller Foundation, No. 122, June 1, 1950."
13. Ibid.
14. L. J. Henderson, *Fitness of the Environment. An Inquiry into the Biological Significance of the Properties of Matter* (New York: Macmillan, 1913; reprinted, The American Naturalist, 1947).
15. W. B. Cannon, "Lawrence J. Henderson," *Biographical Memoirs of the National Academy of Sciences* 23 (1943): 31–58.
16. L. J. Henderson and E. J. Cohn, "The Equilibrium between Acids and Bases in Sea Water," *Proceedings of the National Academy of Sciences* 2 (1916): 618–622.
17. R. M. Ferry, "Obituary. Lawrence Joseph Henderson, 1878–1942," *Science* 95 (1942): 316–318.
18. G. W. Gray, "Confidential Monthly Report for the Information of the Trustees of the Rockefeller Foundation, No. 122, June 1, 1950."
19. E. J. Cohn, "Recent Advances in the Electrochemistry of the Proteins," *Transactions of the Electrochemical Society* 71 (1936): 127–133.
20. E. J. Cohn, "Studies in the Physiology of Spermatozoa," *Biological Bulletin* 34 (1918): 167–218.
21. G. W. Gray, "Confidential Monthly Report for the Information of the Trustees of the Rockefeller Foundation, No. 122, June 1, 1950."
22. Edwin J. Cohn, Jr., interview with the author, Washington, D.C., April 19, 1993.
23. E. J. Cohn, S.B. Wolbach, L. J. Henderson and P. H. Cathcart, "On the Control of Rope in Bread," *The Journal of General Physiology* 1 (1918): 221–230; E. J. Cohn and L. J. Henderson, "The Physical Chemistry of Bread Making," *Science* 48 (1918): 7; E. J. Cohn, P.H. Cathcart and L. J. Henderson, "The Measurement of the Acidity of Bread," *Journal of Biological Chemistry* 36 (1918): 581.
24. A. Baird Hastings, *Crossing Boundaries, Biological, Disciplinary, Human;* Halvor N. Christiensen, Editor (Grand Rapids, Michigan: The Four Corners Press, 1989).
25. E. J. Cohn, "Soren Peter Lauritz Sorensen, 1868–1939," *Journal of the American Chemical Society* 61 (1939): 573–574.

26. G. W. Gray, "Confidential Monthly Report for the Information of the Trustees of the Rockefeller Foundation, No. 122, June 1, 1950."

Chapter 2:

1. G. W. Gray, "Confidential Monthly Report for the Information of the Trustees," January 26, 1944," Rockefeller Foundation Archives, Record Group 1.1, Series 200, Box 141, Folder 174, Rockefeller Archives Center, Tarrytown, New York, cited hereafter as RAC.
2. Information on the Harvard Medical School's De Lamar Fund, and its donor, Captain Joseph De Lamar, can be found in E. L. Wolfe, A. C. Barger and S. Benison, *Walter B. Cannon, Science and Society* (Boston: Boston Medical Library, 2000): 70–73.
3. George Scatchard, "Edwin J. Cohn and Protein Chemistry."
4. J. T. Edsall and W.H. Stockmeyer, "George Scatchard, 1892–1973."
5. George Scatchard, "Edwin J. Cohn and Protein Chemistry."
6. Margaret W. Rossiter, "Philanthropy, Structure, and Personality or, the Interplay of Outside Money and Inside Influence," in C.A. Elliott and M. W. Rossiter, eds., *Science at Harvard University, Historical Perspectives* (Bethlehem, Pa.: Lehigh University Press, 1992), 13–22.
7. Oglesby Paul, *Take Heart; the Life and Prescription for Living of Paul Dudley White, the World's Premier Cardiologist* (Boston: The Francis A. Countway Library of Medicine, 1986), 30.
8. S. Benison, A. C. Barger and E. L. Wolfe, "Walter B. Cannon and the Mystery of Shock: A Study of Anglo-American Cooperation in World War I," *Medical History* 35 (1991): 217–249; E.L. Wolfe, A.C. Barger and S. Benison, *Walter B. Cannon, Science and Society*, p. 234 et seq.
9. E. J. Cohn, "Research in the Medical Sciences," in *Medicine Today: March of Medicine, 1946* (New York: Columbia University Press, 1947), 70–121; printed also in *American Scientist* 37 (1947): 69–90, 243–254. "Research in the Medical Sciences" was one of the New York Academy of Medicine's "Lectures for the Laity," 1945–1946.
10. J. T. Edsall, "Edwin Joseph Cohn, 1892–1953," *Biographical Memoirs of the National Academy of Sciences* 35 (1961): 46–84.
11. E. J. Cohn, "The Physical Chemistry of Proteins," *Physiological Reviews* 5 (1925): 349–437.
12. E. J. Cohn, "Interactions of Proteins and Other Body Constituents," *The Nucleus* 25 (1948): 263–276.

13. E. J. Cohn, "The Physical Chemistry of Proteins."
14. Emil Fischer, "Synthesen in der Purine und Zuckergruppe," *Bericht der Deutschen Chemischen Gesellschaft* 35 (1902): 2660.
15. R. H. McCoy, C. E. Meyer and W. C. Rose, "Feeding Experiments with Mixtures of Highly Purified Amino Acids," *Journal of Biochemistry* 112 (1935): 283–302.
16. E. J. Cohn, "The Physical Chemistry of Proteins."
17. Ibid.
18. Thomas Graham, "Liquid Diffusion Applied to Analysis," *Philosophical Transactions of the Royal Society* 151 (1861): 183–224.
19. E. J. Cohn, "The Physical Chemistry of Proteins."
20. G. Hüfner, "Neue Versuche zur Bestimmung der Sauercapacitüt des Blutstoffs," *Archiv für Physiologie. Physiologische Abhandlung des Archives für Anatomie und Physiologie* (1894): 130–176.
21. E. J. Cohn, "Studies in the Physical Chemistry of Proteins. I. The Solubility of Certain Proteins at Their Isoelectric Points," *Journal of General Physiology* 4 (1921–1922): 697.
22. George Scatchard, "Edwin J. Cohn and Protein Chemistry."
23. N. Bjerrum, "Die Konstitution in der Ampholyte, Besonders der Aminosäuren, und Ihre Dissoziationskonstanten," *Zeitschrift für Physikalische Chemie* 104 (1923): 147–173.
24. George Scatchard, "Edwin J. Cohn and Protein Chemistry."
25. P. Debye and E. Hückel "Zur Theorie der Elektrolyte. I. Gefrierpunktserniedrigung und Verwandte Erscheinungen," *Physikalische Zeitchrift* 24 (1923): 185–206.
26. R. M. Ferry, "Studies in the Chemistry of Hemoglobin. I. The Preparation of Hemoglobin," *Journal of Biological Chemistry* 47 (1923): 819.
27. R. M. Ferry and A. A. Green, "Studies in the Chemistry of Hemoglobin. III. The Equilibrium between Oxygen and Hemoglobin and Its Relation to Changing Hydrogen Ion Activity," *Journal of Biological Chemistry* 81 (1929): 175–203.
28. E. J. Cohn and J. B. Conant, "The Molecular Weights of Proteins in Phenol," *Proceedings of the National Academy of Sciences* 12 (1926): 433–438.
29. Theodor Svedberg and Robin Fåhraeus, "A New Method for the Determination of the Molecular Weight of the Proteins," *Journal of the American Chemical Society* 48 (1926): 430–438.
30. J. T. Edsall, "Some Personal History and Reflections from the Life of a Biochemist," *Annual Review of Biochemistry* 30 (1971): 1–28.
31. W.B. Cannon, "Lawrence Joseph Henderson, 1878–1942."

Chapter 3:

1. G. R. Minot and W.P. Murphy, "Treatment of Pernicious Anemia by a Special Diet," *Journal of the American Medical Association* 87 (1926): 470–476.
2. E. J. Cohn, "George Richards Minot, 1885-1950." *Yearbook of the American Philosophical Society* (1950): 313–319.
3. G. W. Corner, *George Hoyt Whipple and His Friends* (Philadelphia: J. B. Lippincott, 1963).
4. G. R. Minot and W. P. Murphy, "Treatment of Pernicious Anemia by a Special Diet."
5. E. J.Cohn, G. R. Minot, J. F. Fulton, H.F . Ulrichs, F. C. Sargent, J. H. Weare and W. P. Murphy, "The Nature of the Material in Liver Effective in Pernicious Anemia," *Journal of Biological Chemistry* 74 (1927): 60–72.
6. F. M. Rackeman, *The Inquisitive Physician. The Life and Times of George Richards Minot, AB/MD/DSc.* (Cambridge, Mass.: Harvard University Press, 1956), 151.
7. E. J. Cohn, *History of the Development of a Patent Policy, Based on Experiences in Connection with Liver Extracts and Blood Derivatives* (Cambridge, Mass.: Harvard University Printing Office, 1951). Fifth Publication of the University Laboratory of Physical Chemistry Related to Medicine and Public Health, April 1951.
8. A. L. Lowell to D. L. Edsall, March 14, 1947, E. J. Cohn Papers, Countway Library of Medicine, cited hereafter as CLM.
9. E. J. Cohn to Odin Roberts, November 5, 1927, E. J. Cohn Papers, CLM.
10. E. J. Cohn, *History of the Development of a Patent Policy, Based on Experiences in Connection with Liver Extracts and Blood Derivatives.*
11. E. J. Cohn to E. M. Berolzheimer, January 31, 1929, cited in references 7 and 10 above.
12. G. R. Minot and W. P. Murphy, "A Diet Rich in Liver in the Treatment of Pernicious Anemia. Story of One Hundred Cases," *Journal of the American Medical Association* 89 (1927): 729–766.
13. E. J. Cohn, T. L. McMeekin and G. R. Minot, "The Nature of the Substances Effective in Pernicious Anemia," *Transactions of the Association of American Physicians* 45 (1930): 343–349.
14. F. M. Rackemann, *The Inquisitive Physician. The Life of George Richards Minot, AB/MD/DSc.*
15. R. West and E. G. Nichols, "Liver Fractionats in Pernicious Anemia," *Journal of the American Medical Association* 89 (1927): 867–868.

16. G. R. Minot, "The Development of Liver Therapy in Pernicious Anemia," *Lancet* 1 (1935): 361–364. The Nobel Lecture.
17. A. B. Hastings, *Crossing Boundaries, Biological, Disciplinary, Human.*
18. E.J . Cohn, Jr. to the author, October 30, 1993, May 11, 1995, author's files.
19. XIIIth International Physiological Congress, *Program and List of Members, Boston, August 19–23, 1929* (Boston: 1929).
20. Ibid.

Chapter 4:

1. N. F. Thompson to A. L. Lowell, April 16, 1930, Record Group 1.1, Series 200A, Box 89, Folder 1069, RAC.
2. E. J. Cohn, "Die Physikalische Chemie der Eiweisskorper," *Ergebnisse der Physiologie* 33 (1931): 781–882.
3. E. J. Cohn to D. L. Edsall, May 1931, E. J. Cohn Papers, CLM.
4. Ibid.
5. Edwin J. Cohn, Jr., interview with the author, June 12, 1996, author's files.
6. J.T. Edsall, "Edwin Joseph Cohn, 1892–1953."
7. Edwin J. Cohn, Jr., interview with the author, December 24, 1995, author's files.
8. Edwin J. Cohn, Jr., interview with the author, October 30, 1995, author's files.
9. J. T. Edsall, "Edwin Joseph Cohn, 1892–1953."
10. Edwin J. Cohn, Jr., interview with the author, October 30, 1995.
11. A. Baird Hastings, *Crossing Boundaries, Biological, Disciplinary, Human.*
12. J. P. Greenstein, "Studies of the Peptides of Trivalent Amino Acids," *Journal of Biological Chemistry* 93 (1931): 479–494.
13. G. W. Gray, "Confidential Monthly Report for the Information of the Trustees, January 26, 1944."
14. George Scatchard, "Edwin J. Cohn and Protein Chemistry."
15. Jeffries Wyman, Jr., and T. L. McMeekin, "The Dielectric Constant of Solutions of Amino Acids and Peptides," *Journal of the American Chemical Society* 55 (1933): 908–914; "The Dipole Moments of Esters of Amino Acids and Peptides," *Journal of the American Chemical Society* 55 (1933): 915–922.
16. E. J. Cohn to D. L. Edsall, January 16, 1934, E. J. Cohn Papers, CLM.
17. J. T. Edsall and Jeffries Wyman, Jr., "The Polarity of Certain Betaines," *Journal of Biological Chemistry* 105 (1934): proc. 24.

18. P. W. Bridgman and R. B. Dow, "The Compressibility of Solutions of Three Amino Acids," *Journal of Chemical Physics* 3 (1935): 35–41.
19. J.T. Edsall, "Edwin Joseph Cohn, 1892–1953."
20. Edwin J. Cohn, Jr., interview with the author, Washington, D.C., April 19, 1993.
21. Edwin J. Cohn, Jr., to the author, May 11, 1995.
22. A. Baird Hastings, *Crossing Boundaries, Biological, Disciplinary, Human.*
23. Ibid.
24. E. J. Cohn and J.T. Edsall, *Proteins, Amino Acids and Peptides* (New York: Reinhold, 1943).
25. E. J. Cohn to S. C. Burwell, April 21, 1936, Record Group 1.1, Series 200A, Box 89, Folder 1072, RAC.
26. William L. Lawrence, "Vast Network of 'Living Dynamos' Supplies Electric Power to the Body," *New York Times*, April 13, 1936, p. 1, col. 4.
27. "Living Dynamos to the Body U.S. Scientist's Theory," *The Times* (London), April 14, 1936, in Marianne Brettauer Cohn Scrapbook, ACBR.
28. E. J. Cohn to C. S. Burwell, April 21, 1936, Record Group 1.1, Series 200A, Box 89, Folder 1072, RAC.
29. E. J. Cohn to Allan Gregg, April 30, 1936, Record Group 1.1, Series 200A, Box 89, Folder 1072, RAC.

Chapter 5:

1. J. L. Oncley, "Studies of the Dielectric Properties of Protein Solutions. I. Carboxyhemoglobin," *Journal of the American Chemical Society* 60 (1938): 1115–1123.
2. E. J. Cohn to J.B. Conant, March 26, 1938, E.J. Cohn Papers, CLM.
3. E. J. Cohn to C. S. Burwell, July 24, 1940, E. J. Cohn Papers, CLM.
4. J. B. Conant to Warren Weaver, March 19, 1937, Record Group 1.1, Series 200A, Box 141, Folder 1741, RAC.
5. C. S.Burwell to Warren Weaver, March 17, 1937, Record Group 1.2, Series 200A, Box 141, Folder 1741, RAC.
6. J. H. Northrop to Frank Blair Hanson, October 26, 1937, Record Group 1.2, Series 200A, Box 141, Folder 1741, RAC.
7. Michael Heidelberger to Frank Blair Hanson, October 29, 1937, Record Group 1.2, Series 200A, Box 141, Folder 1741, RAC.
8. Warren Weaver, "Diary," November 1, 1937, Record Group 1.2, Series 200A, Box 141, Folder 1741, RAC.

9. W. T. Astbury to Frank Blair Hanson, November 4, 1937, Record Group 1.2, Series 200A, Box 141, Folder 1741, RAC.
10. Max Bergmann to Frank Blair Hanson, November 1, 1937, Record Group 1.2, Series 200A, Box 141, Folder 1741, RAC.
11. C. R. Harington to Frank Blair Hanson, November 12, 1937, Record Group 1.2, Series 200A, Box 141, Folder 1741, RAC.
12. K. Linderstrom-Lang to Frank Blair Hanson, November 22, 1937, Record Group 1.2, Series 200A, Box 141, Folder 1741, RAC.
13. A. von Muralt to F. B. Hanson, February 11, 1938, Record Group 1.2, Series 200A, Box 141, Folder 1742, RAC.
14. A. C. Chibnall, excerpt from letter to Warren Weaver, undated, Record Group 1.2, Series 200A, Box 141, Folder 1741, RAC.
15. Warren Weaver, "Notes," December 7, 1937, Record Group 1.2, Series 200A, Box 141, Folder 1741, RAC.
16. Rockefeller Foundation, Board of Scientific Advisers, "Minutes," [Spring] 1938, Record Group 1.1, Series 100, Box 141, Folder 1741, RAC.
17. E. J. Cohn, "A Brief History of the Support of the Department of Physical Chemistry, Harvard Medical School 1920–1950."
18. E. J. Cohn, "Number and Distribution of the Electrically Charged Group of Proteins," *Cold Spring Harbor Symposia on Quantitative Biology* 6 (1938): 8–20.
19. E. J. Cohn, "Proteins as Chemical Substances and as Biological Components," *Bulletin of the New York Academy of Medicine* 15 (1939): 639–667.
20. E. J. Cohn to J. B.Conant, March 26, 1938, E. J. Cohn Papers, CLM.
21. W. T. Astbury and R. Lomax, "X-Ray Photographs of Crystalline Pepsin," *Nature* 133 (1934): 795.
22. J. D. Bernal and D. Crowfoot, "X-Ray Photographs of Crystalline Pepsin," *Nature* 133 (1934): 794–795.
23. J. T. Edsall, "Some Personal History and Reflections upon the Life of a Biochemist."
24. E. J. Cohn to Warren Weaver, February 26, 1940, Record Group 1.2, Series 200A, Box 141, Folder 1741, RAC. Cohn quoted from a letter he had earlier received from J.B. Bernal.
25. Warren Weaver, "Note re Boston and Cambridge Visit," October 24–26, 1939, Record Group 1.2, Series 200A, Box 141, Folder 1742, RAC.
26. Warren Weaver to E. J.Cohn, February 15, 1940, E. J. Cohn Papers, CLM.
27. E. J. Cohn to Warren Weaver, February 26, 1940, E. J. Cohn Papers, CLM.
28. E. J. Cohn to W. B. Donham, May 10, 1940, E. J. Cohn Papers, CLM.

29. Morris Sayre, President, Corn Products Refining Co., to E. J . Cohn, December 27, 1946, quoted in E. J. Cohn, "A Brief History of the Support of the Department of Physical Chemistry, 1920–1950."
30. J. B. Conant to E. J. Cohn, January 22, 1940, E. J. Cohn Papers, CLM.
31. E. K. Bolton to E. J. Cohn, March 21, 1940, E. J. Cohn Papers, CLM.
32. J. B. Conant to G.H. Chase, February 27, 1940, E. J. Cohn Papers, CLM.
33. E. J. Cohn to J. B. Conant, June 8, 1940, E. J. Cohn Papers, CLM.
34. J. B. Conant to E. J. Cohn, June 15, 1940, E. J. Cohn Papers, CLM.

Chapter 6:

1. Samuel E. Morison, *The Oxford History of the English People* (New York: Oxford University Press, 1965), 992–995.
2. D. W. Stetten, "The Blood Plasma for Great Britain Project," *Bulletin of the New York Academy of Medicine* 28 (1941): 37.
3. "The National Research Council and Medical Preparedness," *Journal of the American Medical Association* 115 (1940): 1640–1643.
4. Ibid.
5. National Research Council, Division of Medical Sciences, Committee on Transfusions, "Minutes," May 31, 1940, p. 1, Archives, National Academy of Sciences, cited hereafter as ANAS.
6. S. Benison, A.C. Barger and E. L. Wolfe, "Walter B. Cannon and the Mystery of Shock. A Study of Anglo-American Cooperation in World War I."
7. W. B. Cannon, *Wound Shock* (New York: Appleton & Co., 1923).
8. O. H. Wangensteen, H. Hall, A. Kremen and B. Stevens, "Intravenous Administration of Bovine and Human Plasma to Man: Proof of Utilization," *Proceedings of the Society for Experimental Biology and Medicine* 43 (1940): 616–621.
9. National Research Council, Division of Medical Sciences, Committee on Transfusions, "Minutes," May 31, 1940, p. 8.
10. D. B. Kendrick, *Blood Program in World War II* (Washington: U.S. Government Printing Office, 1964): 15–16. This was issued by the Office of the Surgeon General, Department of the Army.
11. R. J. S. McDowall, "The Circulation in Relation to Shock," *British Medical Journal* 1 (June 8, 1940): 919–924.
12. A.S. Minot and A. Blalock, "Plasma Loss in Severe Dehydration, Shock and Other Conditions as Affected by Therapy," *Annals of Surgery* 12 (1940): 557–567.

13. D. B. Kendrick, *Blood Program in World War II*, p. 30.
14. Ibid., p. 51.
15. D. B. Kendrick, "Prevention and Treatment of Shock in the Combat Zone," *The Military Surgeon* 88 (1941): 97–113.
16. American National Red Cross, "Memorandum. Chronology of Development of Blood Donor Service," May 30, 1991, of which a copy is in ACBR.
17. L. K. Diamond, "History of Blood Transfusion," in M. M. Wintrobe, *Blood, Pure and Elegant. A Story of Discovery, of People, and of Ideas* (New York: McGraw-Hill, 1980), 659.
18. E. J. Cohn, "The History of Plasma Fractionation," in E. C. Andrus et al., *Advances in Military Medicine* (Boston: Little, Brown, 1948), 364.
19. Ibid.
20. American Medical Association, *New and Nonofficial Remedies* (Chicago: American Medical Association, 1940).
21. National Research Council, Division of Medical Sciences, Subcommittee on Blood Substitutes, "Minutes," November 30, 1940, pp. 9–13, ANAS.
22. E. J. Cohn, T. L. McMeekin, J. L. Oncley, J.M. Newell and W. L. Hughes, Jr., "Preparation and Properties of Serum and Plasma Proteins. I. Size and Charge of Proteins Separating upon Equilibrium across Membranes with Ammonium Sulfate Solutions of Controlled pH, Ionic Strength and Temperature," *Journal of the American Chemical Society* 62 (1940): 3386–3393.
23. T. L. McMeekin, "Preparation and Properties of Serum Plasma Proteins. II. Crystallization of a Carbohydrate-Containing Albumin from Horse Serum," *Journal of the American Chemical Society* 62 (1940): 3393–3396.
24. E. J. Cohn, T. L. McMeekin, J. P. Greenstein and J. H. Weare, "Studies in the Physical Chemistry of Amino Acids, Peptides and Related Substances. VIII. The Relation between Activity Coefficients of Peptides and Their Dipole Moments," *Journal of the American Chemical Society* 58 (1936): 2365–2370.
25. R. M. Ferry, E. J. Cohn and E. S. Newman, "Studies on the Physical Chemistry of the Proteins. XIII. The Solvent Action of Sodium Chloride on Egg Albumin in 25% Ethanol at −5 Deg.," *Journal of the American Chemical Society* 58 (1936): 2370–2375.
26. John Mellanby, "Diphtheria Antitoxin," *Proceedings of the Royal Society of London. Series B, Containing Papers of a Biological Character* 80 (1908): 399–413.
27. W. B. Hardy and Mrs. Stanley Gardiner, "Proteins of Blood Plasma," *The Journal of Physiology* 40 (1910): lxviii–lxxi.

28. E. J. Cohn to C. S. Burwell, July 24, 1940, ACBR.
29. A. J. Kremin, H. L. Taylor and H. Hall, "Skin Sensitivity of Man to Bovine Plasma and Its Albumin and Globulin Fractions," *Proceedings of the Society of Experimental Biology and Medicine* 43 (1940): 532–533.
30. E. J. Cohn, J. A. Leutscher, Jr., J. L. Oncley, S. H. Armstrong, Jr., and B. D. Davis, "Preparation and Properties of Serum and Plasma Proteins. III. Size and Charge of Proteins Separating upon Equilibring across Membranes with Ethanol-water Mixtures of Controlled pH, Ionic Strength and Temperature," *Journal of the American Chemical Society* 62 (1940): 3396–3400.
31. J. D. Porsche, interview with the author, July 14, 1994, author's files.
32. E. J. Cohn to J.B. Conant, September 25, 1940, Harvard Archives, Pusey Library, Conant files—Med. School General.
33. C. S. Stephenson to Rear Admiral Ross T. McIntyre, Surgeon General of the Navy, December 21, 1940. Quoted from E. J. Cohn, "History of Blood Plasma," Appendix, in E. J. Cohn, Manuscript book, 1940-1946, Vol. I, p. 108, ACBR.

Chapter 7:

1. Samuel E. Morison, *Oxford History of the American People*, p. 998.
2. "Red Cross to Start Defense Blood Bank," *New York Times*, February 3, 1941, p. 12, col. 4.
3. L. H. Weed to E. J. Cohn, February 11, 1941, ACBR.
4. National Research Council, Division of Medical Sciences, Subcommittee on Blood Substitutes, "Minutes," April 19, 1941, p. 15, ANAS.
5. L. H. Weed to J. B. Conant, April 5, 1941, ACBR.
6. "Use of Animal Blood for Human Transfusion Studied by Group of Chemists at Harvard," *Boston Herald*, April 22, 1941, p. 1, col. 5.
7. Lawrence E. Strong, interview with the author, March 31, 1992, author's files.
8. E. J. Cohn, "The University and the Biological Laboratories of the State of Massachusetts," *Harvard Medical Alumni Bulletin* 24 (April 1950): 75–79.
9. American National Red Cross, "Memorandum. Chronology of Development of Blood Donor Service."
10. National Research Council, Division of Medical Sciences, Subcommittee on Blood Substitutes, "Minutes," May 8, 1941, p. 37, ANAS.
11. National Research Council, Division of Medical Sciences, Subcommittee on Blood Substitutes, "Minutes," July 18, 1941, p. 89, ANAS.

12. National Research Council, Division of Medical Sciences, Subcommittee on Blood Substitutes, "Minutes," September 19, 1941, p. 121, ANAS.
13. National Research Council, Division of Medical Sciences, Subcommittee on Blood Substitutes, "Minutes," November 3, 1941, p. 130, ANAS.
14. C. A. Janeway, E. A. Stead, Jr., and R.V. Ebert, "Immunological and Clinical Studies on Purified Proteins of Human and Animal Plasma," in S. Mudd and W. Thalheimer, *Blood Substitutes and Blood Transfusions* (Springfield, Ill.: C.C. Thomas, 1942), p. 198.
15. E. J. Cohn, "History of Plasma Fractionation," p. 392.
16. National Research Council, Division of Medical Sciences, Subcommittee on Blood Substitutes, "Minutes," May 8, 1941, p. 50, ANAS.
17. Ibid., p. 47.
18. National Research Council, Division of Medical Sciences, Subcommittee on Blood Substitutes, "Minutes," May 23, 1941, p. 66, ANAS.
19. National Research Council, Division of Medical Sciences, Subcommittee on Blood Substitutes, "Minutes," July 18, 1941, p. 93, ANAS.
20. Ibid., p. 94.
21. E.J. Cohn, K. L. Oncley, L. E.Strong, S. H. Armstrong, Jr., and W. L. Hughes, Jr., "Properties and Functions of the Purified Proteins of Animal and Human Plasma," in S. Mudd and W. Thalheimer, *Blood Substitutes and Blood Transfusions*, pp. 173–183.
22. C. A. Janeway, "War Medicine, with Special Emphasis on the Use of Blood Substitutes," *New England Journal of Medicine* 225 (1941): 371–381.
23. National Research Council, Division of Medical Sciences, Subcommittee on Blood Substitutes, "Minutes," September 19, 1941, pp. 121–126, ANAS.
24. American Medical Association, *New and Nonofficial Remedies*, 1940 ed.
25. "Navy 'Utterly Lacking' Blood Plasma for War," *Boston Herald*, July 20, 1943, Marianne Brettauer Cohn Scrapbook, ACBR.
26. Samuel T. Gibson, M.D., interview with the author, July 20, 1945, author's files.

Chapter 8:

1. J. Pfeiffer, "The Story of Plasma; Blood Transfusion by Remote Control," *Harpers Magazine* (October 1942): 518–525.
2. National Research Council, Division of Medical Sciences, Committee on Transfusions, "Report to Conference on Albumin," Bulletin on Blood

Substitutes (hereafter designated as BBS), January 5, 1942, pp. 151–163, ANAS.

3. National Research Council, Division of Medical Sciences, Committee on Transfusions, "Report to Conference on Albumin," BBS, February 11, 1942, p. 156, ANAS.
4. E. J. Cohn, "The History of Plasma Fractionation," p. 375.
5. National Research Council, Division of Medical Sciences, Committee on Transfusions, "Report to Conference on Albumin," BBS, May 26, 1942, pp. 242–253, ANAS.
6. Ibid., p. 242.
7. National Research Council, Division of Medical Sciences, Committee on Transfusions, "Report on Conference on the Preparation of Normal Human Serum Albumin," BBS, June 5–6, 1942, pp. 254–272, ANAS.
8. J. T. Heyl and C. A. Janeway, "The Use of Human Albumin in Military Medicine. Part I. The Theoretical and Experimental Basis for Its Use," *US Naval Medical Bulletin* 40 (1942): 785–791.
9. G. Scatchard, A. C. Batchelder and A. Brown, "The Osmotic Pressure of Plasma and of Human Serum Albumin," *Journal of Clinical Investigation* 23 (1944): 458–464.
10. L. M. Woodruff and S. T. Gibson, "The Use of Human Albumin in Military Medicine. Part II. The Clinical Evaluation of Human Albumin," *US Naval Medical Bulletin* 40 (1942): 791–796.
11. J. T. Heyl, J. G. Gibson, 2nd, and C. A. Janeway, "Studies on the Plasma Proteins. V. The Effect of Concentrated Solutions of Human and Bovine Serum Albumin on Blood Volume after Acute Blood Loss in Man," *Journal of Clinical Investigation* 21 (1942): 763–777.
12. National Research Council, Division of Medical Sciences, Conference on Albumin, "Minutes," BBS, July 9, 1942, p. 301, ANAS.
13. E. J. Cohn, "Memoranda and Communications on the Preparation of Normal Human Serum Albumin," contained in "Department of Physical Chemistry, Harvard Medical School, Manuscript Volume II, February 11, 1942 to August 18, 1943, pp. 9–13, ACBR.
14. E. J. Cohn, "The History of Blood Fractionation," pp. 375–379.
15. Ibid., pp. 380–385.
16. Ibid., p. 375.
17. L. R. Newhouser and L. L. Lozner, "The Standard Army-Navy Package of Serum Albumin Human (Concentrated)," *US Naval Medical Bulletin* 40 (1942): 796–799.

18. J. F. Enders, "Concentration of Certain Antibodies in Globulin Fractions Derived from Human Blood Plasma," *Journal of Clinical Investigation* 23 (1944): 510–530.
19. R. M. Ferry, "Lawrence Joseph Henderson, 1878–1942."
20. E. J. Cohn, "History of Plasma Fractionation," original text, Department of Physical Chemistry, Harvard Medical School, Manuscript Volume I, Appendix D, p. 114, ACBR.
21. George Scatchard, "Edwin J. Cohn and Protein Chemistry."
22. *Department of Physical Chemistry, Harvard Medical School, 1920–1950. First Publication of the University Laboratory of Physical Chemistry Related to Medicine and Public Health. List of Those Associated, and Bibliography of Published Work* (Cambridge, Mass.: Harvard Printing Office, 1950).
23. E. J. Cohn, *History of the Development of the Scientific Policies of the University Laboratory of Physical Chemistry Related to Medicine and Public Health, Harvard University,* Ninth Publication, (Cambridge, Mass.: Harvard Printing Office, 1952). Also included in A Collection of Pamphlets published under the same name (see chapter 13, note 32).
24. E. J. Cohn to F.B. Hanson, February 2, 1942, Record Group 1.1, Series 200, Box 141, Folder 1744, RAC.
25. E. J. Cohn, "The History of Plasma Fractionation," pp. 379–380.
26. National Research Council, Division of Medical Sciences, Committee on Transfusions, "Report to Conference on Albumin," BBS, June 23, 1942, p. 350, ANAS.
27. D. W. Richards, Jr., "Shock," in E. C. Andrus et al., *Advances in Military Medicine,* p. 335.
28. National Research Council, Division of Medical Sciences, Committee on Transfusions, "Report to Conference on Albumin," BBS, October 20, 1942, p. 385, ANAS.
29. Alfred B. Cohn to the author, September 2, 1998, author's files.

Chapter 9:

1. E. J. Cohn, J. A. Leutscher, Jr., J. L. Oncley, S. H. Armstrong, Jr., and B. D. Davis, "Preparation and Properties of Serum and Plasma Proteins. III. Size and Charge of Proteins Separating upon Equilibrating across Membranes with Ethanol-water Mixtures of Controlled pH, Ionic Strength and Temperature," *Journal of the American Chemical Society* 32 (1940): 3396–3400.

2. C. A. Janeway and J. L. Oncley, "Blood Substitutes," in E. C. Andrus et al. *Advances in Military Medicine*, p. 444.
3. E. J. Cohn, "The History of Plasma Fractionation," p. 392.
4. National Research Council, Division of Medical Sciences, Subcommittee on Blood Substitutes, "Minutes," BBS, January 2, 1942, p. 157, ANAS.
5. National Research Council, Division of Medical Sciences, Subcommittee on Blood Substitutes, "Minutes," BBS, March 10, 1942, p. 183, NAS.
6. National Research Council, Division of Medical Sciences, Subcommittee on Blood Substitutes, "Minutes," BBS, April 15, 1942, p. 187, ANAS.
7. E. J. Cohn, "History of Blood Fractionation," original text, Department of Physical Chemistry, Harvard Medical School, Manuscript Volume I, Appendix G, "Crystallization of Bovine Serum Albumin," p. 126, ACBR.
8. National Research Council, Division of Medical Sciences, Subcommittee on Blood Substitutes, "Minutes," BBS, May 12, 1942, pp. 205–216, ANAS.
9. National Research Council, Division of Medical Sciences, Subcommittee on Blood Substances, "Minutes," BBS, July 16, 1942, p. 311, ANAS.
10. National Research Council, Division of Medical Sciences, Subcommittee on Blood Substitutes, "Minutes," BBS, June 23, 1942, p. 279, ANAS.
11. *Boston Herald*, July 7, 1942, Marianne Brettauer Cohn Scrapbook, ACBR.
12. Vannevar Bush to J. B. Conant, August 29, 1942, National Archives, R.G. 227, OSRD, MR Genl Records, 1940–1946, Bush, Vannevar (Dr) 1941–42 folder.
13. National Research Council, Division of Medical Sciences, Subcommittee on Blood Substitutes, "Minutes of the Conference on Albumin Testing," BBS, October 19, 1942, pp. 350–379, ANAS.
14. *Boston Herald*, December 7, 1942, Marianne Brettauer Cohn Scrapbook, ACBR.
15. *New York Times*, "Bomber Named for Hero Convict," *New York Times*, November 13, 1943, p. 13, col. 1.
16. *Boston Herald*, December 16, 1943, Marianne Brettauer Cohn Scrapbook, ACBR.
17. E. J. Cohn, "The Plasma Proteins: Their Properties and Functions," *Transactions and Studies of the College of Physicians of Philadelphia*, Ser. 3, 10 (December 1942): 149–162.

Chapter 10:

1. D. B. Kendrick, *Blood Program in World War II*, p. 476.
2. Ibid., p. 31.

3. Ibid., p. 398.
4. National Research Council, Division of Medical Sciences, Subcommittee on Blood Substitutes, "Minutes," BBS, April 1943, p. 687, ANAS.
5. Ibid., pp. 689–690.
6. D.B. Kendrick, *Blood Program in World War II*, pp. 395–396.
7. National Research Council, Division of Medical Sciences, Subcommittee on Blood Substitutes, "Minutes," BBS, May 25, 1943, p. 716, ANAS.
8. National Research Council, Division of Medical Sciences, Subcommittee on Blood Substitutes, "Minutes," BBS, August 10, 1943, p. 803, ANAS.
9. National Research Council, Division of Medical Sciences, Subcommittee on Blood Substitutes, "Minutes," BBS, September 24, 1943, p. 866.
10. *New York Times*, August 27, 1943, p. 4, col. 2.
11. E. D. Churchill, *Surgeon to Soldiers* (Philadelphia: Lippincott, 1972), 41–49.
12. D. B. Kendrick, *Blood Program in World War II*, p. 395.
13. S. E. Morison, *The Oxford History of the American People*, pp. 1021–1022.
14. D. B. Kendrick, *Blood Program in World War II*, p. 398.
15. National Research Council, Division of Medical Sciences, Subcommittee on Blood Substitutes, "Minutes," BBS, November 17, 1943, p. 925, ANAS.
16. National Research Council, Division of Medical Sciences, Subcommittee on Blood Substitutes, "Minutes," BBS, March 22, 1943, p. 648, ANAS.
17. Ibid., p. 637.
18. M. V. Veldee to L. H. Weed, July 5, 1943, ANAS.
19. R. T. McIntire to L. H. Weed, July 16, 1943, ANAS.
20. R. T. McIntire to E. J. Cohn, quoted in National Research Council, Division of Medical Sciences, Subcommittee on Blood Substitutes, "Minutes," BBS, November 17, 1943, p. 758, ANAS.
21. National Research Council, Division of Medical Sciences, Subcommittee on Blood Substitutes, "Minutes," BBS, July 19, 1943, p. 757, ANAS.
22. C. A. Janeway et al., "Concentrated Human Serum Albumin: Albumin in Treatment of Shock; Safety of Albumin; Albumin in Treatment of Hypoproteinemia," *Journal of Clinical Investigation* 23 (1944): 465–490.
23. A. Cournand et al., "Clinical Use of Concentrated Human Albumin in Shock, and Comparison with Whole Blood and with Rapid Saline Infusion," *Journal of Clinical Investigation* 23 (1944): 491–505.
24. J.V. Warren et al., "The Treatment of Shock with Concentrated Human Serum Albumin: A Preliminary Report," *Journal of Clinical Investigation* 23 (1944): 506–509.
25. G. Scatchard, "Edwin J. Cohn and Protein Chemistry," pp. 37–44.

26. National Research Council, Division of Medical Sciences, Subcommittee on Blood Substitutes, "Minutes," BBS, November 10, 1942, p. 394, ANAS.
27. National Research Council, Division of Medical Sciences, Subcommittee on Blood Substitutes, "Minutes," BBS, February 23, 1943, pp. 512–572, ANAS.
28. National Research Council, Division of Medical Sciences, Subcommittee on Blood Substitutes, "Minutes," BBS, February 24, 1943, pp. 575–602, ANAS.
29. National Research Council, Division of Medical Sciences, Subcommittee on Blood Substitutes, "Minutes," BBS, September 24, 1943, pp. 810–864, ANAS.
30. National Research Council, Division of Medical Sciences, Subcommittee on Blood Substitutes, "Minutes," BBS, November 17, 1943, p. 927, ANAS.
31. J. F. Enders, "The Concentration of Certain Antibodies in Globulin Fractions Derived from Human Blood Plasma," *Journal of Clinical Investigation* 23 (1944): 510–530.
32. J. L. Oncley, M. Melin, D. A. Richert, J. W. Cameron and P. M. Gross, "Separation of the Antibodies, Isoagglutinins, Prothrombin, Plasminogen, and β_1-Lipoproteins into Subfractions of Human Plasma," *Journal of the American Chemical Society* 71 (1949): 541–550.
33. National Research Council, Division of Medical Sciences, Subcommittee on Blood Substitutes, "Minutes," BBS, February 8, 1943, pp. 620–625, ANAS.
34. J. Stokes, Jr., E. P. Maris and S. S. Gellis, "Chemical, Clinical and Immunological Studies on the Products of Human Plasma Fractionation. XI. The Use of Concentrated Normal Serum Gamma Globulin (Human Immune Serum Globulin) in the Prophylaxis and Treatment of Measles," *Journal of Clinical Investigation* 23 (1944): 531–540.
35. National Research Council, Division of Medical Sciences, Subcommittee on Blood Substitutes, "Minutes," BBS, March 22, 1943, pp. 655–664, ANAS.
36. C. W. Ordman, G. C. Jennings, Jr., and C. A. Janeway, "Chemical, Clinical and Immunological Studies on the Products of Human Plasma Fractionation. XII. The Use of Concentrated Normal Human Serum Gamma Globulin (Human Immune Serum Globulin) in the Prevention and Attenuation of Measles," *Journal of Clinical Investigation* 23 (1944): 541–553.

37. National Research Council, Committee on Medical Research, "Monthly Progress Report," July 1, 1943, ACBR.
38. E. J. Cohn to L. H. Weed, April 6, 1943, in E. J. Cohn, "A Collection of Papers Recording an Institutional Growth and Metamorphosis," ACBR.
39. J. T. Edsall, R. M. Ferry and S. H. Armstrong, Jr., "The Proteins Concerned in the Blood Coagulation Mechanism," *Journal of Clinical Investigation* 23 (1944): 557–565.
40. E. J. Cohn, "Monthly Progress Reports to the National Research Council, Committee on Medical Research, Office of Emergency Management, Contract no. 139," August 1, 1942, ACBR.
41. O. T. Bailey and F. D. Ingraham, "Clinical Use of Products of Human Plasma Fractionation; Use of Products of Fibrinogen and Thrombin in Surgery," *Journal of the American Medical Association* 126 (1944): 680–685.
42. T. J. Putnam, "The Use of Thrombin on Soluble Cellulose in Neurosurgery: Clinical Applications," *Annals of Surgery* 118 (1943): 127.
43. O. T. Bailey and F. D. Ingraham, "Use of Fibrin Foam as Hemostatic Agent in Neurosurgery, Clinical and Pathological Studies," *Journal of Clinical Investigation* 23 (1944): 591–596.
44. J. E. Dees, "Fibrinogen Coagulation as an Aid in the Operative Removal of Renal Calculi," *Journal of Clinical Investigation* 23 (1944): 576–579.
45. National Research Council, Division of Medical Sciences, Subcommittee on Blood Substitutes, "Minutes," BBS, January 28, 1943, pp. 429–510, ANAS.
46. National Research Council, Division of Medical Sciences, Subcommittee on Blood Substitutes, "Minutes," BBS, March 23, 1943, pp. 665–682, ANAS.
47. L. Pillemer et al., "The Separation and Concentration of the Isohemaglutins from Human Serum," *Science* 97 (1943): 75–76.
48. L. Pillemer et al., "Chemical, Clinical and Immunological Studies on the Products of Human Plasma Fractionation. XIII. The Separation and Concentration of Isohemaglutinins from Group-Specific Human Plasma," *Journal of Clinical Investigation* 23 (1944): 550–553.
49. H. K. Beecher and Mark D. Altschule, *Medicine at Harvard, the First Three Hundred Years* (Hanover, N.H.: University Press of New England, 1977), 505–508.
50. "Science Progress Report," *Time Magazine* (November 29, 1943): 40, col. 2.

Chapter 11:

1. D. B. Kendrick, *Blood Program in World War II*, p. 465.
2. Ibid., p. 478.
3. Ibid., pp. 479–480.
4. Ibid., pp. 35–37.
5. Ibid., p. 480 et seq.
6. D. B. Kendrick, "Prevention and Treatment of Shock in the Combat Zone."
7. J. F. Loutit and P. L. Mollison, "Advantages of Disodium-citrate-Glucose Mixture as a Blood Preservative," *British Medical Journal* 2 (1943): 744–745.
8. D. B. Kendrick, *Blood Program in World War II*, p. 481.
9. Ibid., p. 482.
10. Ibid., p. 485.
11. Ibid., p. 488.
12. Ibid., p. 490.
13. Ibid., p. 488.
14. Ibid., p. 489.
15. Ibid., p. 494.
16. National Research Council, Division of Medical Sciences, Subcommittee on Blood Substitutes, "Minutes," BBS, August 30, 1944, pp. 1260–1264, ANAS.
17. National Research Council, Division of Medical Sciences, Committee on Medical Research, 4th Conference on Blood Transfusion, October 4, 1944, pp. 1275–1288.
18. D. B. Kendrick, *Blood Program in World War II*, pp. 610–613.
19. E. J. Cohn, "Monthly Progress Reports, to the National Research Council, Office of Emergency Management, Contract no. 139, No. 16, May 1, 1944, ACBR.
20. C. A. Janeway, "Plasma, the Transport Fluid for Blood Cells and Humors," in M. Wintrobe, *Blood, Pure and Elegant*, pp. 573–599.
21. National Research Council, Division of Medical Sciences, Subcommittee on Blood Substitutes, "Minutes," BBS, May 14, 1945, p. 1341, ACBR.
22. Ibid., p. 1170.
23. Ibid., p. 1054.
24. Ibid., p. 1174.
25. Ibid., p. 1178.
26. J. Stokes, Jr., and J. R. Neefe, "Prevention and Attenuation of Infectious

Hepatitis by Gamma Globulin; Primary Notes," *Journal of the American Medical Association* 127 (1945): 144–145.

27. W. P. Havens and J.R. Paul, "Prevention of Infectious Hepatitis with Gamma Globulin," *Journal of the American Medical Association* 129 (1945): 270–272.
28. E. J. Cohn, "The History of Plasma Fractionation," p. 428.
29. E. J. Cohn, Monthly Progress Reports, to the National Research Council, Office of Emergency Management, Contract no. 139, November 1, 1944, ACBR.
30. Warren Weaver to Frank Blair Hanson, January 26, 1944, Record Group 1.1, Series 200, Box 141, Folder 1744, RAC.
31. National Research Council, Division of Medical Sciences, Subcommittee on Blood Substitutes, "Minutes," BBS, January 1944, p. 984, ANAS.
32. Ibid., p. 989.
33. National Research Council, Division of Medical Sciences, Subcommittee on Blood Substitutes, "Minutes," BBS, March 1944, p. 1052, ANAS.
34. Ibid., p. 1156 et seq.
35. E. J. Cohn, *History of the Development of a Patent Policy Based on Experiences in Connection with Liver Extracts and Blood Derivatives.*
36. Ibid., p. 17.
37. National Research Council, Division of Medical Sciences, Subcommittee on Blood Substitutes, "Minutes," BBS, November 17, 1943, p. 946 et seq., ANAS.
38. E. J. Cohn, J. L. Oncley, L. E. Strong, W. L. Hughes, Jr., and S. H. Armstrong, "The Characterization of the Protein Fractions of Human Plasma," *Journal of Clinical Investigation* 23 (1944): 417–432.
39. H. K. Beecher and M. D. Altschule, *Medicine at Harvard, the First 300 Years*, p. 499.
40. "Work on Synthesis of Penicillin Gains," *New York Times*, March 14, 1944, p. 21, col. 5.
41. William L. Laurence, "War Speeds Study of Blood System," *New York Times*, September 13, 1944, p. 36, cols. 1-2.
42. *New York Herald Tribune*, September 13, 1944. Marianne Brettauer Cohn Scrapbook, ACBR.
43. *Time Magazine*, v. 43, no. 23, June 5, 1944, p. 4.
44. E. J. Cohn, "Blood Proteins and Their Therapeutic Value," *Science* 101 (1945): 51–56.

Chapter 12:

1. Samuel E. Morison, *The Oxford History of the American People*, pp. 1038–1040.
2. D. B. Kendrick, *Blood Program in World War II*, p. 310 et seq.
3. Ibid., p. 215.
4. C. Sidney Burwell to Frank Blair Hanson, February 24, 1945, ACBR.
5. Frank Blair Hanson, "Notes. Harvard, 7–9 March 1945," Record Group 1.1, Series 200, Box 141, Folder 1745, RAC.
6. Secretary, Rockefeller Foundation, to J. B. Conant, May 25, 1945, Dean's Papers, Harvard Medical School, CLM.
7. P. B. Beeson, G. Chesney and A. M. McFarlan, "Hepatitis following Injection of Mumps Convalescent Plasma; Reports from the American Red Cross-Harvard Field Hospital Unit; Use of Plasma in Mumps Epidemic," *Lancet* 1 (1944): 814–815.
8. P. B. Beeson, "Jaundice Occurring One to Four Months after Transfusion of Blood or Plasma; Report of Seven Cases," *Journal of the American Medical Association* 121 (1943): 1332–1334.
9. D. B. Kendrick, *Blood Programs in World War II*, p. 764 et seq.
10. "Army Discovers Jaundice Cause," *New York Times*, January 15, 1945, p. 21, col. 1.
11. National Research Council, Division of Medical Sciences, Subcommittee on Blood Substitutes, Conference on Plasma Fractionation, "Minutes," BBS, March 14, 1945, p. 1330, ANAC.
12. E. B. Grossman, S.G. Stewart and J. Stokes, Jr., "Post-Transfusion Hepatitis in Battle Casualties," *Journal of the American Medical Association* 129 (1945): 991–994.
13. National Research Council, Division of Medical Sciences, Subcommittee on Blood Substitutes, Conference on Plasma Fractionation, "Minutes," BBS, March 14, 1945, p. 1379–1382, ANAS.
14. E. J. Cohn, "Blood Proteins and Their Therapeutic Value."
15. National Research Council, Division of Medical Sciences, Subcommittee on Blood Substitutes, Conference on Plasma Fractionation, "Minutes," BBS, March 14, 1945, p. 1324, ANAS.
16. E. J. Cohn, "Monthly Progress Reports to the National Research Council, Committee on Medical Research, Office of Emergency Management, Contract no. 139," May 1, 1945, ACBR.
17. Ibid., November 1, 1944.

18. E. J. Cohn, "History of Plasma Fractionation," p. 409.
19. G. Scatchard, L.E. Strong, W. L. Hughes, Jr., J. N. Ashworth and A. H. Sparrow, "The Properties of Solutions of Human Serum Albumin of Low Salt Content," *Journal of Clinical Investigation* 24 (1945): 671–676.
20. S. S. Gellis, J. R. Neefe, J. Stokes, Jr., C. A. Janeway and G. Scatchard, "Chemical, Clinical, Immunological Studies on the Products of Human Fractionation. XXXVI. Inactivation of the Virus of Homologous Serum Hepatitis in Solutions of Normal Human Serum Albumin by Means of Heat," *Journal of Clinical Investigation* 27 (1948): 239–244.
21. National Research Council, Division of Medical Sciences, Subcommittee on Blood Substitutes, Conference on Plasma Fractionation, "Minutes," BBS, March 14, 1945, p. 1346, ANAS.
22. G. R. Minot, C. S. Davidson, J. H. Lewis, H. J. Tagnon and F. H. L. Taylor, "Coagulation Defect in Hemophilia; Effect in Hemophilia, of Parenteral Administration, of Fraction in Plasma Globulins Rich in Fibrinogen," *Journal of Clinical Investigation* 24 (1945): 704–707.
23. Samuel E. Morison, *Oxford History of the American People*, p. 1041.
24. Ibid., p. 1045.
25. National Research Council, Division of Medical Sciences, Subcommittee on Blood Substitutes, "Minutes," BBS, May 18, 1945, p. 1555, ANAS.
26. D. B. Kendrick, *Blood Program in World War II*, pp. 490–496.
27. "American Red Cross Distribution of Surplus Dried Blood Plasma," *Journal of the American Medical Association* 129 (1945): 1275.
28. "Harvard Leader Wins Passano Award," *The Evening Sun*, Baltimore, April 16, 1945, copy of award in author's file.
29. "Doctor Is Busy," *Newsweek Magazine*, May 14, 1945, pp. 102, 104, 106.
30. E. J. Cohn, *History of the Development of Scientific Policies of the University Laboratory of Physical Chemistry Related to Medicine and Public Health, Harvard University*, p. 6.
31. E. J. Cohn, "Research in the Medical Sciences," pp. 70–121.
32. "Globulin in the Control of Measles," *Lancet* 1 (1945): 405–406.
33. "Therapeutic Value of Blood Proteins," *Lancet* 1 (1945): 727–728.
34. E. J. Cohn, *History of the Development of Scientific Policies of the University Laboratory of Physical Chemistry Related to Medicine and Public Health, Harvard University*, pp. 7–9.
35. E. J. Cohn to A.N. Richards, October 16, 1945, E. J. Cohn Papers, CLM.
36. E. J. Cohn to Paul Buck, October 27, 1945, ACBR.
37. D. B. Kendrick, *Blood Program in World War II*, p. 699.

Chapter 13:

1. Protein Foundation, Board of Directors, "Minutes," December 11, 1953, ACBR.
2. C. A. Janeway, "Edwin Joseph Cohn, B.S., Ph.D.," *Harvard Medical Alumni Bulletin* 28 (January 1954): 23.
3. Edwin J. Cohn, Jr., interview with the author, Washington, D.C., September 15, 1998, author's files.
4. J. T. Edsall and W. H. Stockmeyer, "George Scatchard, 1892–1973."
5. E. J. Cohn, L. E. Strong, W. L. Hughes, Jr., D. L. Mulford, J. N. Ashworth, M. Melin and H. L. Taylor, "Preparation and Properties of Serum and Plasma Protein. IV. A System for the Separation into Fractions of the Protein and Lipoprotein Components of Biological Tissues and Fluids," *Journal of the American Chemical Society* 68 (1946): 459–475.
6. E. J. Cohn, T. L. Meekin, J. L. Oncley, J. M. Newell and W. L. Hughes, Jr., "Preparation and Properties of Serum and Plasma Proteins. I. Size and Charge of Proteins Separating upon Equilibrium across Membranes with Ammonium Sulfate Solutions of Controlled pH, Ionic Strength and Temperature."
7. E. J. Cohn, J. A. Leutscher, Jr., J. L. Oncley, S. H. Armstrong, Jr., and B. D. Davis, "Preparations and Properties of Serum and Plasma Proteins. III. Size and Charge of Proteins Separating upon Equilibrating across Membranes with Ethanol-water Mixtures of Controlled pH, Ionic Strength and Temperature."
8. E. J. Cohn, L. E. Strong, W. L. Hughes, Jr., D. J. Mulford, J. N. Ashworth, M. Melin and H. L. Taylor, "Preparations and Properties of Serum and Plasma Proteins. IV. A System for the Separation into Fractions of the Protein and Lipoprotein Components of Biological Tissues and Fluids."
9. E. J. Cohn, W. L. Hughes and J. H. Wearn, "Preparations and Properties of Serum and Plasma Proteins. XIII. Crystallization of Serum Albumins from Ethanol-water Mixtures," *Journal of the American Chemical Society* 69 (1947): 1753–1761.
10. E. J. Cohn, J. L. Oncley, L. E. Strong, W. L. Hughes, Jr., and S. H. Armstrong, Jr., "The Characterization of the Protein Fractions of Human Plasma," *Journal of Clinical Investigation* 23 (1944): 417–432.
11. American National Red Cross, Committee on Blood and Blood Derivatives of the Advisory Board of Health, "Minutes," December 14, 1945, Janeway Box, ACBR.

12. D. B. Kendrick, *Blood Program in World War II*, p. 776.
13. American National Red Cross, Committee on Blood and Blood Derivatives of the Advisory Board of Health, "Minutes," March 2, 1946, Janeway Box, ACBR.
14. Ibid., October 22, 1946, Janeway Box, ACBR.
15. Ibid., April 30, 1947, Janeway Box, ACBR.
16. E. J. Cohn to G.F. McGinnes, September 16, 1948, in "A Collection of Papers Recording an Institutional Growth and Metamorphosis" (cited hereafter as IGAM, ACBR), X2, 23.
17. Ibid., July 23, 1947, X2–19.
18. American National Red Cross, "Chronology of Development of Blood Donor Service 1936–1947," American Red Cross Plasma Operations, Washington, DC., pp. 1–12, ACBR.
19. L. K. Diamond, "Edwin J. Cohn Memorial Lecture, The Fulfillment of His Prophecy," *Vox Sanguinis* 20 (1971): 433–440.
20. E. J. Cohn to E. G. Bigwood, December 26, 1946, IGAM, ACBR, X2, 17.
21. S. S. Gellis, J. R. Neefe, J. Stokes, Jr., L. E. Strong, C. A. Janeway and G. Scatchard, "Inactivation of the Virus of Homologous Serum Jaundice in Solutions of Normal Human Serum Albumin by Means of Heat," *Journal of Clinical Investigation* 27 (1947): 239–244.
22. E. J . Cohn to C. J. Van Slyke, June 1946, quoted in E. J. Cohn, *A Memorandum of the Unwisdom of Projects and Reports* (Cambridge, Mass.: Harvard Printing Office, 1952), p. 10. Publication No. 9 of the University Laboratory of Physical Chemistry Related to Medicine and Public Health.
23. C. S. Burwell to Rollo Dyer, October 18, 1946, ACBR.
24. Ernest Allen to C. S. Burwell, October 18, 1946, ACBR.
25. E. J. Cohn, "The Separation of Blood into Fractions of Therapeutic Value," *Annals of Internal Medicine* 26 (1947): 341–352.
26. J. Stokes, Jr., "The Use of Gamma Globulin from Large Pools of Adult Blood Plasma in Certain Infectious Diseases," *Annals of Internal Medicine* 26 (1947): 353–362.
27. G. R. Minot and F. H. L. Taylor, "Hemophilia: The Clinical Use of Antihemophilic Globulin," *Annals of Internal Medicine* 26 (1947): 363–367.
28. C. A. Janeway, "Other Uses of Plasma Fractions with Particular Reference to Serum Albumin," *Annals of Internal Medicine* 26 (1947): 368–376.
29. E. J. Cohn, "Chemical, Physiological and Immunological Properties and Clinical Uses of Blood Derivatives," *Experientia* 3 (1947): 125–136.
30. A. L. Schade and L. Caroline, "Raw Egg White and the Role of Iron in

Growth Inhibition of Shigella dysenteria, Escherichia coli and Saccharomyces cervisia," *Science* 100 (1944): 14.

31. A.L. Schade and L. Caroline, "An Iron Binding Protein in Human Blood Plasma," *Science* 104 (1946): 340.
32. E. J. Cohn, Division of Medical Sciences of the Faculty of Arts and Sciences, Harvard University, *A Collection of Pamphlets Published to Record an Experiment in Organisation for Research, Training and Development in Science Basic to Medicine and Public Health within the Framework of a University* [Cambridge, Mass.: Harvard University Printing Office, 1952]. This contains seven pamphlets published between September 1949 and January 1952, detailing the history of the Division of Medical Sciences, the University Laboratory of Physical Chemistry Related to Medicine and Public Health, the Department of Physical Chemistry, Harvard Medical School, 1920–1950, Harvard's patent policy and scientific policies, etc., ending with A Memorandum on the Unwisdom of Projects and Reports.

Chapter 14:

1. National Research Council, Division of Medical Sciences, Committee on Blood and Blood Derivatives, "Minutes," BBS, April 4, 1948, p. 1, ANAS.
2. Ibid., p. 14.
3. E. J. Cohn to G. F. McGinnes, March 17, 1948, IGAM, ACBR, X2, 22.
4. E. J. Cohn to G. F. McGinnes, April 13, 1948, IGAM, ACBR, X2, 22.
5. National Research Council, Division of Medical Sciences, Committee on Blood and Blood Derivatives, "Minutes," BBS, June 28, 1949, ANAS.
6. S. S. Gellis, J. R. Neefe, J. Stokes, Jr., L. E. Strong and C. A. Janeway, "Inactivation of the Virus of Homologous Serum Hepatitis in Solutions of Normal Human Serum Albumin by Means of Heat."
7. E. J. Cohn to G. F. McGinnes, September 16, 1948, IGAM, ACBR, X2, 23.
8. Lt. General Leslie R. Groves to Vannevar Bush, Chairman of the Research Development Board of the National Military Department, Spring 1948, Quoted by Francis G. Blake in *Blood Characterization and Preservation Laboratory, University Laboratory of Physical Chemistry Related to Medicine and Public Health* (Cambridge, Mass.: Harvard Printing Office, 1951). The pamphlet recorded the dedication exercises of the laboratory on January 8, 1951.
9. James E. McCormack, Exec. Director, Committee on Medical Sciences to Lewis H. Weed, July 16, 1948. IGAM, ACBR, X2, 25.

10. Ibid.
11. Douglas MacN. Surgenor, "Personal notes," author's files.
12. J. E. McCormack to E. J. Cohn, December 16, 1948, IGAM, ACBR, X2, 27 to X2, 28.
13. Copy of the citation in the award of the Medal for Merit to Dr. Edwin J. Cohn, author's files.
14. G. Scatchard, "The Scientific Work of Edwin Joseph Cohn," *The Nucleus* 25 (1948): 263–268.
15. Edwin J. Cohn, "Interactions of Proteins and Other Body Constituents," *The Nucleus* 25 (1948): 269–276.
16. Harvard Medical School, Department of Physical Chemistry, "Minutes, luncheon meeting, September 1948," ACBR.
17. E. J. Cohn, L. E. Strong, W.L. Hughes, Jr., D. Mulford, J. N. Ashworth, M. Melin and H. L. Taylor, "Preparation and Properties of Serum and Plasma Proteins. IV. A System for the Separation into Fractions of the Proteins and Lipoprotein Components of Biological Tissues and Fluids."
18. D. M. Surgenor, L. E. Strong, H. L. Taylor, R. S. Gordon, Jr., and D. M. Gibson, "Preparation and Properties of Serum Plasma Proteins. XX. The Separation of Choline Esterase, Mucoprotein, and Metal-combining Protein into Subfractions of Human Plasma," *Journal of the American Chemical Society* 71 (1949): 1223–1229.
19. E. J. Cohn and B. A. Koechlin, "Crystallization and Characteristics of a Metal-Combining Protein of Human Plasma," *Abstracts, American Chemical Society*, September 1947, p. 30-C.
20. A. L. Schade and L. Caroline, "An Iron Binding Component in Human Blood Plasma," *Science* 104 (1946): 340.
21. D. M. Surgenor, B.A. Koechlin and L.E. Strong, "Chemical, Clinical and Immunological Studies on the Products of Human Plasma Fractionation. XXXVII. The Metal-Combining Globulins of Human Plasma," *Journal of Clinical Investigation* 28 (1949): 73–78.
22. C. E. Rath and C. A. Finch, "Chemical, Clinical and Immunological Studies on the Products of Human Plasma Fractionation. XXXVIII. Serum Iron Transport Measurement of Iron-binding Capacity of Serum in Man," *Journal of Clinical Investigation* 28 (1949): 79–85.
23. G. E. Cartwright and M. M. Wintrobe, "The Anemia of Infection. Studies of the Iron-binding Capacity of Serum," *Journal of Clinical Investigation* 28 (1949): 86–98.
24. E. J. Cohn, D. M. Surgenor and M. J. Hunter, "The State in Nature of Proteins and Protein Enzymes of Blood and Liver." in J. T. Edsall, *Enzymes*

and Enzyme Sytstems. Their State in Nature (Cambridge, Mass.: Harvard University Press, 1951), pp. 107–143.

25. E. J. Cohn, D. M. Surgenor, R. W. Greene, J. D. Porsche and J. B. Lesh, "The State in Nature of the Active Principle in Pernicious Anemia, of Catalase and of other Components of Liver," *Science* 109 (April 1949): 443; and E. J. Cohn, D. M. Surgenor, R.W. Greene, J. D. Porsche, James B. Lesh, and a Committee of the Hematology Study Section, NIH, B. Norberg, G. Derouaux and H. Nitschmann, "The State in Nature of the Active Principle Effective in Pernicious Anemia," in "Manuscripts Presented at a Conference on the Preservation of the Formed Elements and of the Proteins of the Blood, Boston, Mass. January 6, 7, 8, 1949," pp. 257–261, ACBR.

Chapter 15:

1. National Research Council, Division of Medical Sciences, Committee on Blood and Blood Derivatives, "Minutes, 4th Meeting," BBS, January 8, 1949, pp. 44–45, ANAS.
2. National Research Council, Division of Medical Sciences, Committee on Blood and Blood Derivatives, "Minutes," BBS, February 14, 1949, p. 56, ANAS.
3. Ibid., pp. 111–117, ANAS.
4. Ibid., pp. 119–131, ANAS.
5. National Research Council, Division of Medical Sciences, Committee on Blood and Blood Derivatives, "Minutes," BBS, July 8, 1959, p. 136, ANAS.
6. Conant quoted from Foreword to the pamphlet *The University Laboratory of Physical Chemistry Related to Medicine and Public Health* (Cambridge, Mass.: Harvard Printing Office, 1950), 3.
7. D. W. Bailey, Secretary to the President and Fellows of Harvard College, to E. J. Cohn, April 4, 1949, ACBR.
8. G. W. Gray, "Confidential Monthly Report for the Information of the Trustees, the Rockefeller Foundation, No. 122, June 1, 1950," pp. 18–19.
9. E. J. Cohn to P. H. Buck, June 8, 1949, Edwin J. Cohn Papers, CLM.
10. P. H. Buck to E. J. Cohn, May 18, 1949, Edwin J. Cohn Papers, CLM.
11. E. J. Cohn to Carl Cori, May 25, 1949, Edwin J. Cohn Papers, CLM.
12. E. J. Cohn to L. F. Fieser, May 26, 1949; E. J. Cohn to K. V. Thimann, June 2, 1949, Edwin J. Cohn Papers, CLM.
13. P. H. Buck to E. J. Cohn, May 29, 1947, Edwin J. Cohn Papers, CLM.

14. E. J. Cohn, "Report of the Division of Medical Sciences, Faculty of Arts and Sciences, 1909–1949," p. 28, ACBR.
15. C. F. Cori to E. J. Cohn, June 3, 1949, Edwin J. Cohn Papers, CLM.
16. J. H. Welsh to E. J. Cohn, June 10, 1949, Edwin J. Cohn Papers, CLM.
17. L. F. Fieser to E. J. Cohn, June 6, 1949, Edwin J. Cohn Papers, CLM.
18. E. J. Cohn to P. H. Buck, June 8, 1949, Edwin J. Cohn Papers, CLM.
19. Warren Weaver, "Notes" [after June 7, 1949], Record Group 1.1, Series 200, Box 141, Folder 1745, RAC.
20. E. J. Cohn to J. B. Conant, February 5, 1950, ACBR.
21. Frederick Sanger, "Address to the Department of Physical Chemistry, Harvard Medical School, at its luncheon seminar, June 6, 1949," ACBR.
22. National Research Council, Division of Medical Sciences, Committee on Blood and Blood Derivatives, "Minutes," BBS, July 8, 1949, pp. 123–174, ANAS.
23. E. J. Cohn and D. Mittelman, "Protein-Protein Interactions in Alcohol-Water Mixtures of Defined Temperature, pH and Ionic Strength," Abstracts, New York meeting of the American Chemical Society, September 1947, ACBR.
24. E. J. Cohn, F. R. N. Gurd, D. M. Surgenor, B. A. Barnes, R. K. Brown, G. Derouaux, J. M. Gillespie, F. W. Kahnt, W. F. Lever, C. H. Liu, D. Mittelman, R. F. Mouton, K.Schmidt and E.A. Uroma, "System for the Separation of the Components of Human Blood: Quantitative Procedures for the Separation of the Protein Components of Human Plasma," *Journal of the American Chemical Society* 72 (1950): 465–474.
25. E. J. Cohn to V. A. Getting, December 29, 1949, ACBR.
26. E. J. Cohn to G.P . Berry, October 3, 1949, Edwin J. Cohn Papers, CLM.
27. G. P. Berry to E. J. Cohn, Octoer 6, 1949, Edwin J. Cohn Papers, CLM.
28. E. J. Cohn to J. B. Conant, October 12, 1949, E. J. Cohn Papers, CLM; also printed in *The University Laboratory of Physical Chemistry Related to Medicine and Public Health,* 1950, p. 17, ACBR.
29. Commission on Plasma Fractionation and Related Processes, "Minutes of the Annual Meeting," December 27, 1949, p. 4, ACBR.
30. Commission of Plasma Fractionation and Related Processes, "Minutes of Meetings," November 30, 1950, December 12, 1951, January 9, 1952, ACBR.
31. D. M. Surgenor, "Control et Distribution des Gamma Globulins," in *Les Gamma Globulines et la Medecine des Enfants* (Paris: Le Centre International de l'Enfance, 1955), pp. 56–73, ACBR.
32. National Research Council, Division of Medical Sciences, Committee on

Blood and Blood Derivatives, "Minutes," BBS, December 2, 1949, p. 179, ANAS.

33. National Research Council, Division of Medical Sciences, Committee on Blood and Blood Derivatives, "Minutes," BBS, December 3, 1949, p. 240, ANAS.
34. Joint Meeting of the Committee on Blood Banks of the American Medical Association and the Committee on Blood and Blood Derivatives of the Advisory Board on Health Services of the American National Red Cross, "Minutes," December 5, 1949, ACBR.

Chapter 16:

1. *The University Laboratory of Physical Chemistry Related to Medicine and Public Health, Harvard University,* 1950, p. 9, ACBR.
2. Ibid., p. 11.
3. E. J. Cohn to George Kistiakowsky, February 10, 1950, Edwin J. Cohn Papers, CLM.
4. *The University Laboratory of Physical Chemistry Related to Medicine and Public Health, Harvard University,* p. 3.
5. Ibid., p. 5.
6. Ibid., in note on last page.
7. E. Seligman to J. B. Conant, February 17, 1950, Edwin J. Cohn Papers, CLM.
8. Alan Gregg, "Diary," May 23, 1950, Record Group 1.1, Series 2300, Box 141, Folder 1746, RAC.
9. G. C. Marshall to E. J. Cohn, April 20, 1950, ACBR.
10. Samuel E. Morison, *Oxford History of the American People,* pp. 1065–1073.
11. D. B. Kendrick, *Blood Program in World War II,* p. 727.
12. E. J. Cohn to G. C. Marshall, July 24, 1950, IGAM, ACBR, X1, 100.
13. G. C. Marshall to E. J. Cohn, July 26, 1950, filed under "AMCROSS," ACBR.
14. E. J. Cohn to G. C. Marshall, November 30, 1950, IGAM, ACBR, X2, 47.
15. D. B. Kendrick, *Blood Program in World War II,* p. 782.
16. E. J. Cohn, "Demonstration of New Processes of Blood Collection and Separation of Red Blood Cells, White Blood Cells and Platelets, Protein, Glycoprotein, Lipoprotein and Other Components of Plasma," *Science* 112 (1950): 450–451.
17. John Lear, "You May Be Drafted to Give Blood," *Colliers* (March 10, 1951): 11–13, 58–59.

18. E. J. Cohn to G. C. Marshall, November 30, 1950, ACBR.
19. Alfred B. Cohn, interview with the author, September 23, 1998, author's files.

Chapter 17:

1. Harvard University, Laboratory of Physical Chemistry, *Dedication Exercises, Blood Characterization and Preservation Laboratory, January 8, 1951* (Cambridge, Mass.: Harvard University Press, 1951), ACBR.
2. Harvard University, Laboratory of Physical Chemistry, *University Laboratory Conference on the New Mechanical Equipment for Collecting and Processing Human Blood*, June 14–15, 1951 (Cambridge, Mass.: Harvard Printing Office, 1951), ACBR.
3. Harvard University, Laboratory of Physical Chemistry, *The Development of Knowledge of the Blood; an Exhibition* (Cambridge, Mass.: Harvard Printing Office, 1951). The sixth publication of the Laboratory, ACBR.
4. E. J. Cohn, "Implications of New Knowledge About Blood for Transfusion Services and for the Public Health," address to the IV International Congress on Blood Transfusion, Lisbon, Portugal, July 23, 1951, ACBR.
5. E. J. Cohn, "Blood Collection and Preservation," *Proceedings of the Fourth Annual Meeting of the American Association of Blood Banks, Minneapolis, Minnesota, October 22–24, 1951* (1951), pp. 441–444.
6. D. B. Kendrick, *Blood Program in World War II*, p. 735.
7. E. J. Cohn, *History of the Development of Scientific Policies for the University Laboratory of Physical Chemistry Related to Medicine and Public Health, A Memorandum of the Unwisdom of Projects and Reports* (Cambridge, Mass.: Harvard University Printing Office, 1952). Ninth publication of the University Laboratory of Physical Chemistry, ACBR.
8. Commission on Plasma Fractionation and Related Processes, "Minutes and Reports of Meetings," December 12, 1951, ACBR.
9. Edwin J. Cohn, Jr., interview with the author, Washington, D.C., September 15, 1998, author's files.
10. Alfred B. and Barbara N. Cohn, interview with the author, Cambridge, Mass., October 29, 1997, author's files.

Chapter 18:

1. J.B. Conant, "Welcoming Remarks, Seminar on Implications of New Knowledge about Proteins, Protein Enzymes and Cells in Harvard Hall,

in Conference on Implications of New Knowledge about Proteins, Protein Enzymes and Cells," January 8, 1952, ACBR.

2. W.L. Hughes, Jr., "An Albumin Fraction Isolated from Human Plasma as a Crystalline Mercuric Salt," *Journal of the American Chemical Society* 69 (1947): 1753–1761.
3. F. R. N. Gurd and D. S. Goodman, "Preparation and Properties of Serum and Plasma Proteins. XXXII. The Interaction of Human Serum Albumin with Zinc Ions," *Journal of the American Chemical Society* 74 (1952): 670–674.
4. C. A. Janeway, "Clinical and Immunologic Control of Biologic Products," in Conference on Implications of New Knowledge about Proteins, Protein Enzymes and Cells, Commission on Plasma Fractionation, "Minutes," January 7–8, 1952, ACBR.
5. Commission on Plasma Fractionation and Related Processes, "Minutes and Reports of Meetings," 1951, Appendix B., ACBR.
6. E. J. Cohn, D. M. Surgenor, K. Schmid, W.H. Bachelor, H. C. Isliker and E. H. Alameri, "The Physical Chemistry of Proteins," *A General Discussion of the Faraday Society,* No. 13 (1953):176-196, ACBR.
7. E. J. Cohn to J. B. Conant, May 28, 1952, IGAM, ACBR, X1, 77.
8. E. J. Cohn to J. B. Conant, June 4, 1952, ACBR.
9. J. B. Conant to E. J. Cohn, June 4, 1952, IGAM, ACBR, X1, 81.
10. C. I. Barnard to E. J. Cohn, December 6, 1952, ACBR.
11. J. B. Conant to future trustees of the Protein Foundation, January 8, 1953, ACBR.
12. C. I. Barnard to J.W. Barker, March 5, 1953, IGAM, ACBR, X1, 90.
13. Allen Latham, Jr., "The Rewards of Being a Mechanical Engineer," *The Roxbury Latin Newsletter,* 64, no. 12 (July 1991): 12–15, ACBR.
14. J. R. Paul, *Poliomyelitis* (New Haven, Conn.: Yale University Press, 1971), 391.
15. W. McD. Hammon, L. I. Coriell and J. Stokes, Jr., "Evaluation of Red Cross Gamma Globulin as a Prophylactic Agent for Poliomyelitis," *Journal of the American Chemical Society* 150 (1952): 739–760.
16. J. R. Paul, *Poliomyelitis,* p. 393.
17. "Gamma globulin gives hope that polio can be checked," *New York Times,* October 26, 1952, p. 50, col. 3.
18. D. B. Kendrick, *Blood Program in World War II,* p. 784.
19. J. Stokes, Jr., "Immunization in Infants and Children with Particular Reference to Viral Hepatitis," *Johns Hopkins Medical Journal* 121 (1966): 305–328.

20. D. M. Surgenor, "The Nature of New Reagents for Preparation of Stable Protein in Their State of Nature," October 6, 1953, ACBR.
21. J. Stokes, Jr., and J. Smolens, "Report of Conference on Implications of New Knowledge about Proteins, Protein Enzymes and Cells of the Protein Foundation," January 7, 8, 1952; January 15, 1953; May 11, 1953, ACBR.
22. E. J. Cohn to C. I. Barnard, December 6, 1952, February 27, 1953, March 9, 1953, May 11, 1953, ACBR.
23. D. B. Kendrick, *Blood Program in World War II*, pp. 772–776.
24. A. L. Lehninger, *Biochemistry*, 2nd ed. (New York: Worth Publishers, Inc., 1975), Appendix A, p. 1062.
25. J. D. Watson and F. H. C. Crick, "Molecular Structure of Nucleic Acids. A Structure for Deoxyribose Nucleic Acid," *Nature* 171 (1953): 964–967.
26. O. C. Bruton, L. Apt, D. Gitlin and C. A. Janeway, "Absence of Serum Gamma Globulin," *American Journal of Diseases of Children* 84 (1952): 632–636.
27. C. A. Janeway, L. Apt and D. Gitlin, "Agammaglobulinemia," *Transactions of the Association of American Physicians* 66 (1953): 200–202.
28. E. Brand, B. Kassel and L. J. Saidel, "Amino Acid Composition of Plasma Protein," *Journal of Clinical Investigation* 23 (1944): 437–444.
29. E. J. Cohn to J.B. Conant, May 28, 1952, ACBR.
30. E. J. Cohn, "The Fine Structure of Proteins: A Contribution to Order in Biological Systems," Abstract submitted for the American Philosophical Society meeting, November 14, 1952, ACBR.
31. G. Scatchard, "Edwin J. Cohn and Protein Chemistry."
32. J. L. Tullis, "Edwin Cohn: The Man and his Science," Lecture at Harvard Medical School, November 1, 1990, ACBR.
33. W. K. Jordan, "Memorial Minute on Edwin J. Cohn," in "Minutes" of the Protein Foundation, December 11, 1953, ACBR.

Chapter 19:

1. Lasker Foundation Archives, MS C 415, History of Medicine Division, Modern Manuscripts, National Library of Medicine, Bethesda, Maryland.
2. E. J. Cohn, L. E. Strong, W. L. Hughes, Jr., D. J. Mulford, J. N. Ashworth, M. Melin and H. L. Taylor, "Preparation and Properties of Serum and Plasma Proteins. IV. A System for the Separation into Fractions of the Protein and Lipoprotein Components of Biological Tissues and Fluids."

3. E. Tabor, "The Epidemiology of Virus Transmission by Plasma Derivatives: Clinical Studies Verifying the Lack of Transmission of Hepatitis B and C. Viruses and HIV Type I," *Transfusion* 39 (1999): 1160–1168.
4. J. J. Skillman, "Editorial," *Transfusion* 39 (1999): 120–121.
5. J. L. Oncley, M. Melin, D. A. Richert, J. W. Cameron and P. M. Gross, "Preparation and Properties of Serum and Plasma Proteins. XIX. The Separation of Antibodies, Isoglutinens, Prothrombin, Plasminogen and the B1-Lipoproteom into Subfractions of Human Plasma," *Journal of the American Chemical Society* 71 (1949): 541–550.
6. K. F. Austen et al., "Therapeutic Immunology," in R. S. Geha and F.S. Rosen, *Intravenous Immunoglobulin Therapy* (Malden, Mass.: Blackwell Science US, 1996), Chapter 21, pp. 280–296.
7. E. R. Stiehm, "Uses for Intravenous Immune Globulin," Editorial, *New England Journal of Medicine* 325 (1991): 123–125.
8. "Perspectives," *Blood Transfusion Industry* (Boston, Cowen & Co., 1997), p. 46, ACBR.
9. America's Blood Centers *Newsletter*, October 6, 2000, ACBR.
10. J. L. Tullis, D. M. Surgenor, R. Tinch, M. D'Hont, S. Driscoll and W. L. Bachelor, "New Principle of Closed System Centrifugation," *Science* 124 (156): 792–797.
11. A. U. Smith, "Prevention of Haemolysis during Freezing and Thawing of Red Blood Cells," *Lancet* 2 (1956): 910–911.
12. M. M. Ketchel, J. L. Tullis, R. J. Tinch, S. G. Driscoll and D. M. Surgenor, "Use of Biomechanical Equipment for the Long-term Preservation of Erythrocytes," *Journal of the American Medical Association* 168 (1958): 404–408.
13. L. L. Haynes, J. L. Tullis, H. M. Pyle, M. T. Sproul, S. Wallach and W. C. Turville, "Clinical Use of Glycerolized Frozen Blood," *Journal of the American Medical Association* 173 (1960): 1657–1663.
14. S. Bernt, P. Gisholt and E. Smith, "Making Blood Count: the Founding of Haemonetics, Entrepreneurial Management Final Paper," Harvard Business School, April 30, 1993, author's files.
15. Haemonetics, *Annual Report*, 1999, ACBR.
16. Allen Latham, Jr., "The Rewards of Being a Mechanical Engineer."
17. Hans Jörnval, Secretary of the Nobel Assembly, Stockholm, to D. M. Surgenor, June 14, 2000, author's files.

Index

About The Author

After graduating from Williams College in 1939, Douglas M. Surgenor received a master's degree in chemistry from the Massachusetts State College at Amherst and the Ph.D. degree in chemistry from the Massachusetts Institute of Technology in 1946. While at MIT, an interest in studying the chemistry of proteins led him to investigate a course on the Physical Chemistry of Proteins that was being taught by Professor Edwin J. Cohn at the Harvard Medical School. Surgenor went to see Cohn, who explained that his involvement in a project related to the war effort precluded him from offering the course on proteins at the time. A few days later, Cohn called to suggest that Surgenor come to work in his laboratory for the summer, arranging for graduate credits for the time passed there. Surgenor subsequently spent the summer of 1943 in Cohn's laboratory as part of a team that was attempting to isolate human gamma globulin from plasma. Following his graduation from MIT, Surgenor assumed a post in Cohn's Department of Physical Chemistry.

After Cohn's sudden death eight years later, Surgenor's appointment was shifted to the Department of Biological Chemistry under A. Baird Hastings at the Harvard Medical School where he participated in the teaching of the biochemistry course to medical students and continued his research on plasma proteins. In June 1960, Surgenor accepted appointment as Professor and Chairman of the Department of Biochemistry at the University of Buffalo School of Medicine. During his second year at Buffalo, the University of Buffalo was merged with the

State University of New York and Surgenor was appointed Dean of the Buffalo Medical School, a post he held until 1968 when he resigned the deanship to return to teaching. In 1972, while still at Buffalo, he was elected President of the Center for Blood Research, the successor of Cohn's original department; and in 1977 he returned to Boston as Visiting Professor of Pediatrics (Biochemistry) at the Harvard Medical School, serving simultaneously for the next ten years as President of the Center for Blood Research and, between 1977 and 1983, as Director of the American Red Cross Blood Services, Northeast Region.

Surgenor is the author of or contributor to more than 150 scientific and medical papers on the chemistry of proteins and the formed elements of the blood and related topics. He has held numerous advisory positions at the National Heart, Lung and Blood Institute and the National Blood Resource Program. His honors include citations for distinguished service to research by the American Heart Association and the American Association of Blood Banks.